Aperiodic Resonances—Theory and Applications

Jianhua Yang · Miguel A. F. Sanjuan

Aperiodic Resonances—Theory and Applications

Springer

Jianhua Yang
School of Mechanical and Electrical Engineering
China University of Mining and Technology
Xuzhou, Jiangsu, China

Miguel A. F. Sanjuan
Departamento de Física
Universidad Rey Juan Carlos
Madrid, Spain

ISBN 978-981-96-6818-2 ISBN 978-981-96-6819-9 (eBook)
https://doi.org/10.1007/978-981-96-6819-9

This Springer imprint is published by the registered company Springer Nature Singapore Pte Ltd.
The registered company address is: 152 Beach Road, #21-01/04 Gateway East, Singapore 189721, Singapore

To my wife Mengxia Qiu, my son Yaolang Yang, and my daughter Langpeng Qiu

To my wife Céline and my daughters Alicia and Mónica

Preface

Signals are essential for transmitting information across the globe, enriching and energizing our world. In both natural sciences and engineering, aperiodic signals occur far more frequently than periodic ones. Although researchers have examined signals from various perspectives, resonance remains a subject of enduring fascination. The earliest recorded discussion of resonance appears in *Zhuangzi*, an ancient Chinese text from the third to fourth century BC, which describes the resonant phenomenon observed in the strings of the Chinese zither. Despite numerous studies on resonance induced by periodic excitation, no specialized research has focused on resonance caused by aperiodic signals. This gap in knowledge motivates the present monograph.

Aperiodic resonance differs from traditional resonance in several key respects. Unlike periodic signals, aperiodic signals lack distinct frequency components and do not strongly excite a system at its natural frequency. Instead, aperiodic resonance emerges as a synchronization phenomenon, where the system's output tends to align with the input signal, though a phase offset may be present. Even when the input amplitude is weak, the resonance effect amplifies the output while preserving the waveform which is a hallmark of aperiodic resonance.

This monograph is divided into two main parts. Part I focuses on the theoretical study of aperiodic resonances, while the second explores their engineering applications. Chapters 1–10 cover the theoretical foundations of aperiodic resonances, whereas Chaps. 11–13 examine their practical applications in various engineering fields. Aperiodic signals are crucial in inducing resonance within systems, with the mechanism and manifestation of this resonance varying according to the signal's characteristics. Consequently, this monograph begins in Chap. 1 by introducing several typical forms of aperiodic signals. Subsequent chapters delve into the distinct types of aperiodic resonances these signals generate. The signals examined include binary aperiodic signals, M-ary aperiodic signals, linear frequency-modulated signals, and aperiodic pulse signals, among others. Each signal type has been systematically studied and holds clear relevance for various engineering applications. Aperiodic resonances can be classified into several types based on the nature of system excitations. This monograph primarily explores aperiodic system resonance, aperiodic stochastic resonance, aperiodic vibrational resonance, logical resonance, ultrasensitive system resonance, and ultrasensitive vibrational resonance, along with their engineering applications. In Chap. 1, we introduce several key indices to quantify the occurrence and intensity of aperiodic resonance, including the cross-correlation coefficient, spectral amplification factor, bit error rate, and logical accuracy. Each indicator is applied in different research scenarios to analyze various forms of aperiodic resonance. As the field advances, new indices will be developed to further refine the measurement and characterization of these phenomena. Aperiodic system resonance arises when a system is excited by a single aperiodic signal, and it can be induced by tuning the system parameters. In Chaps. 2 and 3, we examine this phenomenon, categorizing it based on the type of excitation signal, namely, binary aperiodic signals, M-ary aperiodic signals, and frequency-modulated signals. Aperiodic stochastic resonance occurs when an aperiodic signal interacts with an appropriate level of noise. This phenomenon is examined in Chaps. 4 and 5, where we first investigate resonance induced by an aperiodic binary signal, followed by resonance caused by different forms of frequency-modulated signals. Aperiodic vibrational resonance occurs in a nonlinear system when it is excited by an aperiodic characteristic signal and an auxiliary signal. Typically, the characteristic signal is slow-varying, while the auxiliary signal is fast-varying. We explore this type of aperiodic resonance in Chaps. 6 and 7, where the characteristic signal is presented in the form of a binary aperiodic signal and a frequency-modulated signal, respectively. Logical resonance is a type of aperiodic resonance in which a nonlinear system's output achieves the highest logical success probability through logical operations. In Chap. 8, we introduce and examine three forms of logical resonance: logical system resonance, logical stochastic resonance, and logical vibrational resonance. Another type of aperiodic resonance, known as ultrasensitive resonance, is examined in Chap. 9 as ultrasensitive system resonance and in Chap. 10 as ultrasensitive vibrational resonance. When a single signal excites an underdamped nonlinear system, it may exhibit ultrasensitive system resonance. In contrast, when the excitation includes both a slow-varying characteristic signal and a fast-varying auxiliary signal, ultrasensitive vibrational resonance can occur. Over time, the resonance pattern may evolve from ultrasensitive to sparse and eventually to steady traditional resonance. This phenomenon is closely linked to chaos and transient chaos, making it a compelling subject in nonlinear dynamics.

Chapters 11–13 showcase various applications of aperiodic resonances. Chapter 11 explores the use of aperiodic stochastic, vibrational, and logical resonances in image processing to extract weak features, enhance image quality, and detect contours. Chapter 12 examines the application of system, vibrational, and stochastic resonances in bearing fault diagnosis, highlighting their effectiveness in extracting fault features from noisy signals under variable operating conditions. Chapter 13 focuses on using aperiodic stochastic resonance in the nondestructive testing of steel wire ropes, where the method extracts damage-related information and assesses damage severity. The monograph's theoretical framework is primarily based on numerical simulations, complemented by experimental studies in the application-focused chapters. Given the inherent complexity of aperiodic resonance, which poses significant challenges for analytical methods, further research is essential. Although this work presents aperiodic resonance phenomena in several typical nonlinear models, many other models remain unexplored. In summary, aperiodic resonance is an evolving research area with numerous open questions, and this monograph highlights several unresolved issues to inspire future investigation.

The content of this monograph is based on over a decade of research achievements by our research group. We express our sincere gratitude to our doctoral and master's students, as well as the undergraduate students who have contributed. We also deeply appreciate the unwavering support of our families, which has enabled us to pursue our research with peace of mind. Additionally, Jianhua Yang extends his heartfelt thanks to his doctoral supervisor, Prof. Xianbin Liu at Nanjing University of Aeronautics and Astronautics, who introduced him to the fascinating field of nonlinear dynamics in 2006 and guided him in exploring the infinite complexities of nonlinear systems.

Finally, we extend our gratitude to the National Natural Science Foundation of China (Grant Nos. 12472036, 12311530053, 12072362, 11672325, and 51305441), the Spanish State Research Agency (AEI) and the European Regional Development Fund (ERDF, EU) under Project No. PID2023-148160NB-I00, and the Priority Academic Program Development of Jiangsu Higher Education Institutions in Jiangsu Province. We also thank the Jiangsu Key Laboratory of Mine Mechanical and Electrical Equipment, the State Key Laboratory of Intelligent Mining Equipment Technology, the School of Mechanical and Electrical Engineering at China University of Mining and Technology, and Universidad Rey Juan Carlos for providing excellent research platforms.

Xuzhou, Jiangsu, China — Jianhua Yang
Móstoles, Madrid, Spain — Miguel A. F. Sanjuan
February 2025

Contents

Part II Applications of Aperiodic Resonances

Acronyms

ABSR	Aperiodic Stochastic Resonance in Bistable System
APSR	Aperiodic Stochastic Resonance in Periodic Potential System
ASR	Aperiodic Stochastic Resonance
BER	Bit Error Rate
BM3D	Block Matching 3D
FM	Frequency-Modulated
LFM	Linear Frequency-Modulated
MSE	Mean Square Error
NLFM	Nonlinear Frequency-Modulated
PCB	Printed Circuit Board
PSNR	Peak Signal-to-Noise Ratio
SNR	Signal-to-Noise Ratio

Part I
Theory of Aperiodic Resonances

Chapter 1
Aperiodic Signals

This chapter introduces several typical types of aperiodic signals, including aperiodic binary signals, aperiodic M-ary signals, various frequency-modulated signals, and aperiodic pulse signals. The aperiodic resonance induced by these signals in different nonlinear systems will be investigated in this monograph. The signals presented in this chapter are simulated signals, with subsequent chapters addressing the practical engineering applications of these signal types and their corresponding experimental counterparts. Aperiodic resonance is characterized by specific indices, such as the cross-correlation coefficient, the spectral amplification factor, the bit error rate, and the logical success probability. This chapter provides a detailed introduction to these indices.

1.1 Aperiodic Binary Signal and M-Ary Signal

The aperiodic binary signal has a wide range of applications, such as vertical cavity surface emitting laser [1, 2], monochrome image processing [3], among others. The expression of the aperiodic binary signal s(t) is [4, 5]

$$s(t) = A \sum_{m=-\infty}^{+\infty} R_m \Gamma(t - mT), \tag{1.1}$$

where A is the amplitude of the signal, $R_m = 1$ or -1 are random numbers that conform to a given statistical distribution, $\Gamma(t) = \begin{cases} 1, & t \in [0, \quad T] \\ 0 & others \end{cases}$, T is the pulse width, i.e., the minimal width of the random pulse of the aperiodic binary signal $s(t)$. In Fig. 1.1, three different waveforms of the aperiodic binary signals are given.

J. Yang and M. A. F. Sanjuan, *Aperiodic Resonances—Theory and Applications*,
https://doi.org/10.1007/978-981-96-6819-9_1

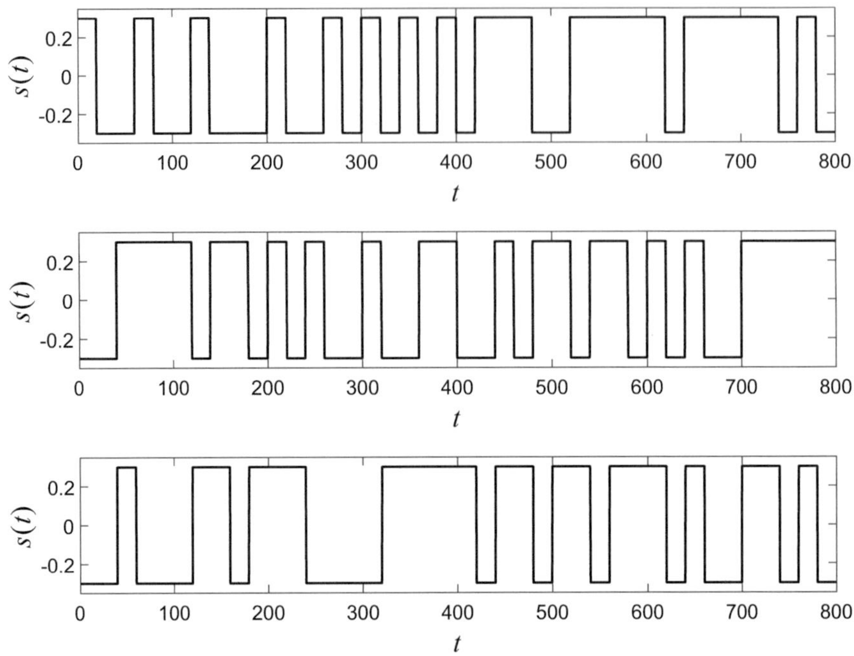

Fig. 1.1 Three different waveforms of the aperiodic binary signal for $A = 0.3$, $T = 20$. From top to bottom, the used random numbers R_m are different

Although the amplitude and pulse width remain unchanged, the random numbers R_m lead to the cause different waveforms.

Another similar aperiodic signal is the M-ary signal $s(t) = A_m, m = 1, 2, M$. As shown in Fig. 1.2, $s(t)$ has an amplitude $A_m = (2m - 1 - M)A$ at each bit with a prior probability. Equation (1.1) can be modified to describe an M-ary signal when R_m is randomly chosen from the integer $2m - 1 - M$, but not only from 1 and -1 [6].

If the amplitude A in Eq. (1.1) is modulated, for instance, by replacing A with $A + F\cos(\Omega t)$, it leads to an amplitude modulated signal, as investigated in our previous work [7]. This modulation transforms the signal into a generalized M-ary signal, where M takes values over a broader range. Examples of the resulting curves are shown in Fig. 1.3.

Existing literature on aperiodic resonance primarily focuses on using aperiodic binary signals as excitation signals, with relatively little research on M-ary signals and amplitude-modulated M-ary signals. The aperiodic resonance induced by these latter two types is more complex, and this will be an open topic explored in the future.

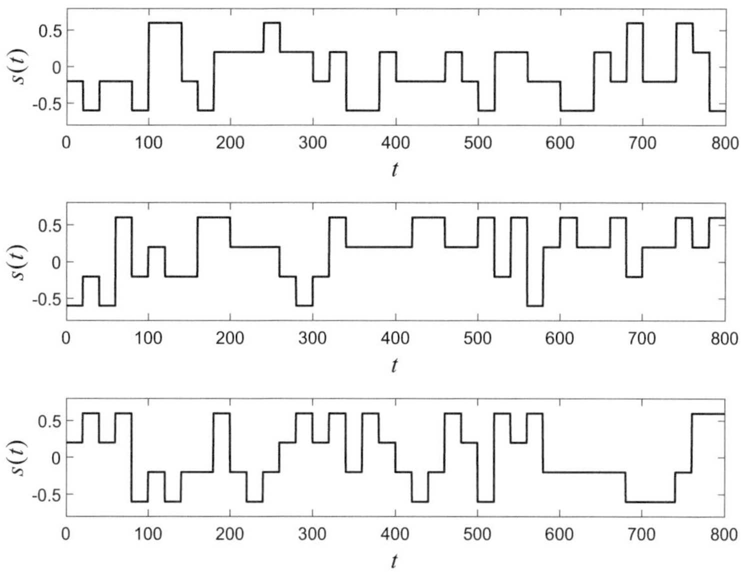

Fig. 1.2 Three different waveforms of the aperiodic M-ary signal for $M = 4$, $A = 0.2$, $T = 20$. From top to bottom, the use random numbers R_m are different

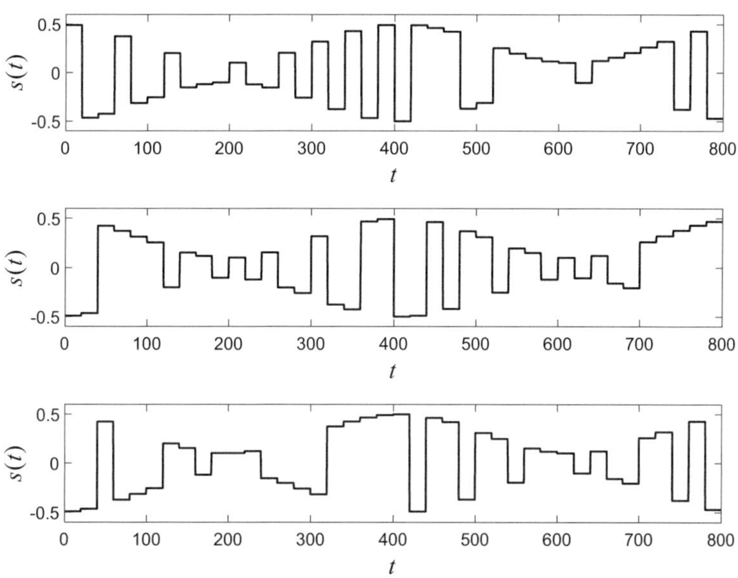

Fig. 1.3 Three different waveforms of the amplitude modulated binary signal for $A = 0.3$, $F = 0.2$, $\Omega = 30$, $T = 20$. From top to bottom, the used random numbers R_m are different

1.2 Linear Frequency-Modulated Signal

A signal whose frequency increases or decreases over time is called a chirp signal. When the frequency increases, it is referred to as an up-chirp, while a decreasing frequency is known as a down-chirp. Chirp signals have found wide applications in radar detection [8, 9], communications [10, 11], and measurement systems [12, 13]. Notably, the linearly frequency-modulated signal introduced below is a typical example of a chirp signal.

Let us first consider the linear frequency-modulated (LFM) signal. Its general mathematical expression is

$$s(t) = A\cos(\pi\gamma t^2 + 2\pi f_0 t + \varphi), \tag{1.2}$$

where A, γ, f_0 and φ are the amplitude, the chirp rate, the starting frequency and the initial phase of the signal, respectively. The phase $\phi(t)$ of the signal $s(t)$ is

$$\phi(t) = \pi\gamma t^2 + 2\pi f_0 t + \varphi. \tag{1.3}$$

Accordingly, the instantaneous frequency $f_{in}(t)$ is

$$f_{in}(t) = \frac{d}{dt}\phi(t) = 2\pi\gamma t + 2\pi f_0. \tag{1.4}$$

The chip rate γ influences the fluctuation and the rate of increase or decrease in the instantaneous frequency. When using radians per second (rad/s) as the unit of frequency, the start frequency is $f_{start} = 2\pi f_0$, and the end frequency is $f_{end} = 2\pi\gamma t_{end} + 2\pi f_0$. Alternatively, when using Hertz (Hz) as the unit of frequency, the start frequency is $f_{start} = f_0$, and the end frequency is $f_{end} = \gamma t_{end} + f_0$, where t_{end} is the signal end time of $s(t)$. The waveform and corresponding spectrum of a linear frequency-modulated signal $s(t)$ are shown in Fig. 1.4. In Fig. 1.4a, b, the instantaneous frequency increases with time t, $f_{start} > f_{end}$. Conversely, in Fig. 1.4c, d, the instantaneous frequency decreases with time t, where $f_{start} < f_{end}$. The spectra in Fig. 1.4b, d are obtained directly via the fast Fourier transform (FFT).

The spectrum in Fig. 1.4**b**, **d** only indicates that the signal is a modulated frequency signal. Nevertheless, it cannot reveal the variation of the frequency with time t. The short time Fourier transform (STFT) provides such an effective tool. Actually, the STFT can not exhibit only the variation of the frequency, but also display the variation of the signal energy, i.e., the signal amplitude with time. Notwithstanding, the STFT has its own shortcoming. As shown in Fig. 1.5, at the beginning and the end of the time axis, some intervals are absent. This is determined by the properties of the STFT. Each missing interval length is half of the length of the Window function divided by the sampling frequency. For example, in Fig. 1.5, we choose the sampling frequency to be 10000, and the length of the Window function 20000. As a result, the missing time interval at the beginning or the end of the time axis is 1 second. In most

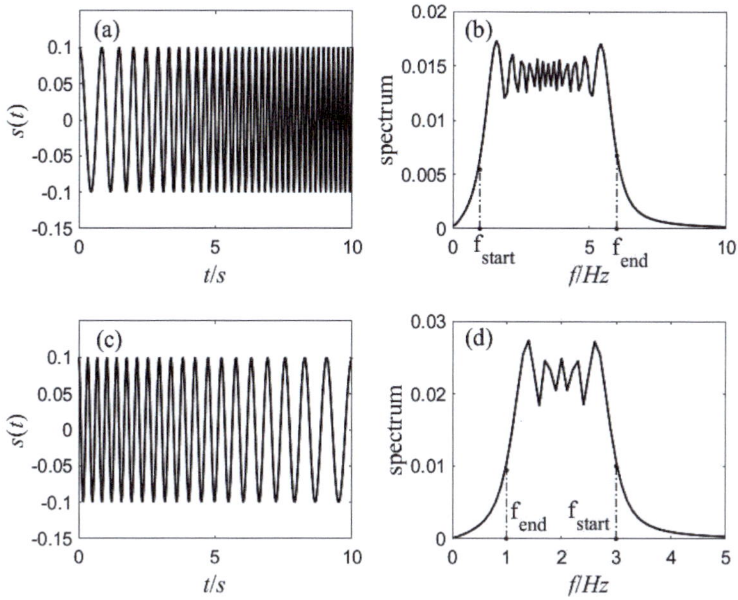

Fig. 1.4 **a** and **c**, the waveform of the linear modulated signal. **b** and **d**, the corresponding spectrum of $s(t)$ corresponding to **a** and **c**, respectively. The simulation parameters are $A = 0.1$, $\varphi = 0$ and $\gamma = 0.5$, $f_0 = 1$ in **a** and **b**, and $\gamma = -0.2$, $f = 3$ in **c** and **d**

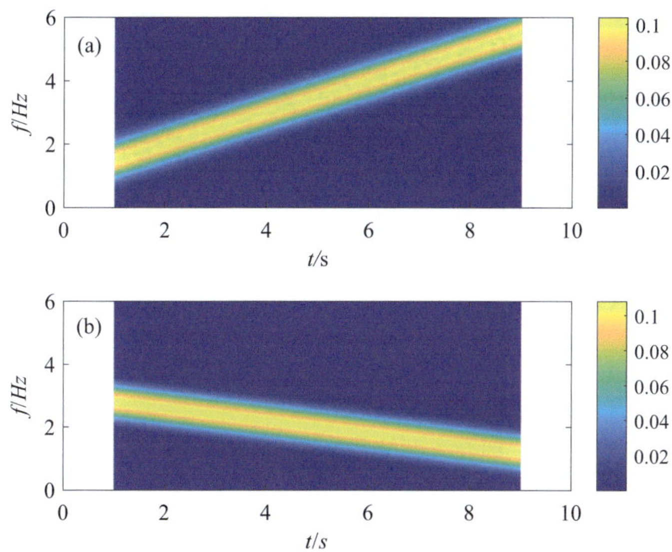

Fig. 1.5 **a** The STFT corresponding to the signal in Fig. 1.4a. **b** The STFT corresponding to the signal in Fig. 1.4c

cases, this shortcoming does not influence the application of the classical transformation method. Certainly, there are also some techniques and improved algorithms to overcome this shortness of the STFT.

1.3 Nonlinear Frequency-Modulated Signal

If the instantaneous frequency of a signal is a nonlinear function of time, it is referred to as a nonlinear frequency-modulated (NLFM) signal. The frequency can be described by various forms of nonlinear functions, depending on the specific application.

1.3.1 Multiple Piecewise Linear Modulated Frequency Case

Herein, we present a nonlinear frequency-modulated (NLFM) signal $s(t)$ in which the time-frequency feature $f_{in}(t)$ is described by a piecewise linear function.

$$f_{in}(t) = \begin{cases} f_0 + \gamma t, & (0 < t \leq t_1) \\ f_0 + \gamma t_1, & (t_1 < t \leq t_2) \\ f_0 + \gamma(t_1 + t_2) - \gamma t, & (t > t_2) \end{cases} \tag{1.5}$$

where f_0 and γ have the same meaning as in Eq. (1.2). At the moments t_1 and t_2, the frequency as a function of time changes. The NLFM signal $s(t)$ corresponding to the cosine function is then described as

$$\begin{aligned} s(t) &= A\cos\left[2\pi \int_0^t f(t')dt' + \varphi\right] \\ &= A\cos\left[\int_0^t 2\pi \begin{cases} f_0 + \gamma t & (0 < t \leq t_1) \\ f_0 + \gamma t_1 & (t_1 < t \leq t_2) \\ f_0 + \gamma(t_1 + t_2) - \gamma t & (t > t_2) \end{cases} dt' + \varphi\right] \\ &= A\cos\left[2\pi \begin{cases} f_0 t + \frac{1}{2}\gamma t^2 & (0 < t \leq t_1) \\ (f_0 t + \gamma t_1 t) & (t_1 < t \leq t_2) \\ [f_0 + \gamma(t_1 + t_2)]t - \frac{1}{2}\gamma t^2 & (t > t_2) \end{cases} + \varphi\right] \end{aligned} \tag{1.6}$$

The frequency variation, linear increase-invariant-linear decrease, is a common variation pattern of NLFM signal frequency.

As an example of Eq. 1.6, Fig. 1.6 shows segmented time-domain waveforms of the NLFM signal $s(t)$. Since the instantaneous frequency varies significantly, only the signals within the time intervals of 0–2 s, $10 - 11$ s and $18 - 20$ s are presented to clearly display the change characteristics of the signal. The characteristics of these

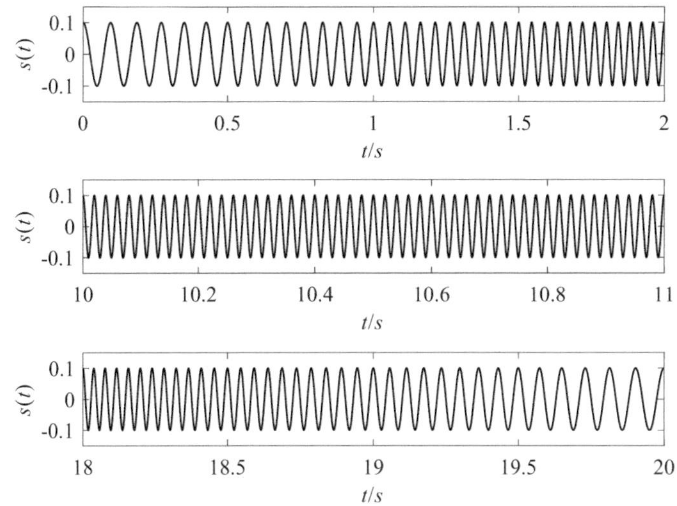

Fig. 1.6 The waveform of the NLFM signal in Eq. (1.6). The simulation parameters are $A = 0.1$, $\gamma = 8$, $f_0 = 10$, $\varphi = 0$, $t_1 = 5$, $t_2 = 15$

three signals include an increase of instantaneous frequency, a constant instantaneous frequency, and a decrease in instantaneous frequency.

The short-time Fourier transform (STFT) spectrum corresponding to the NLFM signal in Fig. 1.6 is presented in Fig. 1.7. The spectrum displays a distinct ridge. The amplitude of the STFT spectrum of the signal is approximately 0.1, and the signal frequency ranges from 10–60 Hz, exhibiting a time-frequency characteristic of linear increase, stationarity, and linear decrease.

In Part 2 of this monograph, we will demonstrate that this frequency variation pattern is commonly observed in rotating machinery. The frequency rise stage corresponds to the startup of the machinery, while the constant frequency stage indicates stable operation. The frequency decrease stage is associated with the deceleration or shutdown of the rotating equipment. Changes in rotational speed are typically controlled by the motor's frequency regulator. Studying NLFM signals is essential for extracting fault features in rotating machinery, highlighting one of the engineering applications of this research. Additionally, there are numerous application cases of NLFM signals in other contexts. Not only is the engineering significance of NLFM signals worthy of exploration, but the theoretical implications of the complex responses induced by these signals in nonlinear systems are also important areas of study.

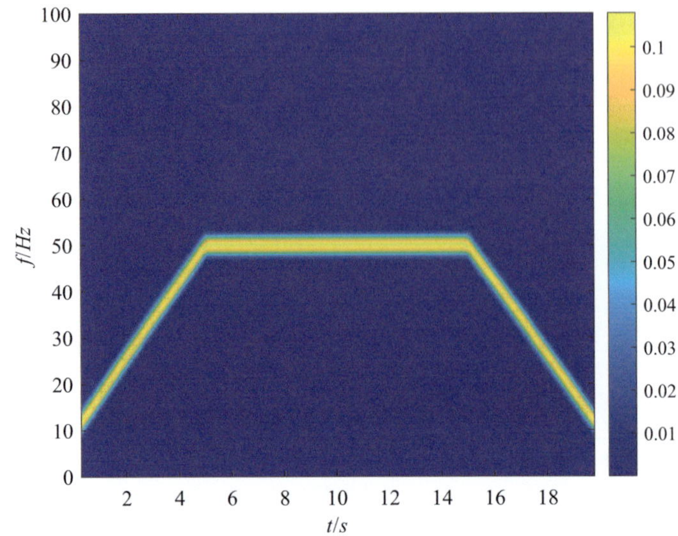

Fig. 1.7 The STFT spectrum of the NLFM signal in Fig. 1.6

1.3.2 Multi-Frequency Modulated Case

According to the reference [15], a type of multi-frequency NLFM signal is expressed by

$$\begin{cases} s(t) = \sum_{h=1}^{m} A_h \exp\{-2\delta\pi f_{\mathrm{n}}[t - v_h(t)]\} \sin\{2\pi f_{\mathrm{n}}[t - v_h(t)] + \varphi\} u[t - v_h(t)] \\ v_h(t) = \sum_{i=1}^{h} T_i \end{cases} \tag{1.7}$$

This signal can be used to simulate the vibration signal of a rolling bearing with outer ring fault under variable speed working condition. In Eq. (1.7), m is the number of pulses, and A_h is the amplitude of the hth pulse, δ is the impact attenuation rate, f_n is the natural frequency of the bearing, $v_h(t)$ is the total time of the occurrence of h pulses, φ is the initial phase, $u(t)$ is the Heaviside step function, and T_i is the repetition period of the ith pulse at the time of occurrence.

The instantaneous frequency is

$$f_{in} = \begin{cases} 5t, & (0 < t \le 3) \\ 15, & (3 < t \le 7) \\ 50 - 5t, & (7 < t \le 10) \end{cases} \tag{1.8}$$

It satisfies

$$\int_0^{v_h} 2\pi f_{in} \mathrm{d}t = 2\pi i \tag{1.9}$$

where i represents the number of revolutions of the bearing.

To simulate the signal, we set the sampling frequency $f_s = 10240\,\mathrm{Hz}$, $\delta = 120000$, $f_n = 2000\,\mathrm{Hz}$, $\varphi = 0$, $A_h = 0.08$. The frequency of the signal increases linearly from 5 Hz to 65 Hz within 0 3 s, remains unchanged within 3−7 s, and decreases linearly from 65 Hz to 5 Hz within 7−10 s. Figure 1.8 shows the segmented time-domain waveform of the NLFM signal, which consists of a series of non-stationary pulse signals and exhibits significant nonlinear time-frequency characteristics. Since the instantaneous frequency changes greatly, in order to show the signal characteristics more clearly, the signals of 0−3 s, 5 5.5 s and 7−10 s are plotted. The pulse interval of the signal decreases first, then stabilizes, and finally increases gradually, which conforms to the law of instantaneous frequency change.

Figure 1.9 shows the STFT spectrum of the NLFM signal corresponding to Fig. 1.8. The STFT spectrum amplitude is approximately 5.5×10^{-4}. Compared with the STFT in Fig. 1.7, there are multiple ridges for the NLFM signal in Eq.(1.7).

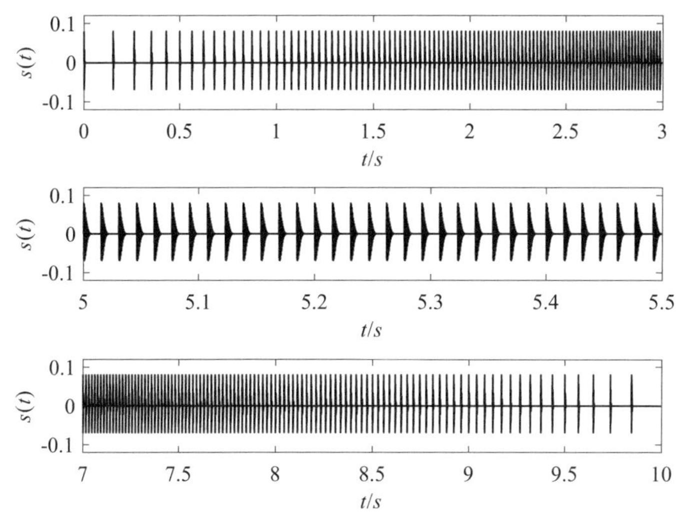

Fig. 1.8 The waveform of the NLFM signal with multi-frequency

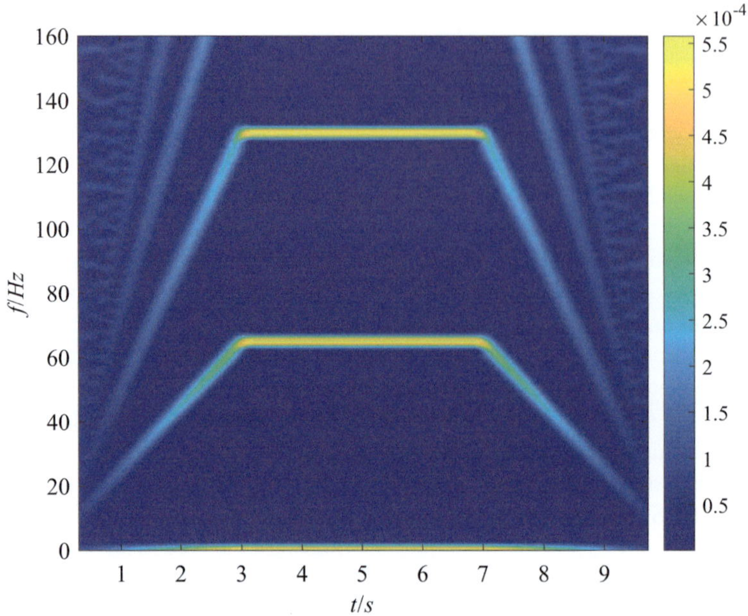

Fig. 1.9 The STFT spectrum of the NLFM signal in Fig. 1.8

1.3.3 Amplitude and Frequency Modulated Simultaneously

Select a common set of amplitude/frequency modulation signals with simple sinusoidal characteristics. The signal $s(t)$ is described as [16]

$$s(t) = 0.08\left(1 + \frac{1}{3}\cos\frac{2\pi t}{9}\right)\cos\left(70\pi t + 30\sin\frac{2\pi t}{3} + 20\cos\frac{5\pi t}{3}\right). \quad (1.10)$$

The signal mainly consists of cosine amplitude/frequency modulation signals. The frequency of simulated signals exhibits complex fluctuations, which is typical in signal processing. Figure 1.10 shows some segmented time-domain waveforms of signal $s(t)$. The signal exhibits obvious amplitude modulated and frequency modulated phenomena, and the amplitude and frequency of each segment fluctuate.

From Fig. 1.11, it can be observed that the frequency of the simulated signal fluctuates in a wide range. The black solid line is the instantaneous frequency curve extracted by phase, which is consistent with the curve in the STFT spectrum. The instantaneous frequency changes in real time, and the frequency characteristics are complex.

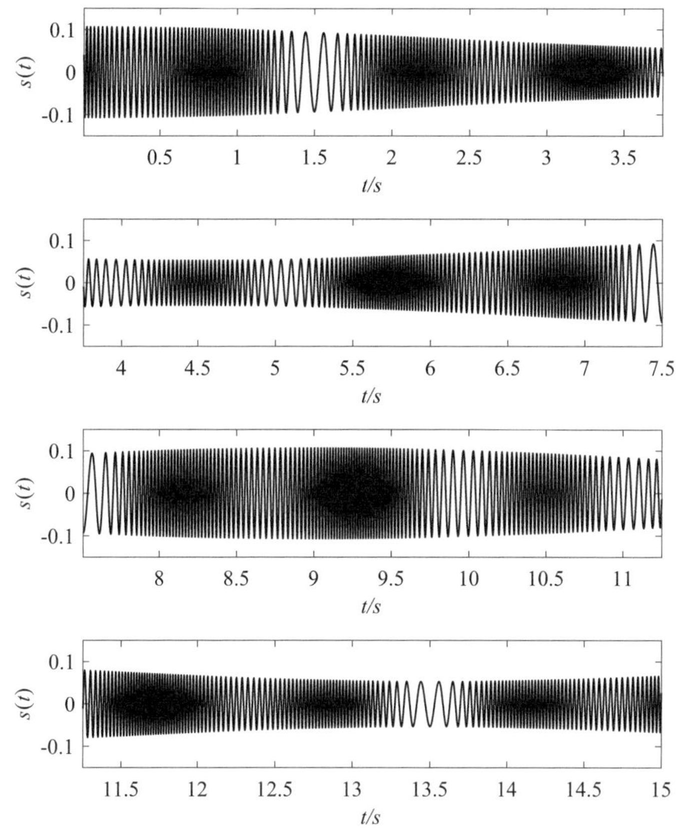

Fig. 1.10 The waveform of the NLFM signal with both amplitude modulation and frequency modulation

1.4 Echo Signal

When the chirp signal is sent out to the target, then reflect back, and the reflected signal is the echo chirp signal, known simplify as echo signal. There are many meaningful works on echo signals, such as range measurement [13], compressed sensing [17, 18], velocity estimation [19], and other fields [20].

To describe echo signal, we re-define the chirp signal $s(t)$ in the complex field

$$s(t) = \begin{cases} e^{j2\pi(B/(2T)t^2+(f_c-B/2)t)}, & 0 \le t \le T \\ 0, & \text{otherwise} \end{cases} \tag{1.11}$$

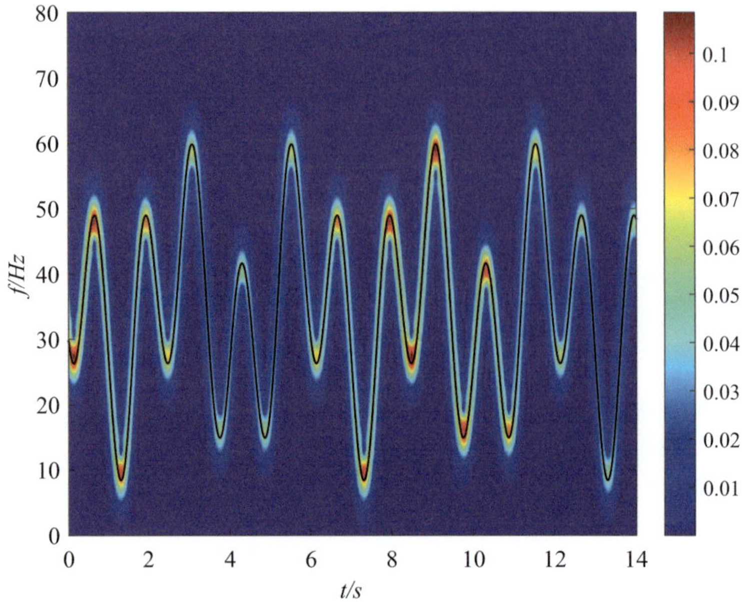

Fig. 1.11 The STFT spectrum of the NLFM signal in Fig. 1.10

where $B > 0$ is the bandwidth, $T > 0$ is the duration, and $f_c > 0$ is the central frequency. The instantaneous frequency f_{in} is

$$f_{in} = \frac{B}{T}t + f_c - \frac{B}{2}. \tag{1.12}$$

The echo signal is affected by the reflection of the object and some other factors, and it is defined as [13, 14]

$$e(t) = \sum_{i=1}^{m} \alpha^{i-1} w(t - id - c)\Re(s(t - id - c)), \tag{1.13}$$

In Eq. (1.10), m is the number of total chirp components, α is the decaying rate, d is the time delay between each chirp component, c is the initial time delay. The symbol $R(\bullet)$ is the real part of the chirp signal $s(t)$. The function $w(t)$ is given as

$$w(t) = \begin{cases} 0.5\left[1 - \cos\left(\frac{2\pi t}{T}\right)\right], & 0 \le t \le T \\ 0, & \text{otherwise} \end{cases} \tag{1.14}$$

We give one component of the echo chirp signal in Fig. 1.12, where $m = 1$.

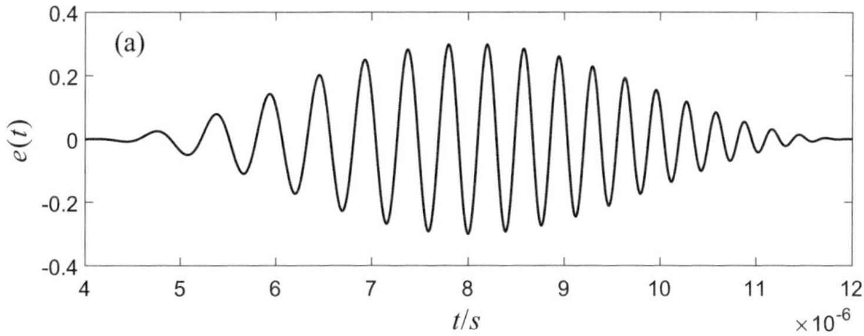

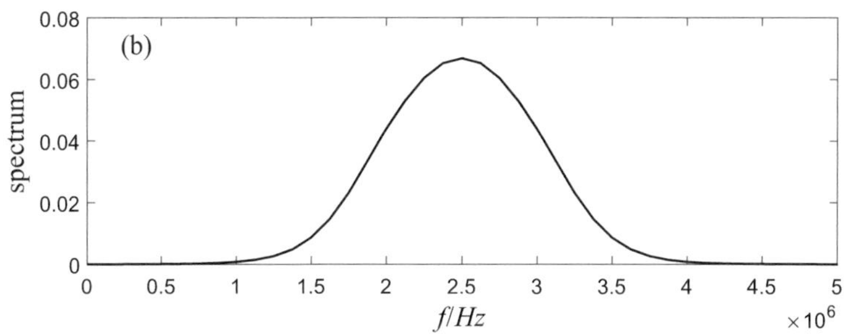

Fig. 1.12 **a** The waveform of the echo signal, **b** the spectrum of the echo signal. The simulation parameters are $B = 2.5 \times 10^6$, $f_c = 2.5 \times 10^6$, $T = 8 \times 10^{-6}$, $c = 1 \times 10^{-6}$, $d = 3 \times 10^{-6}$ and $\alpha = 0.45$

1.5 Aperiodic Pulse Signal

Signals with aperiodic pulse are very common in the engineering fields. They are commonly used in various fields such as magnetic anomaly identification [21–23], weak chromatographic analysis [24, 25], electrochemical battery detection [26, 27], non-destructive testing of the steel wire rope [28, 29] and other fields.

Corresponding to different application backgrounds, there are many different types of aperiodic pulse signals. Herein, we only consider four of the most commonly used types. Their mathematical expressions are described as follows [29, 30]

$$\begin{cases} s_1(t) = A \cdot e^{-2\pi\left(\frac{t-t_0}{W}\right)^2}, \\ s_2(t) = 4\pi A \dfrac{t-t_0}{W^2} \cdot e^{-2\pi\left(\frac{t-t_0}{W}\right)^2}, \\ s_3(t) = -4\pi A \cdot e^{-2\pi \frac{4\pi(t-t_0)^4 - W^2(t-t_0)^2}{W^6}}, \\ s_4(t) = 16\pi^2 A\,(t-t_0) \cdot e^{-2\pi \frac{4\pi(t-t_0)^4 - 3W^2(t-t_0)^2}{W^8}}, \end{cases} \tag{1.15}$$

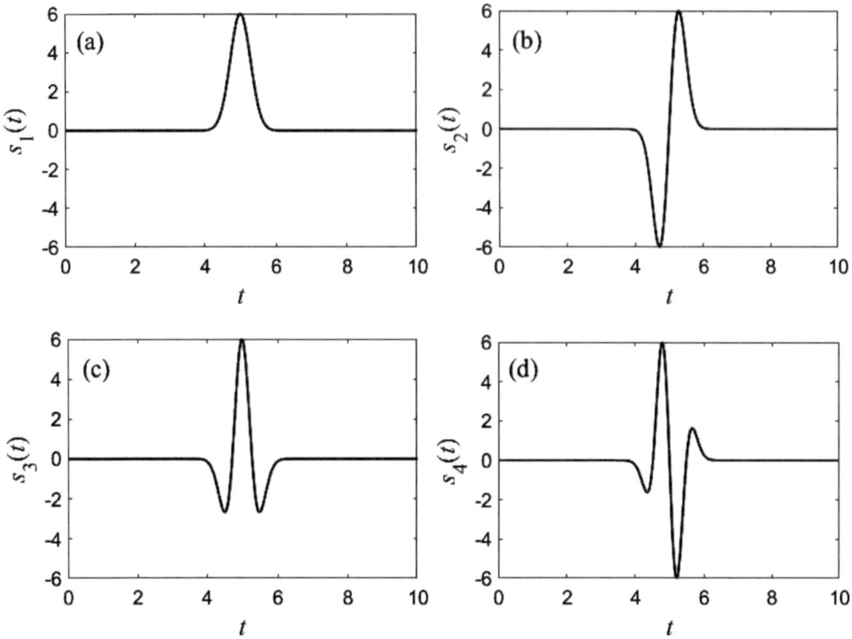

Fig. 1.13 The waveforms from **a–d** corresponding to the four aperiodic pulse signals in Eq. (1.15) in turn. The simulation parameters are $A = 5$, $W = 1$, $t_0 = 5$

where A and W are the peak height and the half-peak width of the pulse signal. The parameters t_0 influences the starting time of the pulse. Figure 1.13 shows the waveforms of the four different aperiodic pulse signals in Eq. (1.15). Apparently, in Fig. 1.13a, c which corresponding to $s_1(t)$ and $s_3(t)$ in Eq. (1.15), the peak of the pulse locates at t_0. Whereas in Fig. 1.13b, d which corresponding to $s_2(t)$ and $s_4(t)$ in Eq. (1.15), pulses flips in two opposite directions, i.e., $s = 0$ at the point t_0.

It is worth noting that in engineering, aperiodic pulse signals may not necessarily be functions of time; they can also be functions of position. For example, in steel wire rope flaw detection, the detector moves along the rope and generates aperiodic pulses at damaged locations, making the flaw detection signal a function of position. Whether the aperiodic pulse signal is a function of time or position does not affect the analysis of its response to a nonlinear system as an excitation. In the second part of this book, we will introduce the aperiodic pulse signal as a non-destructive testing signal for steel wire rope, which is a function of position, and present the related experimental investigations.

Referring to Eq. (1.15), we can also obtain the signal with multiple aperiodic pulses, for example,

$$\begin{cases} s_5(t) = \sum_{i=1}^{4} A(i) \cdot e^{-2\pi\left(\frac{t-t_0(i)}{W}\right)^2}, \\ s_6(t) = \sum_{i=1}^{4} 4\pi A(i) \frac{t-t_0(i)}{W^2} \cdot e^{-2\pi\left(\frac{t-t_0(i)}{W}\right)^2}, \end{cases} \tag{1.16}$$

where i is the number of the pulse. In Fig. 1.14, signal with multiple aperiodic pulses, as described by Eq. (1.16), is presented in sequence. Herein, we only consider the case when $i = 4$, i.e., there are four pulses in each curve. Certainly, we can extend the cases of the aperiodic signals with much more multiple pulses.

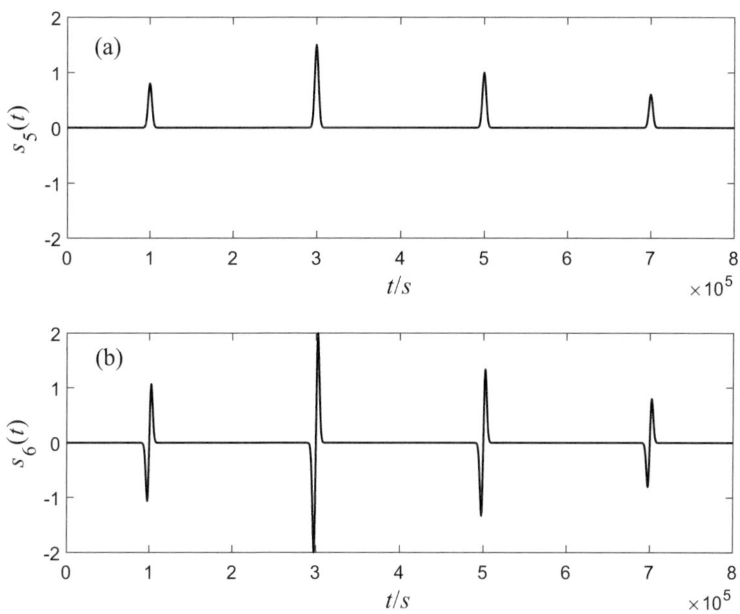

Fig. 1.14 The waveforms from **a–b** correspond, in turn, to the two aperiodic pulse signals described in Eq. (1.16). The simulation parameters are $W = 8000$, $A(i) = 0.8, 1.5, 1, 0.6$, $t_0(i) = 1 \times 10^5$, 3×10^5, 5×10^5, 7×10^5, when i from 1 to 4

1.6 Some Indices for Measuring Aperiodic Signals

In this subsection, we introduce the four most commonly used indices for measuring aperiodic signals in studies.

1.6.1 Cross-Correlation Coefficient

The cross-correlation coefficient is commonly used in measuring the response of the system to an aperiodic binary signal. For two signals $s(t)$ and $x(t)$, we first discretize them. Then, the cross-correlation coefficient between them is calculated by

$$C_{sx} = \frac{\sum_{j=1}^{n}[s(j) - \bar{s}][x(j) - \bar{x}]}{\sqrt{\sum_{j=1}^{n}[s(j) - \bar{s}]^2 \sum_{j=1}^{n}[x(j) - \bar{x}]^2}}. \tag{1.17}$$

Herein, n is the number of the total points of the discrete time series. The symbols $\bar{s}$ and $\bar{x}$ are the average of the two time series.

The cross-correlation coefficient can express the similarity between two time series. When the cross-correlation coefficient is 1, the waveforms formed by the two time series are completely similar, and they are parallel in the graph. When the correlation coefficient is -1, the waveform changes formed by two time series are completely opposite. The closer the correlation coefficient is to 1, the higher the similarity between the waveforms formed by two time series. In the following chapters, we can see that the cross-correlation coefficient close to 1 is a necessary but insufficient condition for the weak aperiodic signal to be amplified after passing through the nonlinear system.

The expression in Eq. (1.17) is the basic formula for calculating the cross-correlation coefficient. In some occasions, we need to improve the formula. For example, when measuring the accuracy of the signal improvement degree by the nonlinear resonance, the modification of the cross-correlation coefficient is proposed to eliminate the negative influence induced by the phase lag between the output and the input [31]. Another example, when studying the aperiodic pulse signals in non-destructive testing signals of steel wire rope, to achieve a better analysis result, we introduced the local cross-correlation coefficient which indicates the feature similarity between the input characteristic signal and the output. Specifically, the local cross-correlation coefficient is calculated as [29],

$$C_{sx}^{w} = \frac{\sum_{j=m}^{n}[s(j) - \bar{s}][x(j) - \bar{x}]}{\sqrt{\sum_{j=m}^{n}[s(j) - \bar{s}]^2 \sum_{j=m}^{n}[x(j) - \bar{x}]^2}}. \tag{1.18}$$

In Eq. (1.18), w denotes the peak width of the pulse, m labels the starting point of the considered pulse, and n is the end point of the pulse. The local cross-correlation coefficient will be used in **Chap.** 13. The local cross-correlation coefficient in Eq. (1.18) is used to calculate on the local feature and can better represent the local characteristic information of the signal than the cross-correlation coefficient in Eq. (1.17). Moreover, to measure the amplification of the signal by the response of a nonlinear system, we also give some other indices such as response amplitude gain in the following chapters.

1.6.2 Spectral Amplification Factor

The spectral amplification factor was once widely used to quantify the response of a harmonic signal passing a nonlinear system [32]. The spectral line of a periodic signal is a discrete line (for harmonic signal) or a series of lines (for anharmonic periodic signal), so the height of the spectral line at a certain frequency can be used to measure the change in energy of the signal at that frequency before and after processing. For the frequency-modulated signal, however, its spectrum is continuous in a frequency interval, as shown in **Sect.** 1.2. It leads to the traditional spectral amplification factor cannot be used directly for the frequency modulated signal. Motivated by the definition of the traditional spectral amplification factor, we proposed the spectral amplification factor for the frequency modulated signal in our former work. It is calculated by

$$\eta = \frac{\frac{1}{f_{end}-f_{start}} \int_{f_{start}}^{f_{end}} S_x(f_{in})d f_{in}}{\frac{1}{f_{end}-f_{start}} \int_{f_{start}}^{f_{end}} S_s(f_{in})d f_{in}} = \frac{\sum_{i=1}^{N} S_x(i)}{\sum_{i=1}^{N} S_s(i)}. \tag{1.19}$$

Here, $S_x(\bullet)$ and $S_s(\bullet)$ represent the discrete spectrum of the two signals being compared. The subscript i denotes the ith point in the discrete spectrum. The other symbols retain the same meanings as mentioned earlier. The spectral amplification factor indicates the average amplification degree of the frequency-modulated signal. Since the instantaneous frequency of the frequency-modulated signal may change significantly with time, we may need to calculate the spectral amplification factor by dividing the complete signal to series of sections.

Considering that the ridge of the instantaneous frequency on the spectrum have a certain width, the spectral amplification factor can be written in the following form [33]

$$\eta=\frac{1}{N}\sum_{i=1}^{N}\left[\frac{\frac{1}{(2m+1)}\sum_{j=1}^{m}\left[S_x(f(t_i)-j)+S_x(f(t_i)+j)+S_x(f(t_i))\right]}{S_s(f(t_i))}\right]. \tag{1.20}$$

In Eq. (1.20), η is the amplification degree of a specific frequency band around the ridge of the instantaneous frequency, m refers the corresponding points on both sides of the frequency curve $f(t)$. For the above two formulas of η, there is the same conclusion, i.e., the signal is amplified when $\eta > 1$.

1.6.3 Bit Error Rate

The bit error rate (BER) is commonly used to quantify the accuracy of the output in relation to the input. BER is calculated by the percentage of the wrong received bits with respect to the total number of transmitted bits. In other words, it is the number of received error bits divided by the total number of the transferred bits. Generally speaking, BER is usually a percentage number. In fact, it is important to define the correct or wrong received bits. BER has been successfully used in the works of stochastic resonance [34–37], and vibrational resonance that induced by aperiodic binary signal [38]. The optimal output-input is achieved when BER reaches the minimum in studying aperiodic resonances. We will give concrete examples in the following chapters.

1.6.4 Logical Success Probability

The logical success probability should be calculated when the logical computation is carried out [39, 40]. Here, the success probability P is defined as

$$P = \frac{\text{The correct turns}}{\text{The total turns}}. \tag{1.21}$$

When the logical response of a nonlinear system is studied, the detailed process for calculating the success probability P is described as follows.

(1) Randomize arrangement of four possible logical states (0, 0), (0, 1), (1, 1), (1, 0) and input it to the nonlinear system.
(2) Exciting the system by each logical state with the same time. The total turns, i.e., all trajectories for the calculation, are set to a large number in simulation.
(3) Sample the response of the nonlinear system within the considered time interval.
(4) Compare the response of the system with the logical relationship truth table to get the success probability of the logical relationship.

Table 1.1 The logical operation truth table

Input set (I_1,I_2)	Value of ($I_1 + I_2$)	**OR**	**AND**	**NOR**	**NAND**	**XOR**	**NXOR**
(0,0)	−1	0	0	1	1	0	1
(0,1)/(1,0)	0	1	0	0	1	1	0
(1,1)	1	1	1	0	0	0	1

In addition, if we give two logical signals I_1 and I_2, for the logical operations **OR**, **AND**, **NOR**, **NAND**, **XOR**, **NXOR**, the logical relationship truth values are shown in Tab. 1.1.

1.7 Concluding Remarks

Besides the aperiodic signals introduced in this section, there are also a large number of some other kinds of aperiodic signals. They appear in different occasions and stand for explicit physical and engineering significance. If we input an aperiodic signal into the nonlinear system as an excitation, it will cause complex dynamic response. The following chapters mainly focus on the research of resonance behavior of several typical nonlinear systems excited by aperiodic signals, and extends from theory to certain applications. As regards to the aperiodic signals corresponding to some engineering occasions, we will give the concrete descriptions in Part 2 of this book. It mainly contains the vibration signal of roll bearing with fault under the time-varying speed condition, magnetic flux leakage detection signal of steel wire rope. The characteristic information will be extracted from the aperiodic signals by aperiodic resonances method.

In addition to the indices introduced in **Sect.** 1.7, there are many other indices that can be used to measure aperiodic signals. For instance, the response amplitude gain, introduced in **Chap.** 2, Eq. (2.2) and Eq. (2.3), quantifies the amplification of a binary signal in the system response. The fractional signal-to-noise ratio, introduced in **Chap.** 5, Eq. (5.6), is used to quantify the response induced by a LFM signal. We should adopt more targeted indices based on specific applications. For example, when we investigate the image processing, the objective assessments of image quality should be chosen [41–43]. The selection of research indices in other fields is also the same.

References

1. V.N. Chizhevsky, E. Smeu, G. Giacomelli, Experimental evidence of vibrational resonances in an optical system. *Phys. Rev. Lett.*, 2003, **91**(22): 220602.
2. V.N. Chizhevsky, G. Giacomelli, Vibrational resonance and the detection of aperiodic binary signals, *Phys. Rev. E*, 2008, **77**(5): 051126.
3. J. Liu, Z. Li, Binary image enhancement based on aperiodic stochastic resonance, *IET Image Processing*, 2015, **9**(12): 1033-1038.
4. J.H. Yang, Q. Ma, C.J. Wu, H.G. Liu, Aperiodic vibrational resonance in the fractional-order bistable system, 2018, *Acta Physica Sinica*, **67**(5): 054501.
5. P.X. Jia, C.J. Wu, J.H. Yang, M.A.F. Sanjuán, G.X. Liu, Improving the weak aperiodic signal by three kinds of vibrational resonance, *Nonlinear Dynamics*, 2018, **9**: 2699-2713.
6. L. Zeng, J. Li, J. Shi, M-ary signal detection via a bistable system in the presence of Lévy noise, *Chaos, Solitons & Fractals*, 2012, **45**(4):378-382.
7. P. Jia, J. Yang, H. Liu, E. Hu, Improving amplitude-modulated signals by re-scaled and twice sampling vibrational resonance methods, *Pramana*, 2018, **91**:38.
8. S.Y. Nusenu, H. Chen, W.Q. Wang, OFDM chirp radar for adaptive target detection in low grazing angle, *IET Signal Processing*, 2018, **12**(5): 613-619.
9. B. Liu, W. Chang, A novel range-spread target detection approach for frequency stepped chirp radar, *Progress in Electromagnetics Research*, 2012, **131**: 275-292.
10. M. Palmese, G. Bertolotto, A. Pescetto, A. Trucco, Experimental validation of a chirp-based underwater acoustic communication method, *J. Acoust. Soc. Am.*, 2008, **123**(5): 3005.
11. P. He, Y. Lv, H. Zhang, Y. Wang, Study of ultra-wideband wireless communication system based on chirp-BOK modulation, *J. Nanjing Univ. Nat. Sci.*, 2006, **26**(2):21-25.
12. Z. Jiang, X. C. Zhang, Electro-optic measurement of THz field pulses with a chirped optical beam, *Appl. Phys. Lett.*, 1998, **72**(16):945-1947.
13. Y. Zhang, C. Zhang, C. Li, C. Li, Z. Zhu, F. Wang, Fractional Fourier transform of ultrasonic chirp signal for range measurement, 2015, *2015 54th Annual Conference of the Society of Instrument and Control Engineers of Japan (SICE)*, IEEE, 140-144.
14. P. Jia, J. Yang, Y. Leng, The echo chirp signal amplification by the vibrational information fusion method, *Int. J. Mod. Phys. B*, 2020, **34**(06): 2050041.
15. Z. Feng, X. Chen, T. Wang, Time-varying demodulation analysis for rolling bearing fault diagnosis under variable speed conditions, *J. Sound Vib.*, 2017, **400**: 71-85.
16. D. Iatsenko, P.V.E. McClintock, A. Stefanovska, Extraction of instantaneous frequencies from ridges in timeCfrequency representations of signals, *Signal Processing*, 2016, **125**: 290-303.
17. M.A. Herman, T. Strohmer, High-resolution radar via compressed sensing, *IEEE Trans. Signal Process.*, 2009, **57**(6): 2275-2284.
18. P. Flandrin, P. Borgnat, Time-frequency energy distributions meet compressed sensing, *IEEE Trans. Signal Process.*, 2010, **58**(6):2974-2982.
19. I. Kuga, N. Tanaka, Alias-free estimation of blood velocity using phase difference and amplitude correlation of chirp echo spectrum, *Acoust. Sci. Technol.*, 2006, **27**(3): 183-186.
20. D.W. Ricker, *Echo Signal Processing*, 2003, Boston: Kluwer Academic Publishers.
21. C. Wan, M. Pan, Q. Zhang, H. Li, M. Pan, Q. Zhang, D. Yang, Magnetic anomaly detection based on stochastic resonance, *Sensors and Actuators A: Physical*, 2018, **278**: 11-17.
22. C. Wan, H. Pang, S. Mou, et al, Magnetic anomaly detection using a parallel stochastic resonance system, *IEEE Trans. Instrum. Meas.*, 2022, **71**: 1-8.
23. T. Qin, L. Zhou, S. Chen, Z. Chen, The novel method of magnetic anomaly recognition based on the fourth order aperiodic stochastic resonance, *IEEE Sensors Journal*, 2022, **22**(17): 17043-17053.
24. W. Zhang, J. Guo, B. Xiang, H. Fan, F. Xu, Improving the detection sensitivity of chromatography by stochastic resonance, *Analyst*, 2014, **139**(9): 2099-2107.
25. R. Li, B. Xiang, H. Deng, S. Xie, The stochastic resonance algorithm with the direct current signal as external force and its application to the detection of weak chromatographic peaks, *J. Sep. Sci.*, 2017, **40**(20): 4091-4096.

26. I. Tiwari, R. Phogat, P. Parmananda, J.L. Ocampo-Espindola, M. Rivera, Intrinsic periodic and aperiodic stochastic resonance in an electrochemical cell, *Phys. Rev. E*, 2016, **94**(2): 022210.
27. P. Parmananda, G.J.E. Santos, M. Rivera, K. Showalter, Stochastic resonance of electrochemical aperiodic spike trains, *Phys. Rev. E*, 2005, **71**(3): 031110.
28. Z. Shan, J. Yang, M.A.F. Sanjuán, C. Wu, H. Liu, A novel adaptive moving average method for signal denoising in strong noise background, *Eur. Phys. J. Plus*, 2022, **137**(1): 50.
29. Z. Xu, Z. Wang, J. Yang, M.A.F. Sanjuán, B Sun, S. Huang, Aperiodic stochastic resonance in a biased monostable system excited by different weak aperiodic pulse signals and strong noise, *Eur. Phys. J. Plus*, 2023, **138**(5): 1-12.
30. B.F. Cao, L. Peng, X.Q. Li, X.Q. Zhang, W.S. Ning, R. Liang, X. Li, M. Hu, Y. Zheng, Detection and parameter estimation of weak pulse signal based on strongly coupled Duffing oscillators, *Acta Physica Sinica*, 2019, **68**(8): 080501.
31. Y. Zhai, J. Yang, S. Zhang, H. Liu, Linear frequency modulated signal induced aperiodic resonance, *Physica Scripta*, 2020, **95**(6): 065213.
32. L. Gammaitoni, P. Hänggi, P. Jung, et al, Stochastic resonance, *Rev. Mod. Phys.*, 1998, **70**(1): 223-287.
33. T. Gong, J. Yang, M.A.F. Sanjuán, H. Liu, Z. Shan, Vibrational resonance by using a real-time scale transformation method, *Physica Scripta*, 2022, **97**(4): 045207.
34. B. Xu, F. Duan, F. Chapeau-Blondeau, Comparison of aperiodic stochastic resonance in a bistable system realized by adding noise and by tuning system parameters, *Phys. Rev. E*, 2004, **69**(6): 061110.
35. S. Sun, B. Lei, On an aperiodic stochastic resonance signal processor and its application in digital watermarking, *Signal Processing*, 2008, **88**(8): 2085-2094.
36. J. Li, L. Zeng, Stochastic resonance in a bistable system subject to multi-time-delayed feedback and aperiodic signal, *J. Phys. A Math. Gen.*, 2010, **43**(49): 495002.
37. L. Zeng, B. Xu, Effects of asymmetric Lévy noise in parameter-induced aperiodic stochastic resonance, *Physica A*, 2010, **389**(22): 5128-5136.
38. V.N. Chizhevsky, G. Giacomelli, Vibrational resonance and the detection of aperiodic binary signals, *Phys. Rev. E*, 2008, **77**(5): 051126.
39. S. Huang, J. Zhang, J. Yang, H. Liu, M.A.F. Sanjuán, Logical vibrational resonance in a symmetric bistable system: Numerical and experimental studies, *Commun. Nonlinear Sci. Numer. Simul.*, 2023, **119**: 107123.
40. M. Hou, J. Yang, S. Shi, H. Liu, Logical stochastic resonance in a nonlinear fractional-order system, *Eur. Phys. J. Plus*, 2020, **135**(9): 747.
41. J.C. Yao, G.Z. Liu, Objective assessment method of image quality based on visual perception of image content, *Acta Physica Sinica*, 2018, **67**(10): 108702.
42. J.C. Yao, J. Shen, Objective assessment of image quality based on image content contrast perception, *Acta Physica Sinica*, 2020, **69**(14): 148702.
43. C.Y., Shi, Y.D. Lin, Objective image quality assessment based on image color appearance and gradient features, *Acta Physica Sinica*, 2020, **69**(22): 228701.

Chapter 2
Aperiodic System Resonance Caused by Single Aperiodic Binary/M-Ary Signal

This chapter presents aperiodic resonance caused by a single excitation. In a nonlinear system, we use the aperiodic binary signal and the aperiodic M-ary signal as excitations, respectively. When the time scale of the signal matches well with the system parameters, aperiodic system resonance can be achieved by adjusting the system parameters. If they do not match well, it is necessary to achieve matching through scale transformation. This chapter introduces the theoretical framework of the general re-scaling method to achieve aperiodic system resonance under aperiodic excitation.

2.1 Aperiodic Binary Signal as the Characteristic Signal

It is well known that a harmonic signal can induce resonance in a linear system when the excitation frequency approaches the system's natural frequency. For an anharmonic periodic signal, resonance is more likely to occur, as each of the series of fundamental and harmonic frequency components in the signal can cause resonance when they are close to the natural frequency. In a nonlinear system, besides primary resonance, subharmonic and superharmonic resonances can also occur. These types of periodic resonance have been extensively studied. However, can an aperiodic excitation induce aperiodic resonance in a nonlinear system? When aperiodic resonance occurs, the output signal waveform exhibits a certain similarity to the input signal waveform, with an amplified amplitude. Moreover, if the output waveform is entirely opposite to the input waveform and the amplitude is amplified, it can be considered that the system has undergone reverse aperiodic resonance. In fact, there are many related studies on aperiodic resonance, such as *aperiodic stochastic resonance* [1–3] and *aperiodic vibrational resonance* [4–6]. To distinguish different types of aperiodic resonance in subsequent chapters, we refer to the phenomenon of aperiodic resonance caused by a single aperiodic excitation as *aperiodic system*

J. Yang and M. A. F. Sanjuan, *Aperiodic Resonances—Theory and Applications*,
https://doi.org/10.1007/978-981-96-6819-9_2

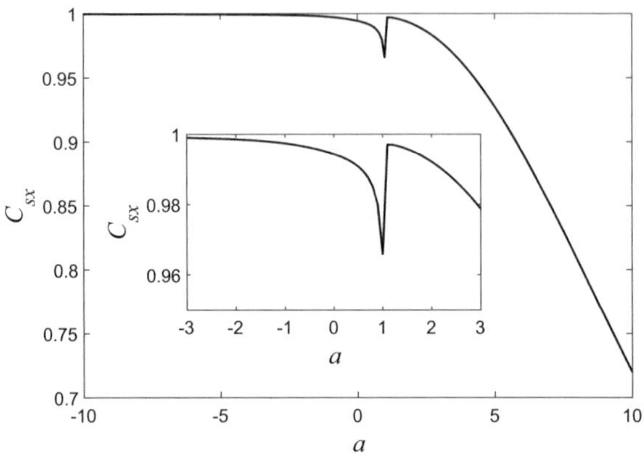

Fig. 2.1 The cross-correlation coefficient versus the system parameter a, here $b = 2$. For the aperiodic binary signal, $A = 0.3$, $T = 200$

resonance, since it is primarily induced by changes in system parameters. There is little literature on the topic of aperiodic system resonance.

We previously investigated aperiodic system resonance in a simple overdamped bistable system excited by an aperiodic binary signal [7]. Here, we continue to use the same nonlinear system

$$\frac{dx}{dt} = ax - bx^3 + s(t), \tag{2.1}$$

where a and b are the system parameters, and $s(t)$ is an aperiodic binary signal that has been introduced in Chap. 1 in Eq. (1.1). The potential function of Eq. (2.1) is $V(x) = -\frac{a}{2}x^2 + \frac{b}{4}x^4$. When $a > 0$ and $b > 0$, the potential function presents bistable configuration, see Fig. A.1 of the Appendix A.1.1.[1] Whereas when $a < 0$ and $b > 0$, the potential function presents a monostable configuration, see Fig. A.5 of the Appendix A.2.1. When the system parameter changes from a positive value to a negative one, the system undergoes pitchfork bifurcation [8, 9].

In Fig. 2.1, based on the cross-correlation coefficient, we study the aperiodic system resonance induced by the system parameters a. When $a = 1$, the cross-correlation coefficient reaches the first valley value. When $a = 1.1$, the cross-correlation coefficient reaches the first peak value. In each interval of a, the trend of change is very obvious. The aperiodic binary signal used here is shown in Fig. 2.2.

The curve of the cross-correlation coefficient only provides a reference for determining aperiodic resonance. While a maximum cross-correlation coefficient is a necessary condition for aperiodic resonance, it is not sufficient. In other words, aperiodic

[1] For the convenience of reference and comparison, the main potential function figures involved in this monograph are included in Appendix 1.

resonance may not occur even when the cross-correlation coefficient reaches its max imum, but when aperiodic resonance does occur, the cross-correlation coefficient should be relatively large.

Corresponding to Fig. 2.1, to directly exhibit the aperiodic resonance, we present the time series of the output compared with the input binary signal in Fig. 2.2. From Fig. 2.2a–d, the amplitude of the output gradually increases. Then, from Fig. 2.2d–f, the amplitude of the output gradually decreases. Additionally, the time series corresponding to the valley of the curve in Fig. 2.1 are shown in Fig. 2.2d, where the signal is significantly improved. Furthermore, the time series corresponding to the peak of the curve in Fig. 2.1 are given in Fig. 2.2e, where the signal is weakened despite the cross-correlation coefficient being larger than that in Fig. 2.2d. This once again indicates that the cross-correlation coefficient cannot determine whether the

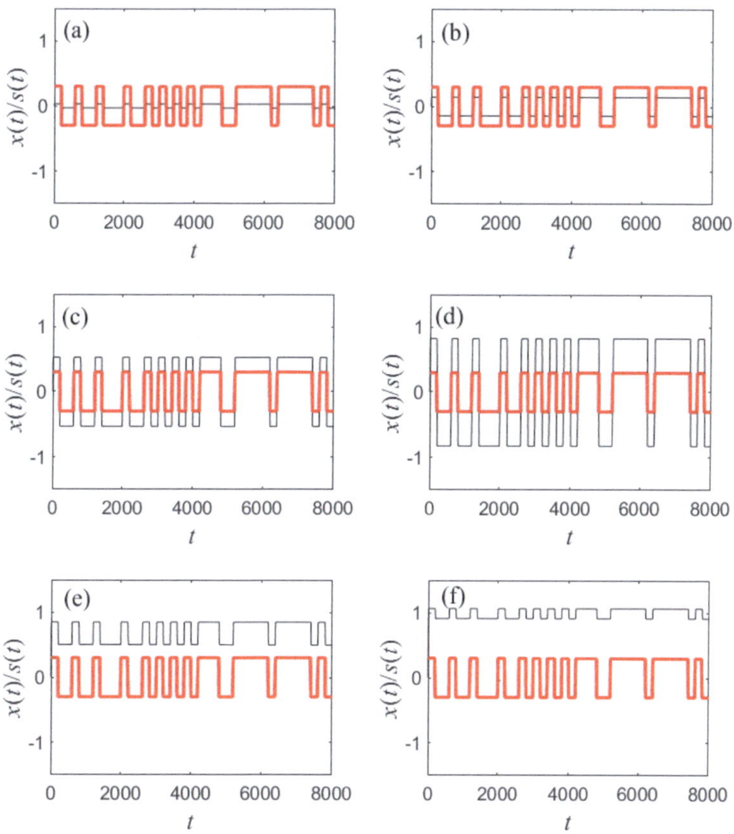

Fig. 2.2 The input binary signal (thin line) and the corresponding output (thick line) of the system, here $b = 2$. **a** $a = -10$, **b** $a = -2$, **c** $a = 0$, **d** $a = 1$, **e** $a = 1.1$, **f** $a = 2$. For the aperiodic binary signal, $A = 0.3$, $T = 200$

signal is amplified. It can only measure the similarity between two time series. Figure 2.2 also illustrates that changes in the cross-correlation coefficient in Fig. 2.1 are related not only to the amplitude of the output but also to signal drift.

To measure the amplification of an aperiodic signal through a nonlinear system, a new index called the response amplitude gain was proposed in [7]. It is approximately calculated by

$$\beta = \frac{1}{2}\frac{\max x(t) - \min x(t)}{A}. \tag{2.2}$$

When calculating the response amplitude gain using Eq. (2.2), the impact of initial value $x(0)$ on the system must be considered, and the transient response caused by the initial values should be removed. Naturally, this formula has a limited scope of application, and it is no longer applicable to the time series shown in Fig. 2.11.

Using the same simulation parameters as in Fig. 2.1, the response amplitude gain versus the system parameter a is shown in Fig. 2.3. The amplification of the binary signal by aperiodic resonance is clearly illustrated in the curve. Specifically, when $a = 1$, $\beta = 2.75$, indicating that the signal is amplified 2.75 times by the nonlinear system when aperiodic system resonance occurs. For all cases where $\beta > 1$, the signal is amplified.

Besides the cross-correlation coefficient and the response amplitude gain, the bit error rate (BER) can also be used to measure aperiodic resonance. The definition of a correct or incorrect received bit of the system output influences the value of BER. As is well known, the response amplitude does not usually equal the level of the input signal. Here, we consider the output to be correct if the absolute value of the discrete response point is larger than a given threshold value A_{th}, and incorrect otherwise. The choice of A_{th} is therefore crucial, as different values indicate different physical meanings. We use two typical values for A_{th}: the first is $A_{th} = 0$, and the second is $A_{th} = A$. When $A_{th} = 0$, it means that the output and the input have the same signs, ensuring similarity between the output series and the input binary aperiodic

Fig. 2.3 Curve of the response amplitude gain β versus the system parameter a, here $b = 2$. For the aperiodic binary signal, $A = 0.3$, $T = 200$

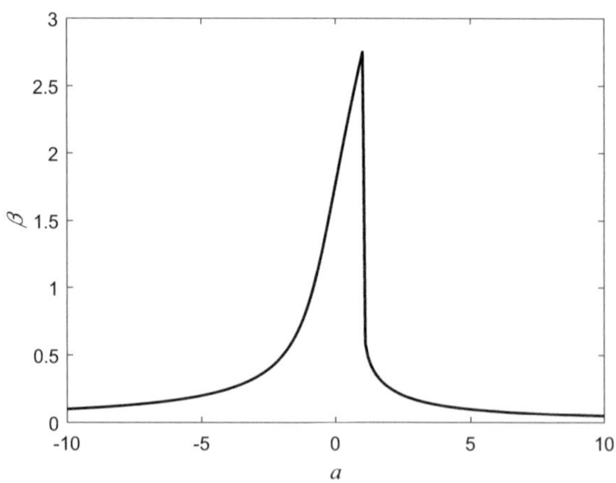

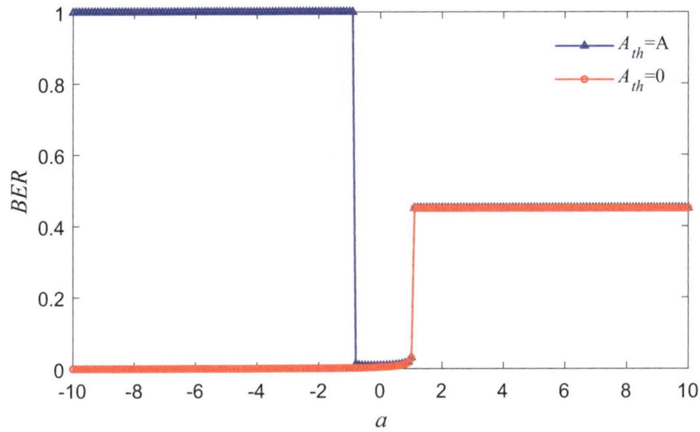

Fig. 2.4 Curve of the BER versus the system parameter a, here $b = 2$. For the aperiodic binary signal, $A = 0.3$, $T = 200$

signal. When $A_{th} = A$, it indicates that the output is amplified compared to the input, ensuring both signal similarity and amplification. Based on this approach, we plot BER versus the system parameter a, choosing a step size of 0.1 for a. From the two curves, we observe that when $a \leq -0.9$, $BER = 0$ for both $A_{th} = A$ and $A_{th} = 0$, indicating that the output and input are completely similar, although the output is not amplified compared to the input. In the range $a \in [-0.8, 1.0]$, the values of BER for both $A_{th} = A$ and $A_{th} = 0$ are consistently low, indicating that the output and input are completely similar, although the output is not amplified compared to the input. When $a \geq 1.1$, the value of BER is 0.45 for both $A_{th} = A$, and $A_{th} = 0$, and the two curves coincide completely. This behavior is primarily due to the signal drifting into a potential well and oscillating within it, causing the output and input symbols to no longer match exactly, resulting in a relatively high BER. Nevertheless, it is promising that, regardless of the index used, whether the cross-correlation coefficient, response amplitude gain, or bit error rate, when $a = 1$, both similarity and amplification are maintained between the output and the input signal.

To demonstrate the influence of pulse width T on aperiodic system resonance, Fig. 2.5, illustrates the dependence of three indices, the cross-correlation coefficient, the response amplitude gain, and the bit error rate, on the value of T. It is evident that when T is small, the system response does not achieve a strong aperiodic resonance state. To address this issue, it is necessary to introduce the re-scaling theory.

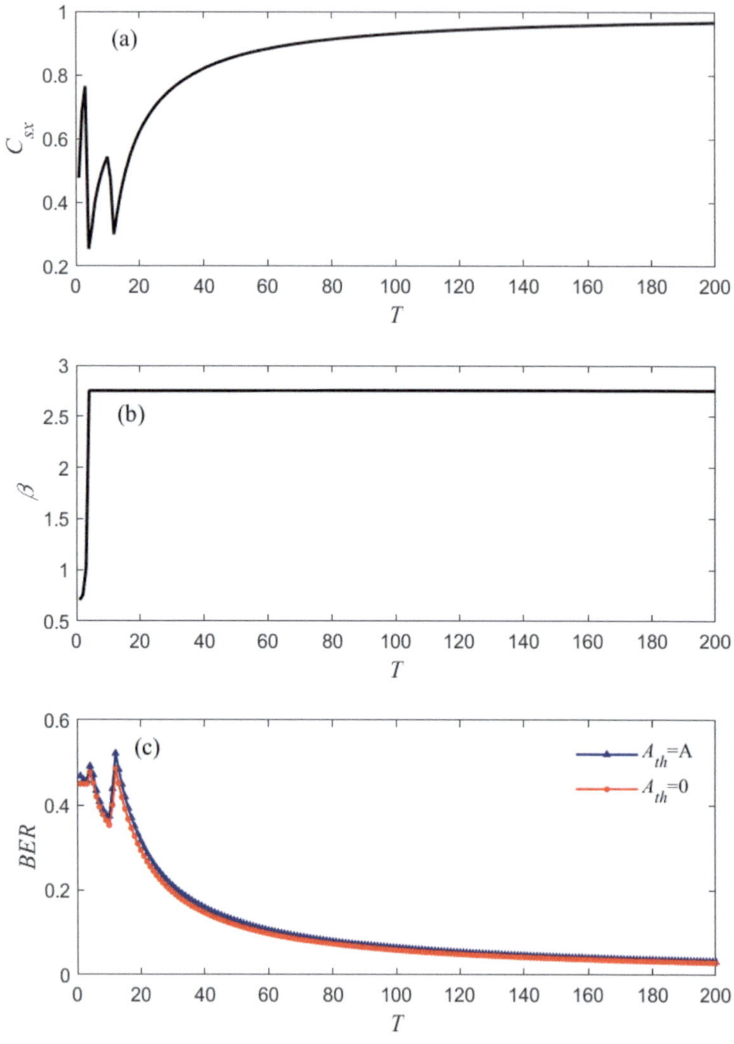

Fig. 2.5 Dependence of the aperiodic system resonance indices on the pulse width T. The simulation parameters are $a = 1, b = 2, A = 0.3$. **a** The dependence of the cross-correlation coefficient C_{sx} on the minimum pulse width T. **b** The dependence of the response amplitude gain β on the minimum pulse width T. **c** The dependence of the bit error rate BER on the minimum pulse width T

2.2 Aperiodic M-Ary Signal as the Characteristic Signal

If we choose a complex signal such as M-ary signal in **Fig.** 1.2, the top one, to replace the binary signal in Eq. (2.1), the cross-correlation coefficient versus the system parameter a is shown in Fig. 2.6. The curve reaches the valley value and the peak value at $a = 1.6$ and 2.4, respectively. In the label, X gives the value of the horizontal axis coordinate, and Y gives the value of the vertical axis coordinate of the labelled point. The labels in the subsequent figures also have the same meanings as these.

To clearly demonstrate the amplification of the M-ary signal, the time series of the output is presented in Fig. 2.7. Due to the multi-level nature of the M-ary signal, achieving perfect aperiodic system resonance is challenging, even when the cross-correlation coefficient is large. In other words, it is difficult to simultaneously achieve waveform similarity and amplification. The output waveform may not maintain the same proportionality at each stage as the original input signal, and it is even challenging to ensure that the output and input signals remain synchronized within the same potential well.

Due to the multi-level of the M-ary signal, the response amplitude gain can no longer be calculated using Eq. (2.2). We amend it to [7]

$$\beta = \frac{\max x(t) - \min x(t)}{\max s(t) - \min s(t)}. \tag{2.3}$$

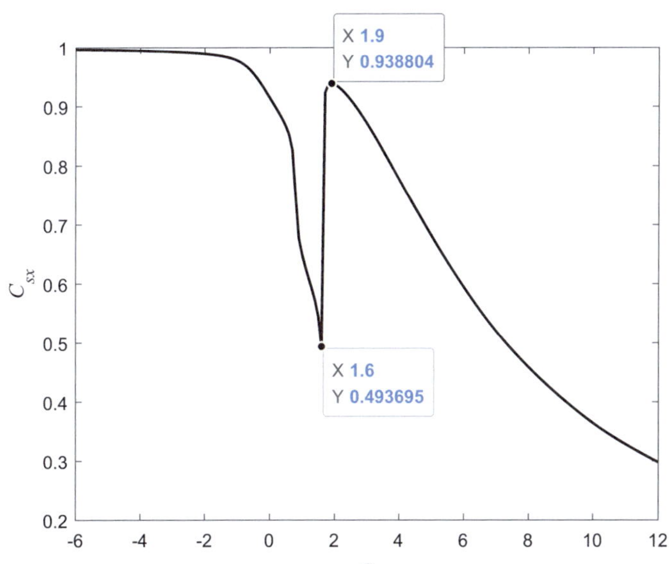

Fig. 2.6 The cross-correlation coefficient versus the system parameter a, here $b = 2$. For the aperiodic M-ary signal, $M = 4$, $A = 0.2$ and $T = 20$

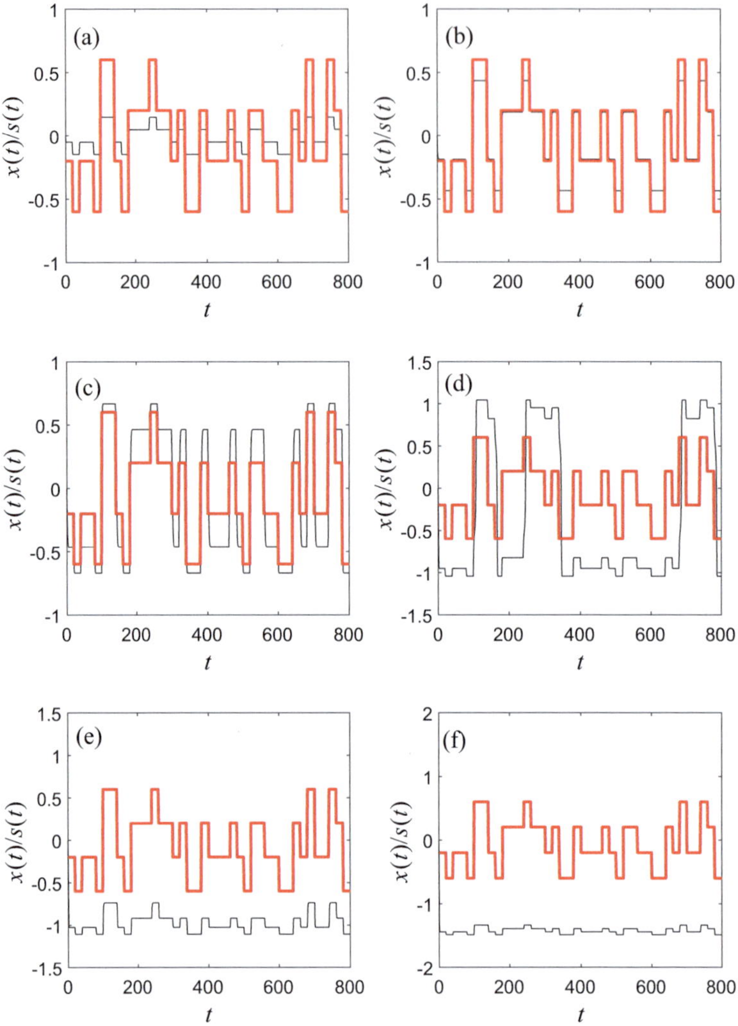

Fig. 2.7 The input M-ary signal (thin line) and the corresponding output (thick line) of the system, here $b = 2$. **a** $a = -4$, **b** $a = -1$, **c** $a = 0$, **d** $a = 1.6$, **e** $a = 1.9$, **f** $a = 4$. For the aperiodic M-ary signal, $M = 4$, $A = 0.2$ and $T = 20$

The response amplitude gain for the same M-ary signal is shown in Fig. 2.8. The response amplitude gain also reaches its peak value at $a = 1.6$, which corresponds to the value of a at the valley in Fig. 2.6.

Using the same process as calculating BER in Fig. 2.4, we present the BER curves in Fig. 2.9. The threshold $A_{th} = 0$ ensures the similarity between the input and output waveforms, while the threshold $A_{th} = A$ ensures the amplification of the

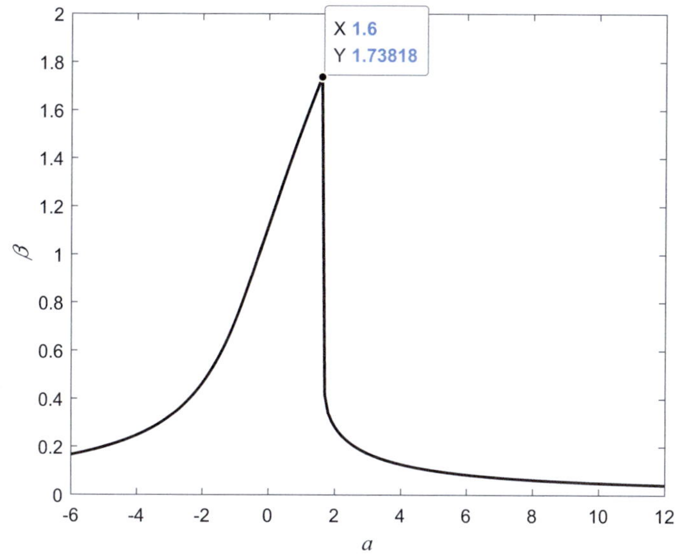

Fig. 2.8 The response amplitude gain versus the system parameter a, here $b = 2$. For the aperiodic M-ary signal, $M = 4$, $A = 0.2$ and $T = 20$

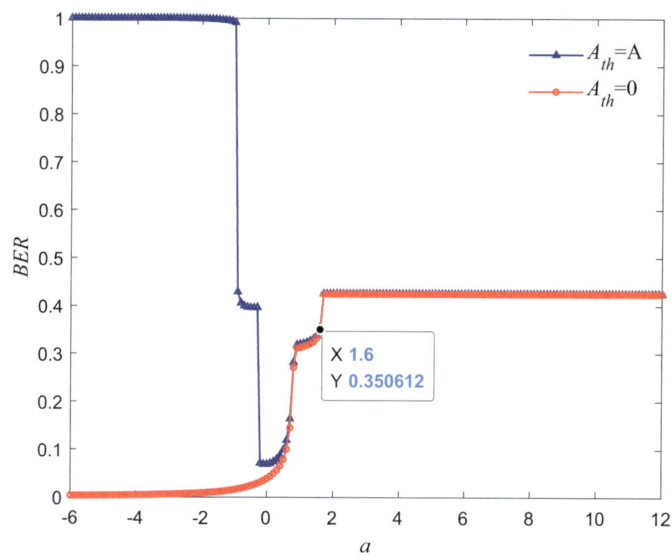

Fig. 2.9 BER versus the system parameter a, here $b = 2$. For the aperiodic M-ary signal, $M = 4$, $A = 0.2$ and $T = 20$

input binary signal. Only in the interval $a \in [-0.2, 0.8]$ do both BER values corresponding to $A_{th} = 0$ and $A_{th} = A$ remain very small, indicating a strong aperiodic system resonance. This interval can be considered the resonance zone.

It must be stated that neither the cross-correlation coefficient nor the response amplitude gain can effectively amplify the M-ary signal. This is due to the complex nature of the changes in the M-ary signal, and both indices are merely approximate evaluations in an average sense. To achieve better results, the signal could be divided into multiple segments for processing. Additionally, new indices or even a fusion of several indices should be proposed and employed. This remains an open issue for further investigation.

2.3 Aperiodic System Resonance: General Re-scaled Theoretical Framework

It is relatively easy to identify resonance regions within small parameter ranges of the system when the excitation signal is an aperiodic signal with a minimum time scale. For example, in Chap. 2.1, the system parameter is on the order of 1, while the pulse width is on the order 100. Small parameter systems can induce resonance phenomena under slowly varying signal excitation. However, there are numerous fast-varying signals in natural and engineering fields. Under the excitation of these signals, resonance requires large system parameters. Achieving the matching of system parameters and signals remains a challenge, which has been addressed by the re-scaled theoretical framework.

The re-scaled theoretical framework of resonance was first applied to solve stochastic resonance in systems excited by high-frequency signals. Based on this theory, Hu et al. introduced the normalized scale transformation to achieve stochastic resonance in such systems [10–12]. Subsequently, many researchers have applied this method to fault diagnosis of rolling bearings under noisy backgrounds [13–16]. Huang et al. [17], proposed the general scale transformation stochastic resonance method, which compared to stochastic resonance realized by normalized scale transformation, allows the system output to achieve an optimal stochastic resonance. Liu et al. [18], then introduced the normalized scale transformation to realize vibrational resonance in systems excited by two high-frequency signals. We extended the re-scaled theory of vibrational resonance from the normalized scale transformation to the general scale transformation [19], and further expanded it from vibrational resonance under harmonic excitations to aperiodic vibrational resonance under aperiodic binary excitations [20, 21].

To further reveal what happens to the system output when the pulse width T is set to a small value, we give an example for the aperiodic response induced by an aperiodic binary signal. As shown in Fig. 2.10, we choose $T = 0.01$ and $T = 0.1$, respectively. Clearly, none of the index curves exhibit a resonance phenomenon. This fully indicates that there is neither synchronization nor amplification between the system output and the input signal.

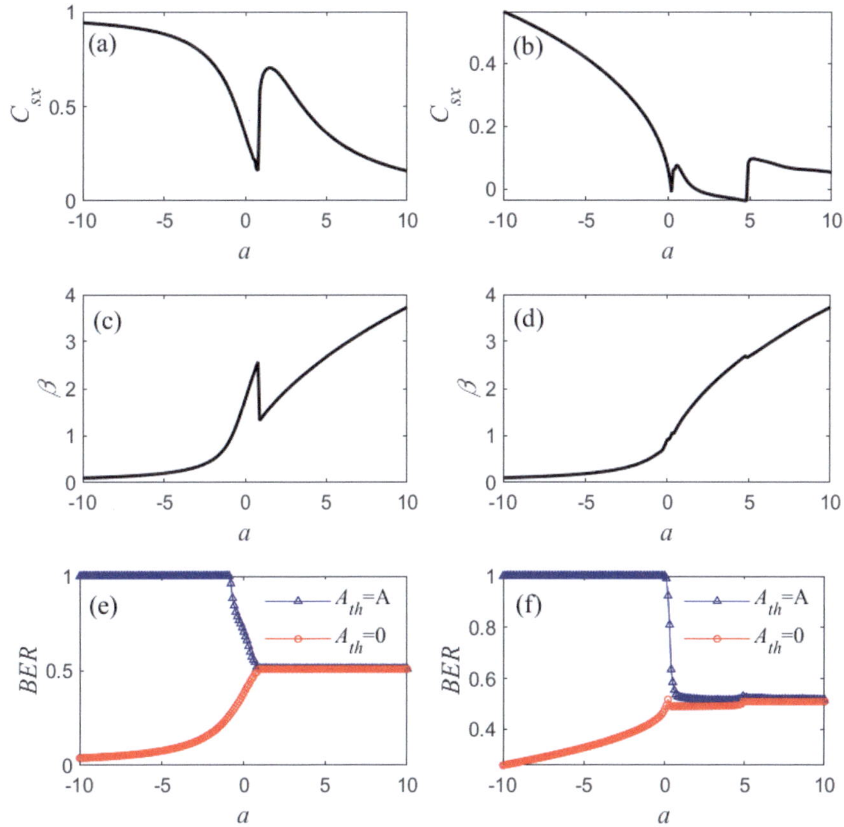

Fig. 2.10 Different indices cannot reflect the occurrence of the aperiodic system resonance. **a** and **b**, the cross-correlation coefficient versus the system parameter a. **c** and **d**, the response amplitude gain versus the system parameter a. **e** and **f**, the bit error rate versus the system parameter a. The simulation parameters are $b = 2$, $A = 0.3$, $T = 1$ for **a**, **c**, **e**, and $T = 0.1$ for **b**, **d**, **f**

In Fig. 2.11, we provide the corresponding time series for the point with the "strongest resonance" in Fig. 2.10a, i.e., the cross-correlation coefficient achieving the peak for the case $T = 1$. We cannot see any direct relationship between the system output and input signal in Fig. 2.11. It is worth noting that from the output time series, the calculation formula for the response amplitude gain in Eq. (2.2) is no longer valid, and we cannot use this formula anymore. As a result, the curves of the response amplitude gain in Fig. 2.10 cannot reflect the actual situation and is an incorrect curve.

Now, we introduce the re-scaled method to achieve aperiodic system resonance. First, we use the scale transformation, i.e.,

$$\tau=\beta t, \quad x(t) = z(\tau), \tag{2.4}$$

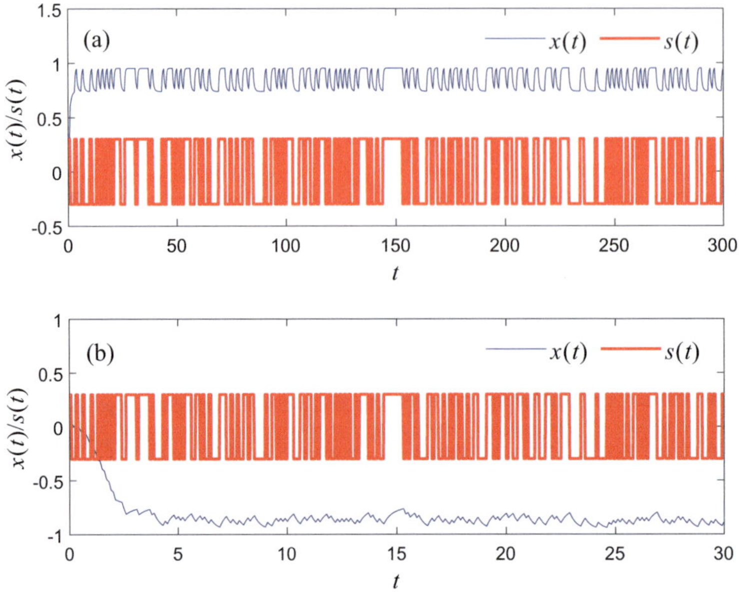

Fig. 2.11 The input binary signal (thin line) and the corresponding output (thick line) of the system, here $a = 1.5$, $b = 2$, $A = 0.3$, **a** $T = 1$, **b** $T = 0.1$

where β is the scale coefficient. Substituting Eq. (2.4) into Eq. (2.1), we obtain an equation for the new time scale τ

$$\frac{dz(\tau)}{d\tau} = \frac{a}{\beta}z(\tau) - \frac{b}{\beta}z^3(\tau) + \frac{A}{\beta}\sum_{m=-\infty}^{+\infty} R_m \Gamma(t - m\beta T). \tag{2.5}$$

At the new time scale, the minimum pulse width becomes β times of the original one. Then, letting

$$a_1 = \frac{a}{\beta}, \quad b_1 = \frac{b}{\beta}, \tag{2.6}$$

then, we have

$$\frac{dz(\tau)}{d\tau} = a_1 z(\tau) - b_1 z^3(\tau) + \frac{A}{\beta}\sum_{m=-\infty}^{+\infty} R_m \Gamma(t - m\beta T). \tag{2.7}$$

In its current form, if the scale coefficient is sufficiently large, the parameters a_1 and a_2 can be kept in small, while the minimal pulse width βT can be made large enough. On the time scale, the potential for resonance is achieved. However, the amplitude of the excitation is reduced to $\frac{1}{\beta}$ of its original value. To restore the amplitude to its original value, the re-scaled equation is obtained in the form

$$\frac{dz(\tau)}{d\tau} = a_1 z(\tau) - b_1 z^3(\tau) + A \sum_{m=-\infty}^{+\infty} R_m \Gamma(t - m\beta T). \tag{2.8}$$

Based on the above analysis, to achieve aperiodic system resonance of fast-varying excitation at the time scale t, the following equation must be solved

$$\frac{dx(t)}{dt} = a_1 \beta x(t) - b_1 \beta x^3(t) + \beta A \sum_{m=-\infty}^{+\infty} R_m \Gamma(t - mT). \tag{2.9}$$

In Fig. 2.1, where $T = 200$, according to the re-scaled theory outlined above, to achieve the same resonant effect as shown in Fig. 2.12, we need to choose the scale coefficients $\beta = 200$ (left column) and $\beta = 2000$ (right column), respectively. Clearly, the resonance indicated by the three different indices is consistent and aligns

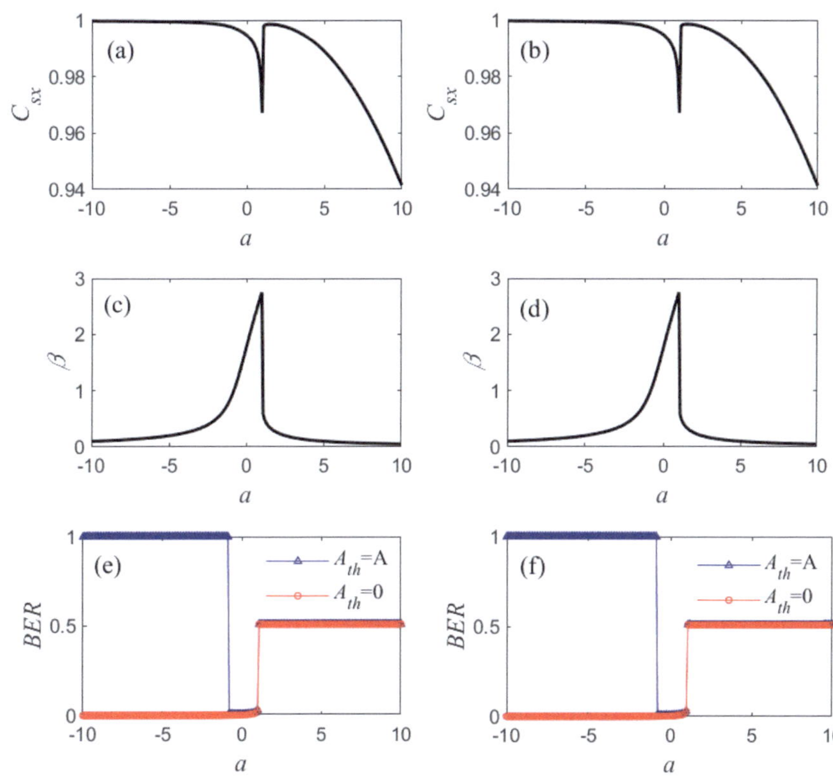

Fig. 2.12 Different indices reflect the occurrence of the aperiodic system resonance in the re-scaled theoretical framework. **a** and **b**, the cross-correlation coefficient versus the system parameter a. **c** and **d**, the response amplitude gain versus the system parameter a. **e** and **f**, the bit error rate versus the system parameter a. The simulation parameters are $b = 2$, $A = 0.3$, and $T = 1$, $\beta = 200$ for **a**, **c**, **e**, and $T = 0.1$, $\beta = 2000$ for **b**, **d**, **f**

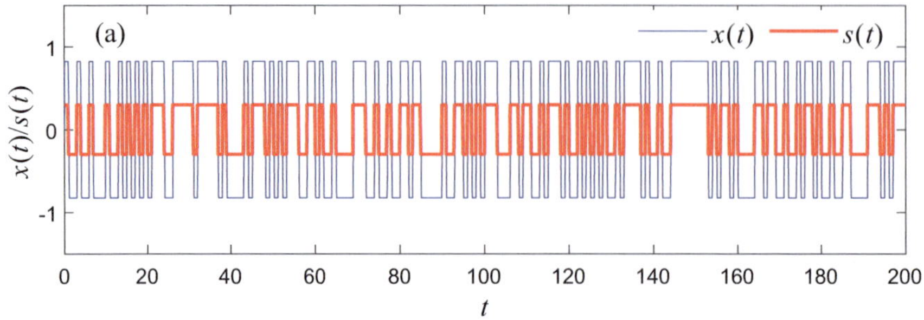

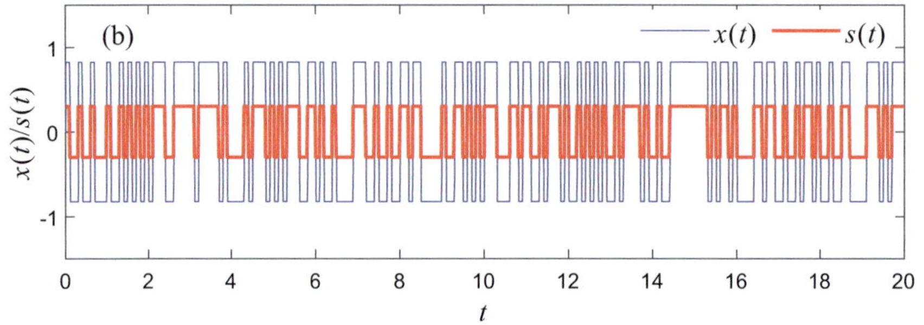

Fig. 2.13 The input binary signal (thin line) and the corresponding output (thick line) of the system under the re-scaled aperiodic resonant state, here $a = 1$, $b = 2$, $A = 0.3$, **a** $T = 1$, $\beta = 200$, **b** $T = 0.1$, $\beta = 2000$

with the resonance observed in Figs. 2.1, 2.3, and 2.4. This consistency demonstrates the effectiveness of the re-scaled resonance theory. Importantly, to achieve a strong aperiodic system resonance effect, the selected sampling rate must be sufficiently high for rapidly changing excitations.

Fig. 2.13 provides a direct illustration of the aperiodic resonance of the system response from a time series perspective. It is evident that the output and input feature signals are synchronized, with the output exhibiting significant amplification. By applying different scale coefficients to input signals with varying time scales, we can convert them to the same time scale within the equivalent system, thereby achieving a consistent resonance effect.

2.4 Concluding Remarks

The pervasive presence of aperiodic signals underscores the importance of studying aperiodic resonance. This chapter explores the aperiodic system resonance induced by binary or M-ary aperiodic signals in a bistable system. Aperiodic system resonance is achieved by adjusting the system parameters. When the external signal is in a slow-varying form, aperiodic resonance can be induced by directly modifying the system parameters. Conversely, for fast-varying external signals, a general re-scaled method must be employed.

In addition to the aperiodic system resonance occurring in the bistable system, it is essential to consider various nonlinear systems based on practical needs. For instance, ordinary differential equations can take on many different typical forms. This chapter focuses exclusively on the overdamped bistable system and does not address other forms of nonlinear terms, nor does it explore underdamped nonlinear systems, coupled systems, and so forth.

References

1. J.J. Collins, C.C. Chow, A.C. Capela, T.T. Imhoff, Aperiodic stochastic resonance, *Phys. Rev. E*, 1996, **54**(5): 5575-5584.
2. J.J. Collins, C.C. Chow, T.T. Imhoff, Aperiodic stochastic resonance in excitable systems, *Phys. Rev. E*, 1995, **52**(4): R3321-R3324.
3. K. Park, Y.C. Lai, Z. Liu, A. Nachman, Aperiodic stochastic resonance and phase synchronization, *Phys. Lett. A*, 2004, **326**(5-6): 391-396.
4. V.N. Chizhevsky, G. Giacomelli, Vibrational resonance and the detection of aperiodic binary signals, *Phys. Rev. E*, 2008, **77**(5): 051126.
5. P.X. Jia, C.J. Wu, J.H. Yang, M.A.F. Sanjuán, G.X. Liu, Improving the weak aperiodic signal by three kinds of vibrational resonance, *Nonlinear Dynamics*, 2018, **91**: 2699-2713.
6. J. Yang, S. Rajasekar, M.A.F. Sanjuán, Vibrational resonance: A review, *Phys. Rep.*, 2024, **1067**: 1-62
7. Z.Q. Wang, J.H. Yang, Aperiodic resonance of a nonlinear system excited by aperiodic binary signal or M-ary signal, *Acta Physica Sinica*, 2023, **72**(22): 222501.
8. A. Medio, M. Lines, Nonlinear Dynamics: A Primer, 2001, Cambridge University Press, pp.142-145.
9. J.M.T. Thompson, H.B. Stewart, Nonlinear Dynamics and Chaos, 2002, John Wiley & Sons, pp.114-115.
10. N. Hu, M. Chen, G. Qin, L. Xia, Z. Pan, Z. Feng, Extended stochastic resonance (SR) and its applications in weak mechanical signal processing, *Front. Mech. Eng. in China*, 2009, **4**: 450-461.
11. X. Zhang, N. Hu, Z. Cheng, L Hu, Enhanced detection of rolling element bearing fault based on stochastic resonance, *Chin. J. Mech. Eng.*, 2012, **25**(6): 1287-1297.
12. X.F. Zhang, N.Q. Hu, L. Hu, Z. Cheng, Multi-scale bistable stochastic resonance array: A novel weak signal detection method and application in machine fault diagnosis, *Sci. China Technol. Sci.*, 2013, **56**: 2115-2123.
13. J. Tan, X. Chen, J. Wang, H. Chen, H. Cao, Y. Zi, Z. He, Study of frequency-shifted and re-scaling stochastic resonance and its application to fault diagnosis, *Mech. Syst. Signal Process.*, 2009, **23**(3): 811-822.
14. Q. He, J. Wang, Y. Liu, D. Dai, Multiscale noise tuning of stochastic resonance for enhanced fault diagnosis in rotating machines, *Mech. Syst. Signal Process.*, 2012, **28**: 443-457.
15. Q. He, J. Wang, Effects of multiscale noise tuning on stochastic resonance for weak signal detection, *Digit. Signal Process.*, 2012, **22**(4): 614-621.
16. X. Liu, H. Liu, J. Yang, G. Litak, G. Cheng, S. Han, Improving the bearing fault diagnosis efficiency by the adaptive stochastic resonance in a new nonlinear system, *Mech. Syst. Signal Process.*, 2017, **96**: 58-76.

17. D. Huang, J. Yang, J. Zhang, H. Liu, An improved adaptive stochastic resonance method for improving the efficiency of bearing faults diagnosis, *Proc. Inst. Mech. Eng., Part C*, 2018, **232**(13): 2352-2368.
18. H.G. Liu, X.L. Liu, J.H. Yang, G. Litak, G. Cheng, S. Han, Detecting the weak high-frequency character signal by vibrational resonance in the Duffing oscillator, *Nonlinear Dynamics*, 2017, **89**: 2621-2628.
19. J.H. Yang, M.A.F. Sanjuán, H.G. Liu, Enhancing the weak signal with arbitrary high-frequency by vibrational resonance in fractional-order Duffing oscillators, *J. Comput. Nonlinear Dyn.*, 2017, **12**(5): 051011.
20. J.H. Yang, Q. Ma, C.J. Wu, H.G. Liu, Aperiodic vibrational resonance in the fractional-order bistable system, *Acta Physica Sinica*, 2018, **67**(5): 054501.
21. P.X. Jia, C.J. Wu, J.H. Yang, M.A.F. Sanjuán, G.X. Liu, Improving the weak aperiodic signal by three kinds of vibrational resonance, *Nonlinear Dynamics*, 2018, **91**: 2699-2713.

Chapter 3
Aperiodic System Resonance Caused by Single Frequency-Modulated Signal

This chapter presents the aperiodic system resonance induced by a frequency-modulated signal. We introduce the theoretical framework of the piecewise re-scaled method and the real-time re-scaled method to achieve aperiodic system resonance under a linear or nonlinear frequency-modulated signal. The improvement of aperiodic system resonance by vibrational resonance, i.e., the effect of an auxiliary fast-varying signal in the excitation, is also clearly demonstrated.

3.1 Aperiodic System Resonance Realized by the Piecewise Re-scaled Method

As discussed in Chap. 2, when the excitation is in a fast-varying form, the re-scaled theoretical framework must be used to match the system parameters with the excitation to achieve aperiodic resonance. For a fixed time scale of the fast-varying excitation, using a constant scaling coefficient can achieve aperiodic system resonance within the general re-scaled theoretical framework introduced in Chap. 2. However, for complex variations in the time scale of the excitation, particularly when a constant scaling coefficient is insufficient to achieve appropriate matching between the system parameters and the excitation over an extended period, the piecewise re-scaled theory must be applied [1–4].

For excitations with frequencies changing over time, a fixed scaling coefficient can make it challenging to match system parameters with the excitation frequencies over an extended period. The piecewise re-scaled theoretical framework can be employed to achieve aperiodic resonance in the system's response. In this context, we consider the following traditional bistable system excited by a linear frequency-modulated signal.

$$\dot{x}(t) = ax(t) - bx^3(t) + A\cos\left(\pi\gamma t^2 + 2\pi f t\right), \tag{3.1}$$

J. Yang and M. A. F. Sanjuan, *Aperiodic Resonances—Theory and Applications*,
https://doi.org/10.1007/978-981-96-6819-9_3

where $a > 0$ and $b > 0$. We use the piecewise re-scaled method to realize the aperiodic resonance in Eq. (3.1). First, we divide the signal into n segments to achieve an approximate scale match between the system parameters and the instantaneous frequency of each segment. We define a time scale for each time series, for example, the time scale for the ith segment is

$$\beta = \beta_0 (f + \gamma t_{ei}). \tag{3.2}$$

Here, β_0 is a constant coefficient, referred to as the base of the scale coefficient. t_{ei} is an arbitrary time point in the time series segment. By achieving aperiodic system resonance in the output of each time series and then concatenating these resonance outputs to form a complete time series, we realize aperiodic system resonance under frequency-modulated signal excitation through the piecewise re-scaled method. This approach can be applied not only to aperiodic system resonance caused by linear frequency-modulated signal excitation but also to resonance induced by other, more complex forms of frequency-modulated signals.

We will continue using the scale transformation in Eq. (2.4) and substitute it into Eq. (3.1), then, we obtain

$$\frac{dz(\tau)}{d\tau} = \frac{a}{\beta} z(\tau) - \frac{b}{\beta} z^3(\tau) + \frac{A}{\beta} \cos\left[\pi\gamma\left(\frac{\tau}{\beta}\right)^2 + 2\pi f \frac{\tau}{\beta}\right]. \tag{3.3}$$

The equation for matching the system parameters and the signal time scale is

$$\frac{dz(\tau)}{d\tau} = a_1 z(\tau) - b_1 z^3(\tau) + \frac{A}{\beta} \cos\left[\pi\gamma\left(\frac{\tau}{\beta}\right)^2 + 2\pi f \frac{\tau}{\beta}\right], \tag{3.4}$$

where a_1 and b_1 are still defined in Eq. (2.6). Moreover, here, a_1 and b_1 are not fixed constants since they depend closely on the time point t_{ei}. By recovering the excitation to its original amplitude, aperiodic system resonance under frequency-modulated excitation will be achieved in the following equation

$$\frac{dx(t)}{dt} = a_1 \beta x(t) - b_1 \beta x^3(t) + \beta A \cos\left(\pi\gamma t^2 + 2\pi f t\right). \tag{3.5}$$

We provide an example to demonstrate the piecewise re-scaled method and the influence of the time point t_{ei} in Fig. 3.1. The spectral amplification factor as a function of the system parameter shows clear evidence of aperiodic system resonance. In the numerical calculations, we set the total time to 12 s and divide the time series into 12 equal parts. We select 5 cases for the time point t_{ei} in each segment. Specifically, t_{ei} corresponds to the initial time, quarter time, middle time, three-quarters time, and end time of the segmented time series, from case 1 to case 5 in sequence. The influence of the time point t_{ei} on the aperiodic resonance effect is minimal. Additionally,

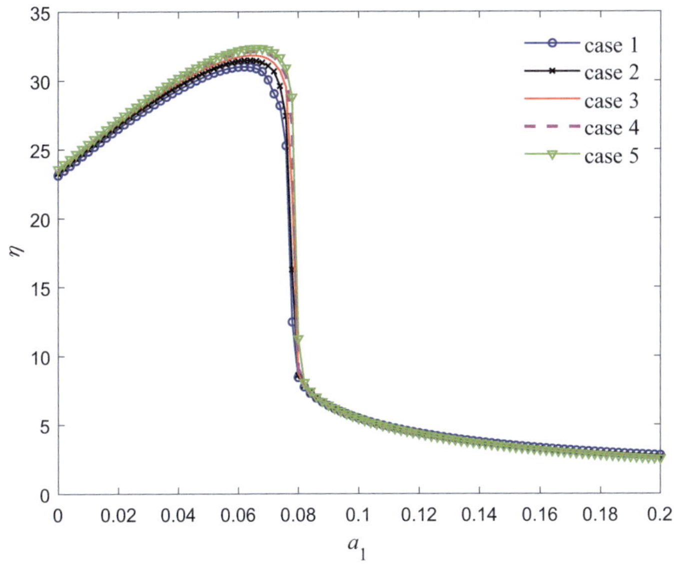

Fig. 3.1 The influence of the selection of t_{ei} on the aperiodic system resonance when the system parameter a_1 is the control variable. The simulation parameter are $A = 0.1, \gamma = 1, f = 1, b_1 = 0.01$ and $\beta_0 = 400$

in Fig. 3.1, the spectral amplification factor is significantly large when $a_1 = 0$, indicating that the linear frequency-modulated signal can also induce aperiodic system resonance in a monostable system.

By selecting the same 5 cases of t_{ei} and using the re-scaled constant β_0 as a control, Fig. 3.2 clearly illustrates the dependence of the spectral amplification factor η on the re-scaled constant β_0. Notably, a re-scaled constant β_0 that is too small cannot produce the ideal aperiodic resonance effect. As β_0 increases, the spectral amplification factor η reaches a saturation point. Therefore, as long as the value of the re-scaled constant β_0 is chosen within a reasonable range, its specific numerical value is not a major concern.

To better illustrate the aperiodic resonance caused by the system parameters, we present Fig. 3.3, which simultaneously shows how the adjustments of parameters a_1 and b_1 affect the aperiodic resonance. It is evident that both parameters a_1 and b_1 exhibit aperiodic system resonance. This further demonstrates the feasibility of the proposed piecewise re-scaled method in inducing aperiodic system resonance. In the three-dimensional plot, X, Y and Z of the label represent the values of the coordinate values of the two horizontal axes and the vertical axis of the labelled point. The coordinate annotation of subsequent three-dimensional plot has the same meaning as this.

To more intuitively illustrate the aperiodic system resonance, we present Fig. 3.4, which shows the input signal and the system output when strong aperiodic system

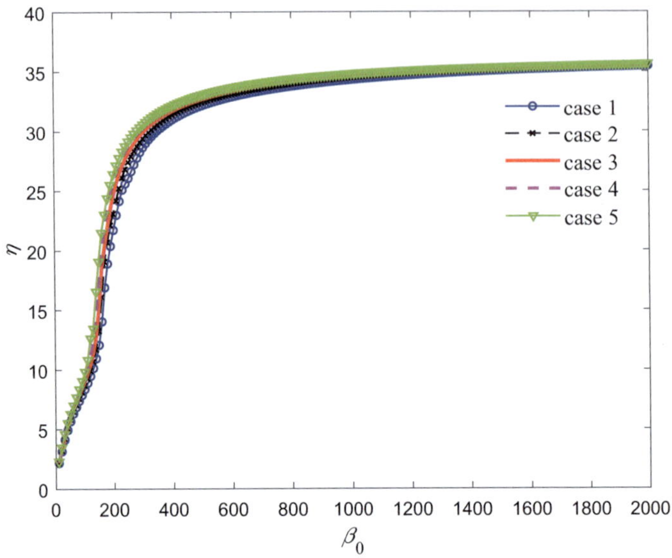

Fig. 3.2 The influence of the selection of t_{ei} on the aperiodic system resonance when the re-scaled constant β_0 is the control variable. The simulation parameter are $A = 0.1, \gamma = 1, f = 1, a_1 = 0.06$ and $b_1 = 0.01$

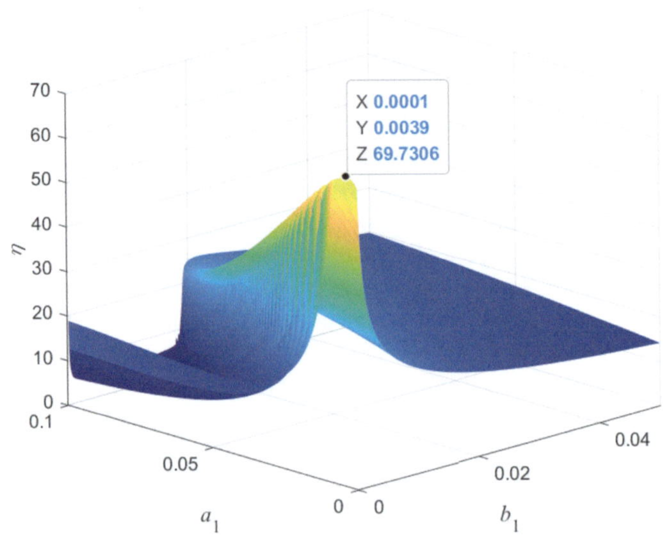

Fig. 3.3 The aperiodic system resonance induced by the parameters a and b. The simulation parameter are $A = 0.1, \gamma = 1, f = 1$ and $\beta_0 = 400$

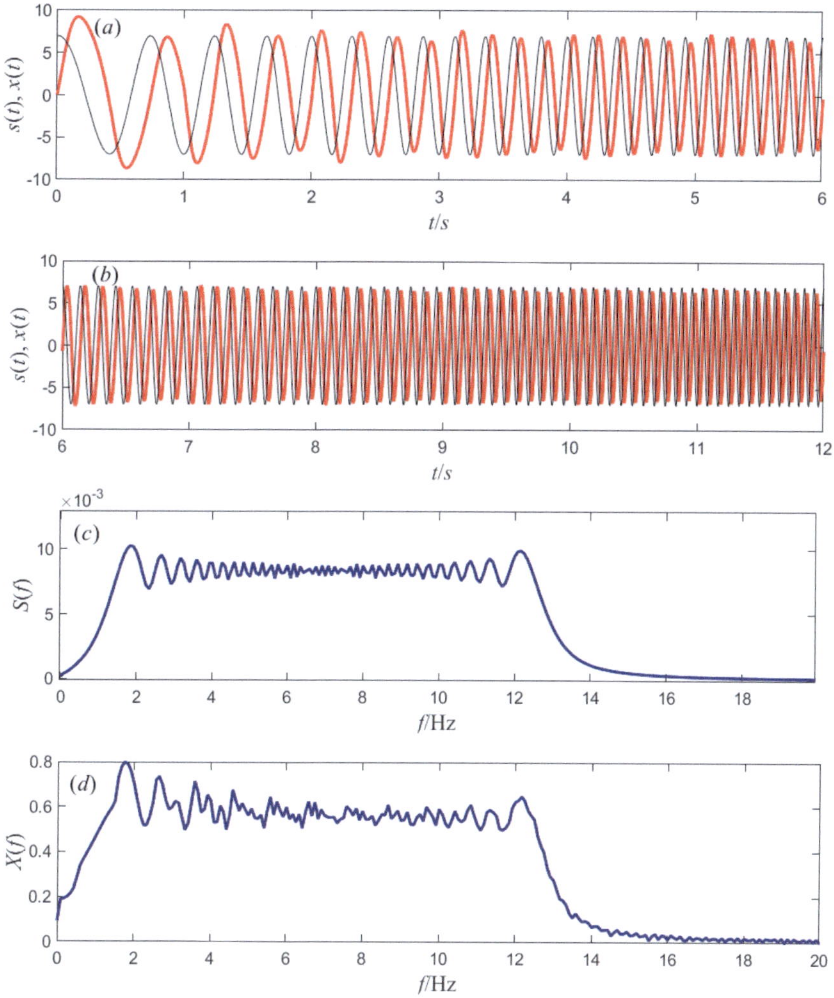

Fig. 3.4 The output of the system when the aperiodic resonance occurs compared with the input signal in the time domain and frequency domain. In **a** and **b**, the thick line is the output and the thin line is 70 times the input signal. In **c** and **d**, the spectrum of the input signal and the output are presented respectively. The simulation parameter are $a_1 = 0.0039$, $b_1 = 0.0001$, $A = 0.1$, $\gamma = 1$, $f = 1$ and $\beta_0 = 400$

resonance occurs. From Fig. 3.4, it is evident that when aperiodic system resonance takes place, the signal is amplified by nearly 70 times. However, due to the influence of system damping, there is a phase lag between the system output and the input signal. Despite this, the system can still experience strong resonance under the excitation of a linear frequency-modulated signal.

In our former work, we also extended the nonlinear system to a general bistable system [4], i.e.,

$$\dot{x}(t) = ax(t) - bx(t)|x(t)|^{\alpha-1} + A\cos\left(\pi\gamma t^2 + 2\pi f t\right). \tag{3.6}$$

The fractional exponent $\alpha > 1$, and its value can be an integer or a fractional number. Here, we consider the dynamical equation as a signal processor. The potential of the equation is

$$V(x) = -\frac{a}{2}x^2 + \frac{b}{\alpha+1}|x|^{\alpha+1}. \tag{3.7}$$

When $\alpha = 3$, this special potential function degenerates into the traditional bistable potential, and Eq. (3.6) also simplifies to Eq. (3.1). The configurations of this type of potential function, featuring a fractional exponent nonlinear term, for different values of α are presented in Fig. A.2 of the Appendix A.1.2. Additionally, the potential function described in Eq. (3.7) has been explored in relation to various dynamical topics. The specific value of α typically depends on the characteristics of the material, and the corresponding physical backgrounds have been discussed in detail in previous works [5–8].

Clearly, the spectral amplification factor closely depends on the fractional exponent α, as shown in Fig. 3.5. For different values of a_1, the spectral amplification factor may monotonically decrease (such as the values of $a_1 = 0.1$ and 0.12), or

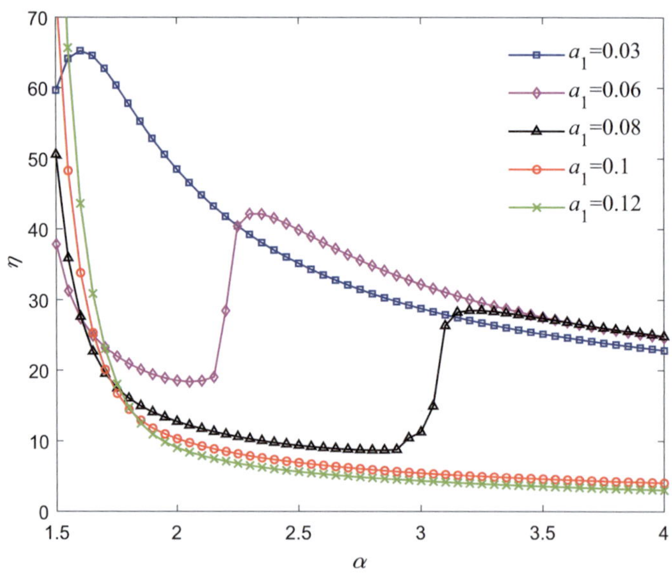

Fig. 3.5 The dependence of the spectral amplification factor on the fractional exponent α under different values of the system parameter a_1. The simulation parameter are $A = 0.1$, $\gamma = 1$, $f = 1$, $b_1 = 0.01$ and $\beta_0 = 400$

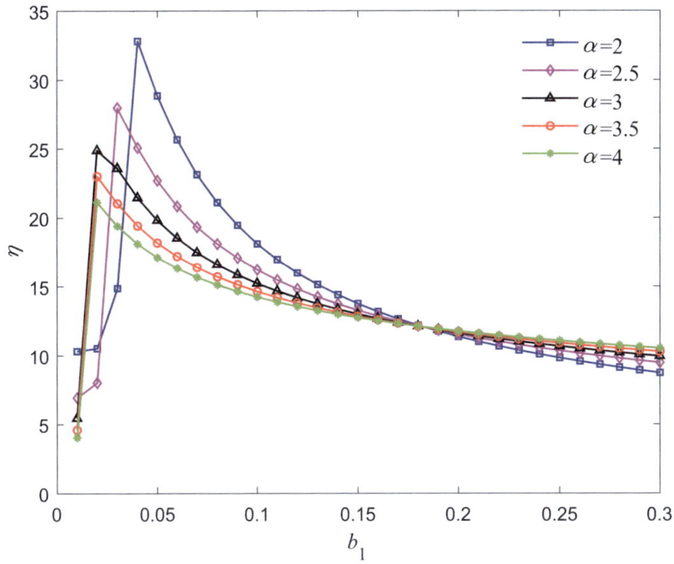

Fig. 3.6 The dependence of the spectral amplification factor on the system parameter b_1 under different values of the fractional exponent α. The simulation parameter are $A = 0.1$, $\gamma = 1$, $f = 1$, $a_1 = 0.1$ and $\beta_0 = 400$

there may be a resonance peak present in the curve (such as the values of $a_1 = 0.03$, 0.06 and 0.08). Thus, it can be seen that aperiodic resonance can also be achieved by adjusting the value of the exponent α when the system parameters a and b are determined. For cases where the fractional exponent α is small, although the value of the spectral amplification factor η may be higher, we also need to consider the stability of the response. The smaller the fractional exponent α, the easier the system response is to diverge.

Considering different cases of the fractional exponent α, in Fig. 3.6, we show the resonance that induced by the system parameter b_1. In this figure, both for the traditional bistable system ($\alpha = 3$) and or the fractional exponent case ($\alpha \neq 3$), there is resonance caused by the system parameter b.

In fact, the piecewise re-scaled theoretical framework not only can process the linear frequency-modulated signal, but it can also process more complex nonlinear frequency-modulated signals [9]. Although nonlinear frequency-modulated signals are presented in various forms, the handling method and process are essentially the same, and will not be repeated here. Readers are advised to practice on their own.

3.2 Aperiodic System Resonance Realized by the Real-Time Re-scaled Method

For frequency-modulated signals with more complex frequency variations, we introduced real-time re-scaled theoretical framework to achieve aperiodic system resonance [10–13]. Consider the overdamped bistable system with time-varying parameters

$$\frac{\mathrm{d}x(t)}{\mathrm{d}t} = a(t)x(t) - b(t)x^3(t) + s(t). \tag{3.8}$$

Herein, $a(t)$ and $b(t)$ are time-varying parameters with $a(t) > 0$ and $b(t) > 0$. $s(t)$ is the frequency-modulated signal with instantaneous frequency $f(t)$ that constantly changes over time. In order to achieve the matching of scale coefficient with the excitation $s(t)$ and the system parameters $a(t)$ and $b(t)$, the real-time scale coefficient $\beta(t)$ is introduced as

$$\beta(t) = \beta_0 f(t), \tag{3.9}$$

where β_0 is the scale coefficient constant, and $\beta(t)$ usually satisfies $\beta(t) \gg 1$. Meanwhile, the system parameters $a(t)$, $b(t)$ and the real-time scale coefficient $\beta(t)$ satisfy the following relationship

$$\frac{a(t)}{\beta(t)} = a_1, \frac{b(t)}{\beta(t)} = b_1. \tag{3.10}$$

Introducing the variable substitutions $\tau = \beta(t)t,\ x(t) = z(\tau)$, Eq. (3.8) can be approximately written as

$$\frac{\mathrm{d}z(\tau)}{\mathrm{d}\tau} = \frac{a(t)}{\beta(t)}z(\tau) - \frac{b(t)}{\beta(t)}z^3(\tau) + \frac{1}{\beta(t)}s\left(\frac{\tau}{\beta(t)}\right). \tag{3.11}$$

Additionally, an approximation has been made during the derivation of Eq. (3.11). Specifically,

$$\frac{dx(t)}{dt} = \frac{dz(\tau)}{d\tau}\frac{d\tau}{dt} = \frac{dz(\tau)}{d\tau}\left(\beta_0\frac{df(t)}{dt} + \beta(t)\right) \approx \beta(t)\frac{dz(\tau)}{d\tau}, \tag{3.12}$$

where the inequality $\beta(t) \gg \beta_0\frac{df(t)}{dt}$ serves as a prerequisite for obtaining Eq. (3.12). This prerequisite can be met in many cases. Then, by substituting Eq. (3.10) into Eq. (3.11), we get

$$\frac{\mathrm{d}z(\tau)}{\mathrm{d}\tau} = a_1 z(\tau) - b_1 z^3(\tau) + \frac{1}{\beta(t)}s\left(\frac{\tau}{\beta(t)}\right). \tag{3.13}$$

Among these, Eq. (3.13) satisfies the condition for system resonance with small parameters. In comparison to Eqs. (3.8), (3.13) reduces the amplitude of the excitation

$s(t)$ to $1/\beta(t)$. By multiplying the amplitude of the excitation in Eq. (3.13) by $\beta(t)$, it becomes

$$\frac{dz(\tau)}{d\tau} = a_1 z(\tau) - b_1 z^3(\tau) + s\left(\frac{\tau}{\beta(t)}\right). \tag{3.14}$$

If the system resonance occurs in Eq. (3.14), then the system resonance must also occur in the following equation

$$\frac{dx(t)}{dt} = a_1\beta(t)x(t) - b_1\beta(t)x^3(t) + \beta(t)s(t). \tag{3.15}$$

Equation (3.15) is a large parameter model that can detect the signal with arbitrary high-frequency.

(i) NLFM Signal with Piecewise Linear Time-Frequency

We first use the NLFM signal with piecewise linear time-frequency defined in Eq. (1.6) as an example. We give its time-domain waveform and time-frequency representation in Fig. 3.7 under a group of simulation parameters. In Fig. 3.7b, the STFT spectrum reflects the frequency characteristic curve of the signal increases 0 Hz 150 Hz within 0–3 s, then remains unchanged over 7 s, and finally drops 0 Hz.

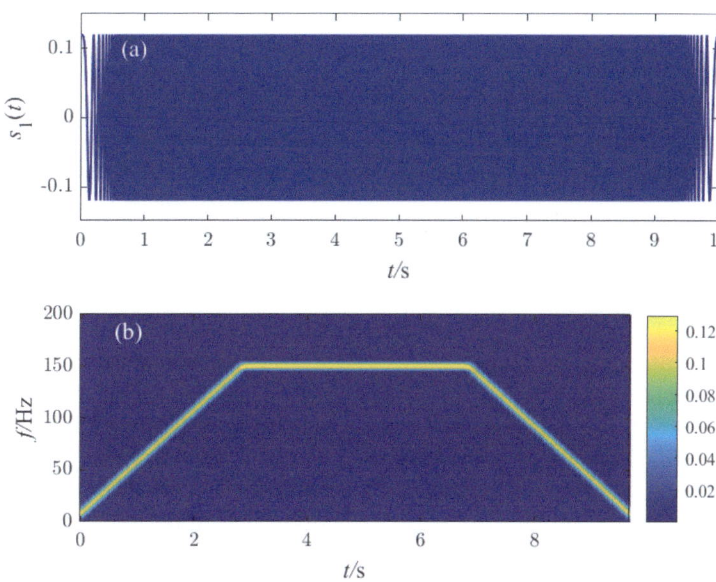

Fig. 3.7 The input NLFM signal with piecewise linear time-frequency. **a** The waveform of the time series in the time domain. **b** The STFT spectrum in the time-frequency domain. The simulation parameters are $A = 0.12$, $f_0 = 0$, $\gamma = 50$, $\phi = 0$, $t_1 = 3$, $t_2 = 10$

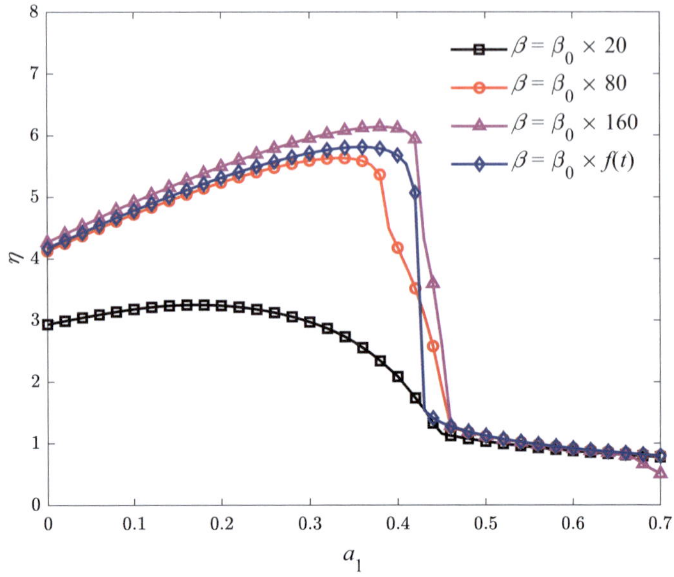

Fig. 3.8 The spectrum amplification factor η as a function of the system parameter a_1 for different values of scale coefficient $\beta = \beta_0 \times 20$, $\beta = \beta_0 \times 80$, $\beta = \beta_0 \times 160$, $\beta = \beta_0 \times f(t)$. The simulation parameters are $A = 0.12$, $f_0 = 0$, $\gamma = 50$, $\phi = 0$, $t_1 = 3$, $t_2 = 10$, $b_1 = 1$, $\beta_0 = 100$, $m = 2$

For the signal shown in Fig. 3.7, we present the spectral amplification factor η of the system output at different values of the scale coefficient in Fig. 3.8. The fixed scale coefficients are $\beta = \beta_0 \times 20$, $\beta_0 \times 80$, $\beta_0 \times 160$, while the real-time scale coefficient is given by $\beta = \beta_0 \times f(t)$. As the system parameter a_1 increases, the real-time spectral amplification factor also increases with the fixed scale coefficient β. The response achieves the strongest aperiodic system resonance at the maximum value of η. Furthermore, the real-time spectral amplification factor of the system output is maximized using the real-time re-scaled method. In comparison to the curve of $\beta_0 \times 160$, the maximum value of $\beta_0 \times f(t)$ is slightly smaller, although the difference is not significant.

To investigate the influence of system parameters on the aperiodic resonance response, the real-time spectral amplification factor η versus the system parameters a_1 and b_1 are shown in Fig. 3.9. The color codes the degree of the aperiodic resonance, while the higher value indicated by the color, the stronger the aperiodic resonance. An obvious resonance region similar to a "ridge" can be easily found in the figure, where the strongest aperiodic resonance of the response can occur, achieving the effect of NLFM signal enhancement.

According to the resonance phenomenon indicated by Fig. 3.10, when $a_1 = 0.146$ and $b_1 = 0.11$, the spectral amplification factor η take the maximum value, $\eta = 10.3635$, the corresponding time series in the time domain and the spectral in the time-frequency domain are given in Fig. 3.10. In comparison to Fig. 3.7, the

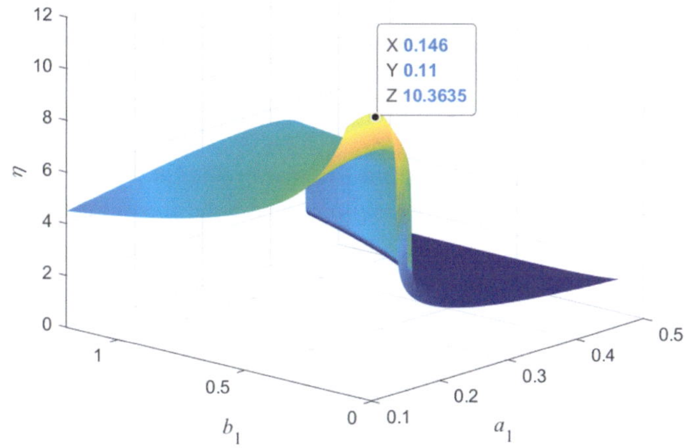

Fig. 3.9 The spectrum amplification factor η versus the system parameters a_1 and b_1 under different values of scale coefficient $\beta = \beta_0 \times 20$, $\beta = \beta_0 \times 80$, $\beta = \beta_0 \times 160$, $\beta = \beta_0 \times f(t)$. The simulation parameters are $A = 0.12$, $f_0 = 0$, $\gamma = 50$, $\phi = 0$, $t_1 = 3$, $t_2 = 10$, $\beta_0 = 100$, $m = 2$

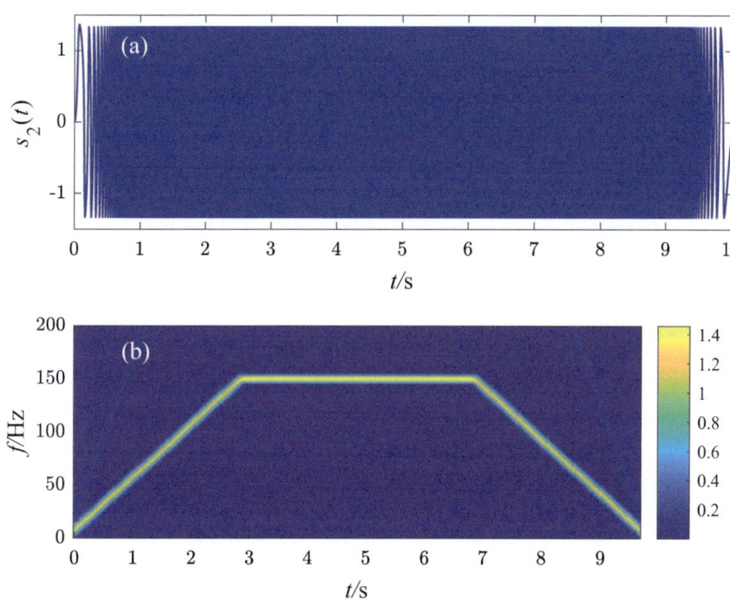

Fig. 3.10 The output in the time domain and time-frequency domain when the aperiodic resonance occurs. **a** The waveform of the time series in the time domain. **b** The STFT spectrum in the time-frequency domain. The simulation parameters are $A = 0.12$, $f_0 = 0$, $\gamma = 50$, $\phi = 0$, $t_1 = 3$, $t_2 = 10$, $a_1 = 0.146$, $b_1 = 0.11$, $\beta_0 = 100$, $m = 2$

signal amplitude is significantly enhanced. Additionally, the continuity of the signal is also very good. In this subsection, we employ both the real-time re-scaled method and the general re-scaled method to achieve aperiodic resonance. While the piece-wise re-scaled method can also accomplish the same effect, the real-time re-scaled method proves to be more advantageous for handling more complex frequency-modulated signals.

(ii) NLFM Signal with Amplitude and Frequency Modulation

Consider a typical NLFM signal with complex amplitude and frequency modulation simultaneously

$$s(t) = 0.12\left(1 + \frac{1}{3}\cos\frac{2\pi t}{9}\right)\cos\left(80\pi t + 30\sin\pi t + 20\cos\frac{5\pi t}{3}\right). \tag{3.16}$$

The amplitude of the signal is modulated by a sine function, while its frequency is simultaneously modulated by sine and cosine functions. Figure 3.11 shows the time-domain waveform and STFT spectrum of the signal. The amplitude and frequency of the signal vary with the time, the frequency distribution ranging from 10 to 75 Hz. The solid line represents the instantaneous frequency curve extracted from the phase of the NLFM signal, which aligns with the time-frequency characteristic curve. For this type of NLFM signal, the general re-scaled method with fixed scale coefficient is not applicable.

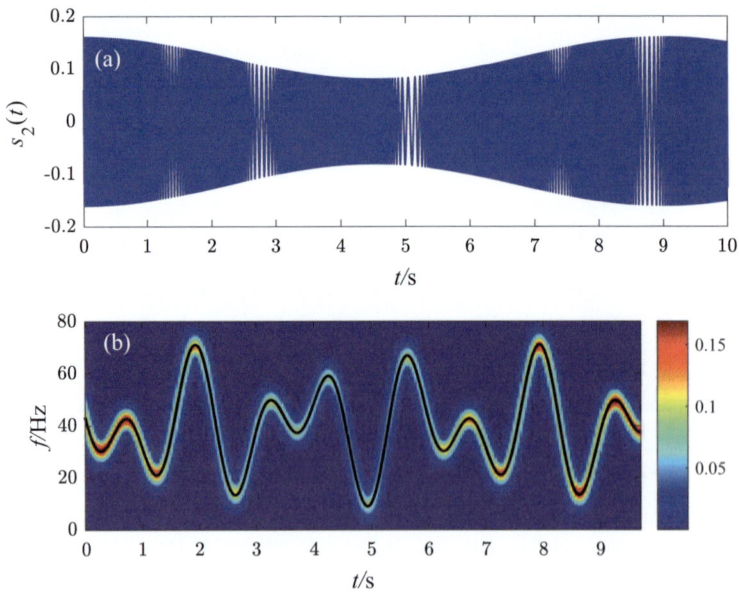

Fig. 3.11 The input NLFM signal of Eq. (3.16). **a** The waveform of the time series in the time domain. **b** The STFT spectrum in the time-frequency domain

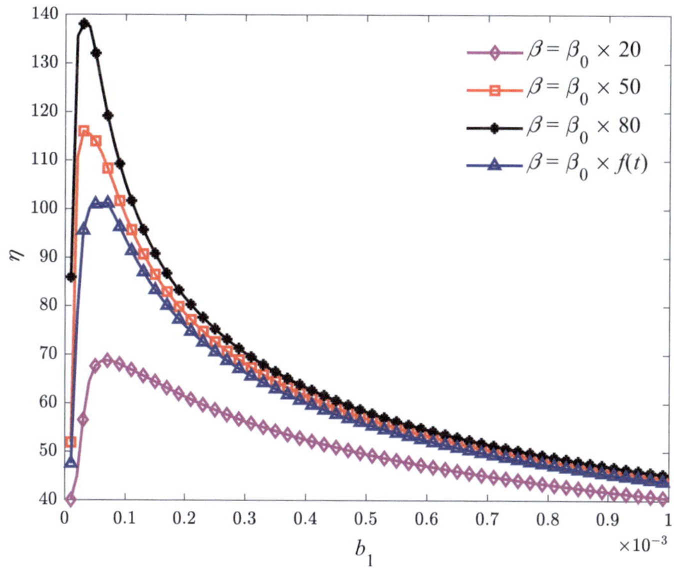

Fig. 3.12 The spectrum amplification factor η as a function of the system parameter b_1 for different values of scale coefficient. The simulation parameters are $a_1 = 0.01$, $m = 5$, $_0 = 1000$

The curves in Fig. 3.12 show the real-time spectral amplification factor at different scale coefficients. The curves indicate that the real-time spectral amplification factor η first increases and then decreases as the system parameter b_1 rises. Corresponding to $\beta = \beta_0 \times 20, \beta_0 \times 50, \beta_0 \times 80$, the resonance peak appears at $b_1 = 7 \times 10^{-5}, 3 \times 10^{-5}, 3 \times 10^{-5}$, respectively, with peak values of $\eta_{max} = 68.736, 116.037, 138.082$. In contrast, for the case where $\beta = \beta_0 f(t)$, the resonance peak appears at $b_1 = 7 \times 10^{-5}$ with $\eta_{max} = 101.075$.

Corresponding to the resonance peak of the case $\beta = \beta_0 \times f(t)$, Fig. 3.13 shows the time-domain waveform and STFT spectrum of the response. The signal amplitude is significantly enhanced while ensuring the continuity of the time-frequency characteristics of the signal. It is an obvious advantage of the real-time re-scaled method.

Although in Fig. 3.12, the real-time spectral amplification factor η corresponding to the fixed scale coefficient may be higher than that obtained by the real-time re-scaled process, we should also further understand the time series of system response when the aperiodic resonance occurs in the time domain, as shown in Fig. 3.14. The poor continuity of the NLFM signal that processed by the general re-scaled process is very obviously. The phenomenon of energy divergence is severe, far less effective than the time series obtained by the real-time re-scaled theoretical framework in Fig. 3.13.

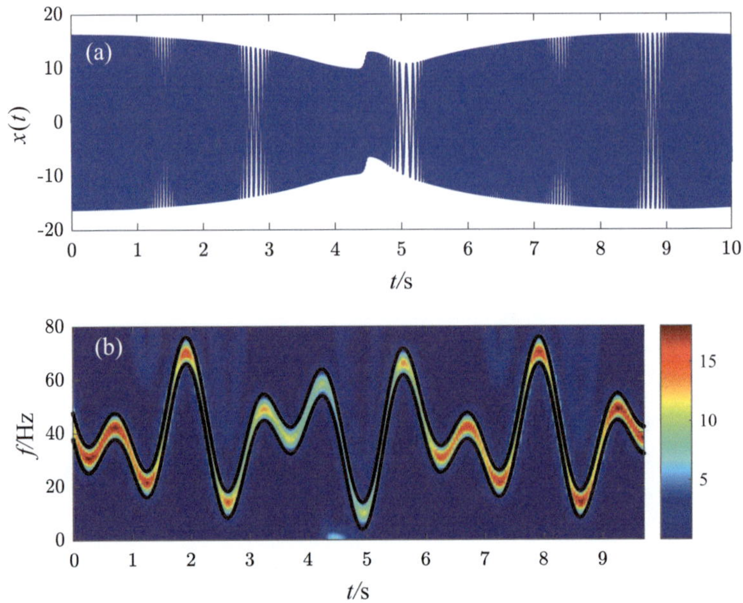

Fig. 3.13 The response corresponding to the aperiodic resonance in the real-time re-scaled theoretical framework. **a** The waveform of the time series in the time domain. **b** The STFT spectrum in the time-frequency domain. The simulation parameters are $m = 5$, $\beta_0 = 1000$

This section addresses the aperiodic system resonance of NLFM signals, offering insights into NLFM signal enhancement and detection. It holds practical significance and value for signal amplification. Additionally, it outlines the relevant influence laws governing the optimal output response of the system under NLFM signal excitation, providing valuable reference points for studying the dynamic behavior of the system under complex signal excitations.

3.3 Aperiodic System Resonance Improved by Vibrational Resonance

Vibrational resonance means that the response of a nonlinear system to a weak slow-varying characteristic signal can be effectively improved by an appropriate fast-varying auxiliary signal [14]. Vibrational resonance is similar to the well-known stochastic resonance [15], but their mechanisms are different. Vibrational resonance uses the effect of the fast-varying auxiliary signal to change the coefficient equivalent nonlinear system [16, 17]. In fact, vibrational resonance is based on the fact that the fast-varying signal can change the linear stiffness of the system [18, 19].

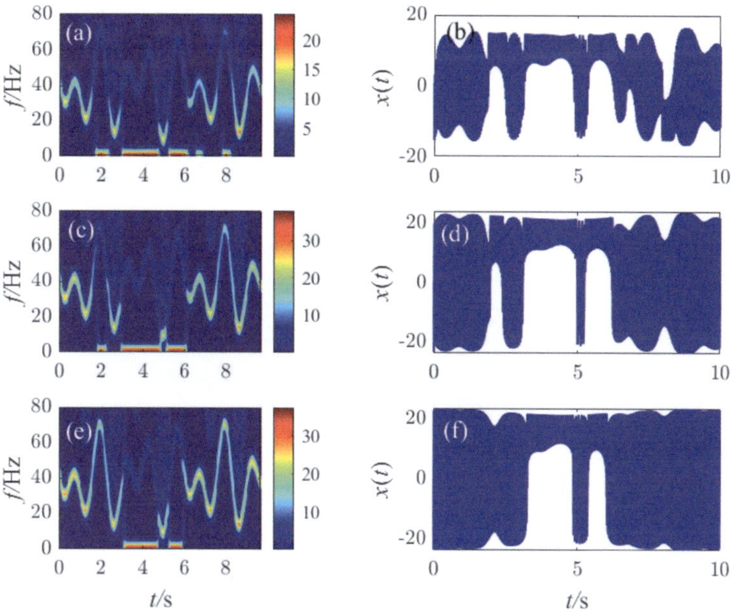

Fig. 3.14 The aperiodic system resonance processed by the general re-scaled method under different values of scale coefficient β. **a** The output STFT spectrum under $\beta = \beta_0 \times 20$. **b** The output time domain waveform under $\beta = \beta_0 \times 20$. **c** The output STFT spectrum of system output under $\beta = \beta_0 \times 50$. **d** The output time domain waveform under $\beta = \beta_0 \times 50$. **e** The output STFT spectrum under $\beta = \beta_0 \times 80$. **f** The output time domain waveform under $\beta = \beta_0 \times 80$

Herein, we study a bistable nonlinear system excited by a linear frequency-modulated characteristic signal and we also add an auxiliary fast-varying signal in the excitation. The specific dynamical equation is described by

$$\frac{\mathrm{d}x(t)}{\mathrm{d}t} = ax(t) - bx^3(t) + s(t) + F(t). \tag{3.17}$$

In addition to adding a fast-varying auxiliary signal $F(t)$, Eq. (3.17) is the same as Eq. (3.1). The form of the auxiliary signal is

$$F(t) = B\cos(\pi k\gamma t^2 + 2\pi k f t). \tag{3.18}$$

As a result, the instantaneous frequency of the fast-varying signal is ten times that of the slow-varying signal. To achieve aperiodic resonance, we continue to employ the piecewise re-scaled method. In the following process, we divide the entire signal into 12 segments. The time scale is defined as in Eq. (3.2), where t_{ei} represents the end time of each time series. The constant coefficient is set to $\beta_0 = 400$.

The effect of the fast-varying auxiliary signal on the aperiodic system resonance is presented in Fig. 3.15. The amplitude of the fast-varying auxiliary signal significantly

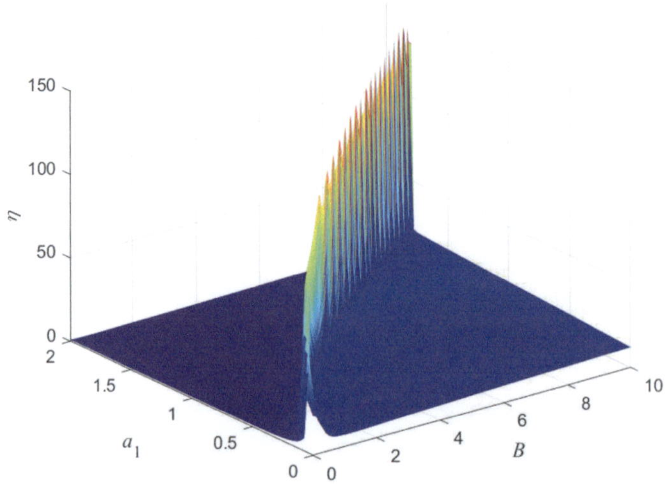

Fig. 3.15 The spectrum amplification factor η versus the system parameters a_1 and the auxiliary signal amplitude B in the three-dimensional surface. The simulation parameters are $A = 0.1, \gamma = 1, f = 1, k = 10, b_1 = 0.01$

affects the spectrum amplification factor of the system response. In fact, it indicates the aperiodic system resonance improved by vibrational resonance. A detailed discussion on aperiodic vibrational resonance will be provided in Chap. 7.

To further illustrate the performance of aperiodic system resonance enhanced by vibrational resonance, we present Fig. 3.16, which again reflects the impact of the auxiliary fast-varying signal. It is evident that the maximum value of the spectral amplification factor η is significantly greater when $B > 0$ compared to the case when $B = 0$. In other words, the aperiodic system resonance is markedly improved by the presence of the auxiliary fast-varying signal.

3.4 Concluding Remarks

In this chapter, we present the piecewise re-scaled method and the real-time re-scaled method to achieve aperiodic resonance in a nonlinear system excited by various types of frequency-modulated signals. Regardless of the complexity of the modulation form, reasonable scale transformations allow for real-time adjustments of the system parameters in accordance with the signal's time scale. This enables effective matching between the signal and system parameters, resulting in the observation of aperiodic resonance phenomena. Additionally, the choice of re-scaled method influences both the degree of signal amplification during aperiodic resonance and the continuity of the signal following the resonance processing. Furthermore, we highlight the beneficial

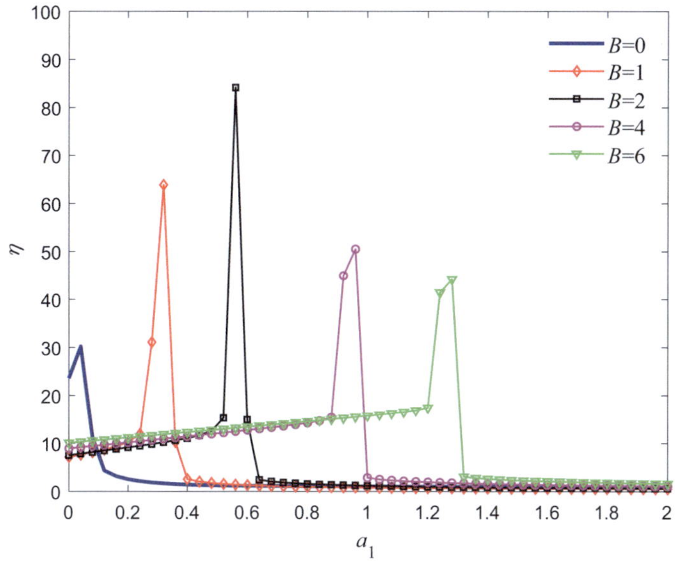

Fig. 3.16 The effect of the fast-varying auxiliary signal on the aperiodic system resonance. The simulation parameters are $A = 0.1$, $\gamma = 1$, $f = 1$, $k = 10$, $b_1 = 0.01$

impact of vibrational resonance on aperiodic system resonance; by incorporating an auxiliary fast-varying signal into the excitation, the performance of aperiodic system resonance can be significantly enhanced.

As potential research directions for aperiodic resonance, we propose two open-ended questions. The first question addresses the behavior of linear systems under aperiodic excitation. While a linear system experiences resonance at the excitation frequency under harmonic excitation, all aperiodic resonances discussed in this chapter occur within nonlinear systems. Can we observe any aperiodic resonance in linear systems? The second question investigates the phenomenon of subharmonic and superharmonic resonance in nonlinear systems under harmonic excitation. Similarly, do nonlinear systems exhibit comparable resonance phenomena when subjected to aperiodic excitation, such as frequency-modulated signals?

References

1. P. Jia, J. Yang, C. Wu, M.A.F. Sanjuán, Amplification of the LFM signal by using piecewise vibrational methods, *Journal of Vibration and Control*, 2019, **25**(1): 141-150.
2. P. Jia, J. Yang, X. Zhang, M.A.F. Sanjuán, On the LFM signal improvement by piecewise vibrational resonance using a new spectral amplification factor, *IET Signal Processing*, 2019, **13**(1): 65-69.
3. C. Wu, Z. Wang, J. Yang, D. Huang, M.A.F. Sanjuán, Adaptive piecewise re-scaled stochastic resonance excited by the LFM signal, *The European Physical Journal Plus*, 2020, **135**: 1-14.

4. Y. Zhai, J. Yang, S. Zhang, H. Liu, Linear frequency modulated signal induced aperiodic resonance, *Physica Scripta*, 2020, **95**(6): 065213.
5. D. Russell, T. Rossing, Testing the nonlinearity of piano hammers using residual shock spectra, *Acta Acustica United with Acustica*, 1998, **84**(5): 967-975.
6. C.K. Kwuimy, B.N. Nbendjo, Active control of horseshoes chaos in a driven Rayleigh oscillator with fractional order deflection, *Physics Letters A*, 2011, **375**(39): 3442-3449.
7. H. Li, X. Liao, S. Ullah, L. Xiao, Analytical proof on the existence of chaos in a generalized Duffing-type oscillator with fractional-order deflection, *Nonlinear Analysis: Real World Applications*, 2012, **13**:(6): 2724-2733.
8. C.K. Kwuimy, G. Litak, C. Nataraj, Nonlinear analysis of energy harvesting systems with fractional order physical properties, *Nonlinear Dynamics*, 2015, **80**:(1): 491-501.
9. J. Yang, S. Zhang, M.A.F. Sanjuán, H. Liu, Time-frequency analysis of a new aperiodic resonance, *Communications in Nonlinear Science and Numerical Simulation*, 2020, **85**: 105258.
10. T. Gong, J. Yang, M.A.F. Sanjuán, H. Liu, Z. Shan, Vibrational resonance by using a real-time scale transformation method, *Physica Scripta*, 2022, **97**(4): 045207.
11. G. Tao, J.H. Yang, Z. Shan, Z.L. Wang, H.G. Liu, Optimal resonance response of nonlinear system excited by nonlinear frequency modulation signal, *Acta Physica Sinica*, 2022, **71**(5): 050503.
12. Z. Shan, Z. Wang, J. Yang, D. Zhou, H. Liu, Extracting non-stationary signal under strong noise background: Time-varying system analysis, *Journal of Vibration and Control*, 2023, **29**(17-18): 4036-4045.
13. T. Gong, J. Yang, Z. Shan, Z. Wang, H. Liu, Optimal resonance response of nonlinear system excited by nonlinear frequency modulation signal, *Acta Physica Sinica*, 2022, **71**(5): 050503.
14. P.S. Landa, P.V.E. McClintock, Vibrational resonance, *Journal of Physics A: Mathematical and General*, 2000, **33**(45): L433-L438.
15. L. Gammaitoni, P. Hänggi, P. Jung, F. Marchesoni, Stochastic resonance, *Reviews of Modern Physics*, 1998, **70**(1): 223-287.
16. I.I. Blekhman, P.S. Landa, Conjugate resonances and bifurcations in nonlinear systems under biharmonical excitation, *International Journal of Non-Linear Mechanics*, 2004, **39**(3): 421-426.
17. J. Yang, S. Rajasekar, M.A.F. Sanjuán, Vibrational resonance: A review, *Physics Reports*, 2024, **1067**: 1-62.
18. J.J. Thomsen, Some general effects of strong high-frequency excitation: stiffening, biasing and smoothening, *Journal of Sound and Vibration*, 2002, **253**(4): 807-831.
19. J.J. Thomsen, Vibrations and Stability, 2003, Berlin: Springer-Verlag.

Chapter 4
Aperiodic Stochastic Resonance Caused by Aperiodic Binary Signal and Noise

This chapter focuses on aperiodic stochastic resonance induced by noise excitation and an aperiodic binary signal. We examine several typical nonlinear models to study this phenomenon, including bistable systems, asymmetric bistable systems, monostable systems, tri-stable systems, and fractional systems. Aperiodic stochastic resonance can be realized through an appropriate level of noise, as measured by the cross-correlation coefficient, and can also be enhanced by adjusting system parameters. Additionally, we address how vibrational resonance improves aperiodic stochastic resonance.

4.1 Aperiodic Stochastic Resonance in Stochastic Ordinary Differential System

Scholars across various fields are familiar with the well-known phenomenon of stochastic resonance [1, 2]. In a nonlinear system subjected simultaneously to a weak characteristic signal and noise, stochastic resonance is a physical phenomenon that leverages noise to create a harmonious interaction among noise, the system, and the weak signal, thereby enhancing the weak signal's characteristics in the response. When the weak characteristic signal is aperiodic, it leads to aperiodic stochastic resonance in the response.

Aperiodic signals come in various forms, and the stochastic resonance induced by different aperiodic signals demonstrates considerable diversity. These variations not only manifest in distinct phenomena but also necessitate different research approaches. Aperiodic stochastic resonance primarily focuses on cases induced by aperiodic binary signals [3–5], linear or nonlinear frequency-modulated signals [6–8], and aperiodic pulse signals [9]. Additionally, some studies discuss methods for effectively recovering signal features using aperiodic stochastic resonance in the presence of strong background noise and completely unknown signals

J. Yang and M. A. F. Sanjuan, *Aperiodic Resonances—Theory and Applications*,
https://doi.org/10.1007/978-981-96-6819-9_4

[10, 11]. In this chapter, we discuss in detail the aperiodic stochastic resonance caused by an aperiodic binary characteristic signal. The phenomena induced by frequency-modulated signals and aperiodic pulse signals will be addressed in the following two chapters.

In this section, we mainly study the stochastic resonance phenomenon of an ordinary bistable system induced by an aperiodic binary signal. The nonlinear system first discussed is governed by

$$\frac{dx}{dt} = ax - bx^3 + s(t) + N(t). \tag{4.1}$$

Equation (4.1) comes from Eq. (2.1) and only an additive noise $N(t)$ is added in the excitation. Herein, $N(t)$ is a Gaussian white noise that satisfies the following statistical properties

$$\begin{cases} \langle N(t) \rangle = 0, \\ \langle N(t), N(0) \rangle = 2D\delta(t), \end{cases} \tag{4.2}$$

where D represents noise intensity, and $\delta(t)$ is the Dirac Delta function. Usually, the Gaussian white noise time series can be generated by the following formula

$$N(i) = \sqrt{\frac{2D}{\Delta t}}\xi(i), \quad i = 1, 2, \ldots n, \tag{4.3}$$

where $\xi(i)$ are random numbers with a standard normal distribution, which can be generated using the "randn" command in MATLAB software. As a result, the discrete equation for the numerical simulation of Eq. (4.1) is given by the form

$$x(i+1) = x(i) + \left[ax(i) - bx^3(i) + s(i) + N(i)\right]\Delta t, \tag{4.4}$$

where Δt is the time step.

We commonly use the cross-correlated coefficient as defined in Eq. (1.17), to quantify aperiodic stochastic resonance. As shown in Fig. 4.1, the effect of noise intensity D on the cross-correlated coefficient C_{sx} is clearly depicted. As noise intensity increases, we observe a pronounced peak in the cross-correlation coefficient, which is generally associated with aperiodic stochastic resonance. However, if the noise intensity continues to increase, the cross-correlation coefficient begins to decrease. This phenomenon is referred to as *aperiodic stochastic resonance*.

Even when the noise intensity is very low, it can still result in a large cross-correlation coefficient. However, this merely causes the output waveform to closely resemble the input signal without amplifying it in the response, as shown in Fig. 4.2a. When we slightly increase the noise intensity, as depicted in Fig. 4.2b, the positive role of noise becomes evident, leading to some amplification of the signal, though complete synchronization between the input and output has not yet been achieved. Figure 4.2c corresponds to the labeled peak of the $C_{sx} - D$ curve of the case $A = 0.35$ in Fig. 4.1. In this scenario, the response exhibits optimal aperiodic stochastic

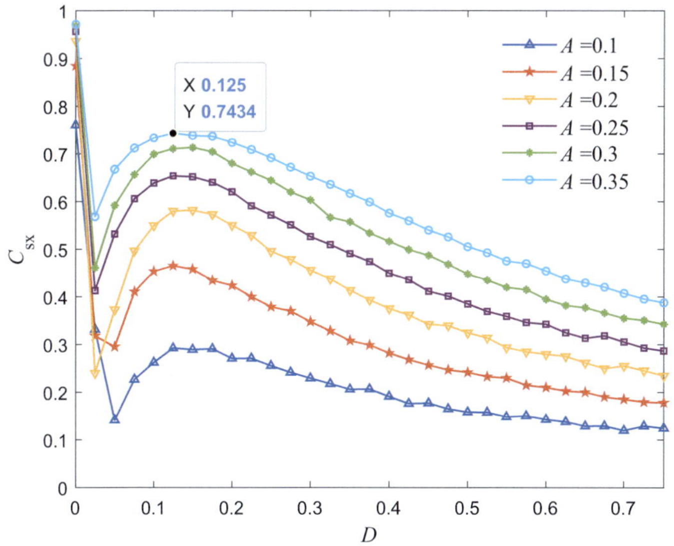

Fig. 4.1 The cross-correlation coefficient C_{sx} is plotted against the noise intensity D. The simulation parameters are $a = 1$, $b = 1$ and $T = 20$

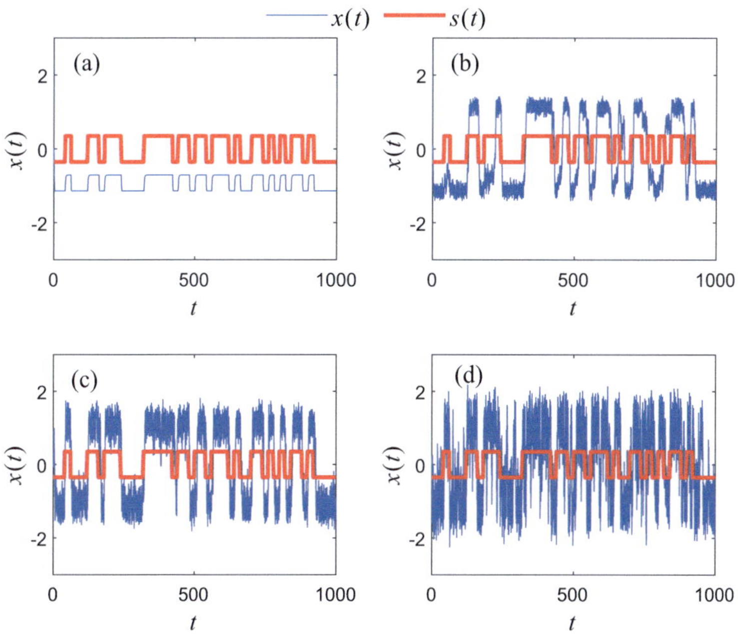

Fig. 4.2 The time series of the output corresponds to different noise intensity D, **a** $D = 0$, **b** $D = 0.025$, **c** $D = 0.125$, **d** $D = 0.4$. The simulation parameters are $a = 1$, $b = 1$, $A = 0.35$ and $T = 20$

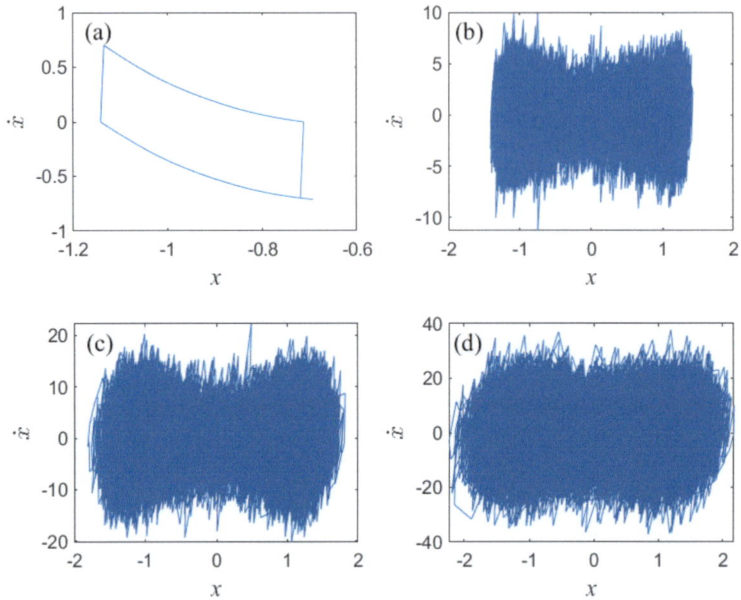

Fig. 4.3 The phase diagrams correspond to the time series presented in Fig. 4.2 sequentially. Each diagram reflects the system's response under varying noise intensities, illustrating how the dynamics change in relation to the time series at different levels of noise

resonance, as illustrated in the time series presented in Fig. 4.2c. However, if we continue to increase the noise intensity beyond this point, it disrupts the order between the system's output and the input, resulting in irregular random fluctuations in the output. To further illustrate the response of the stochastic ordinary bistable system, Fig. 4.3 presents the phase diagrams of Eq. (4.1) for different values of noise intensity. In the absence of noise, the response is confined within a potential well, as shown in Figs. 4.2a and 4.3a. The phase diagrams in Fig. 4.3 vividly reflect the aperiodic characteristics of the system response. As noise intensity increases, the range of the system response gradually expands. While the phase diagrams can illustrate changes in the oscillation range of the system response, they do not directly convey the synchronization characteristics between the system response and the input signal.

In Fig. 4.4, we illustrate the influence of noise on parameter-induced resonance. It is evident that both the location and value of the resonance peak are affected by noise intensity. The nonlinear relationship between noise intensity and the maximum peak further underscores the occurrence of aperiodic stochastic resonance. The stochastic resonance achieved by adjusting system parameters while keeping the signal and noise intensity fixed is referred to as *adaptive stochastic resonance*. This type of stochastic resonance is often realized through optimization algorithms designed to identify the optimal system parameters [12–14].

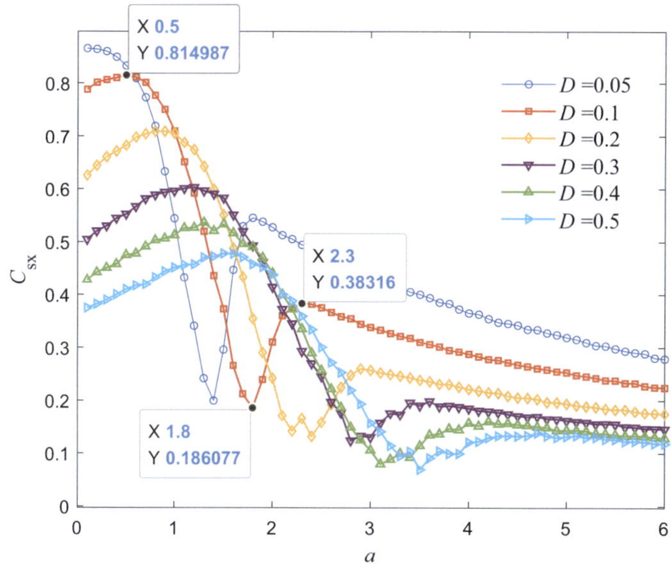

Fig. 4.4 The cross-correlation coefficient C_{sx} is plotted against the system parameter a under different values of noise intensity D. The simulation parameters are $b = 1$, $A = 0.3$ and $T = 20$

In Fig. 4.5, we present the time series of the output corresponding to the curve in Fig. 4.4. The optimal output is shown in Fig. 4.5b, corresponding to the labeled peak in Fig. 4.4. When the parameter a is increased to a larger value, as illustrated in Fig. 4.5c, d, the potential well deepens, preventing the output from crossing the potential barrier, resulting in irregular random fluctuations. The adjustment of the system parameter to achieve aperiodic stochastic resonance is more intuitively reflected in the time series presented in Fig. 4.5.

The potential function of Eq. (4.1) is in symmetric bistable form. In addition to this, we aim to explore aperiodic stochastic resonance in an asymmetric bistable system. The dynamical equation for this system is given by:

$$\frac{dx}{dt} = ax - bx^3 + r + s(t) + N(t), \tag{4.5}$$

where r is the bias constant. The two non-zero equilibrium points are not symmetric. The potential function of Eq. (4.5) is

$$V(x) = -rx - \frac{a}{2}x^2 + \frac{b}{4}x^4. \tag{4.6}$$

It displays asymmetric bistable configuration. See the configuration of this potential function in Fig. A.3 of the Appendix A.1.3.

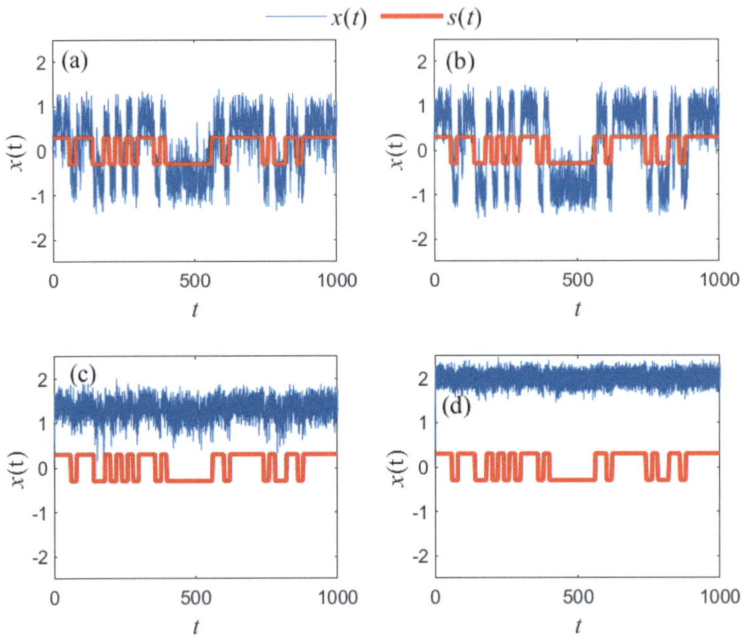

Fig. 4.5 The time series of the output corresponds to different system parameter a, **a** $a = 0.1$, **b** $a = 0.5$, **c** $a = 1.8$, **d** $a = 4$. The simulation parameters are $b = 1$, $A = 0.3$, $T = 20$ and $D = 0.1$

In Fig. 4.6, we illustrate aperiodic stochastic resonance in the asymmetric bistable system. The behavior observed is quite similar to that in the symmetric bistable system, indicating that the asymmetry does not impede the occurrence of aperiodic stochastic resonance.

To illustrate the role of asymmetry in aperiodic stochastic resonance, we present Fig. 4.7, which displays the curves of C_{sx} versus D. Notably, most curves reach their maximum values at $c = 0$, indicating that aperiodic stochastic resonance is more effectively achieved in symmetric bistable systems compared to asymmetric ones. However, there is one curve where, at a relatively low value of D (specifically $D = 0.05$), the asymmetric bistable system demonstrates a certain advantage in achieving aperiodic stochastic resonance.

Certainly, there are various forms of asymmetric bistable systems. We will leave the relevant issues for readers to explore in depth and will not engage in detailed discussions here. In our previous review paper [15], we also provided a summary of the equations governing different asymmetric bistable systems.

Now, we will discuss the existence of aperiodic stochastic resonance in a monostable system. The dynamical equation for this system is given by

$$\frac{dx}{dt} = -ax - bx^3 + s(t) + N(t). \tag{4.7}$$

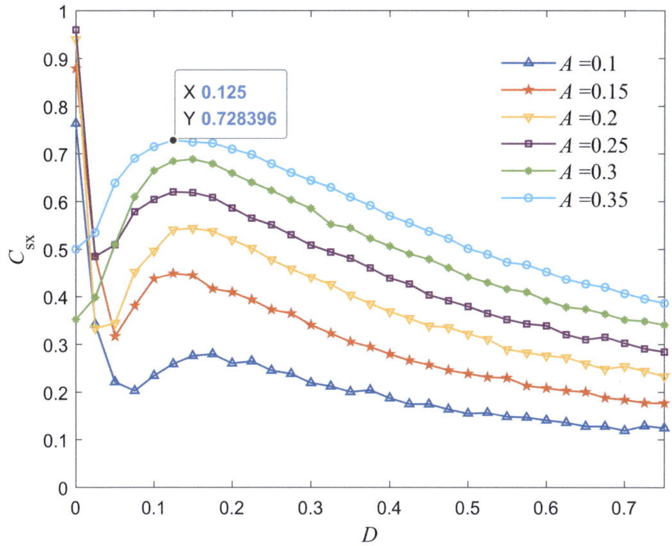

Fig. 4.6 The cross-correlation coefficient C_{sx} is plotted against the noise intensity D in the asymmetric bistable system. The simulation parameters are $a = 1$, $b = 1$, $c = 0.1$ and $T = 20$

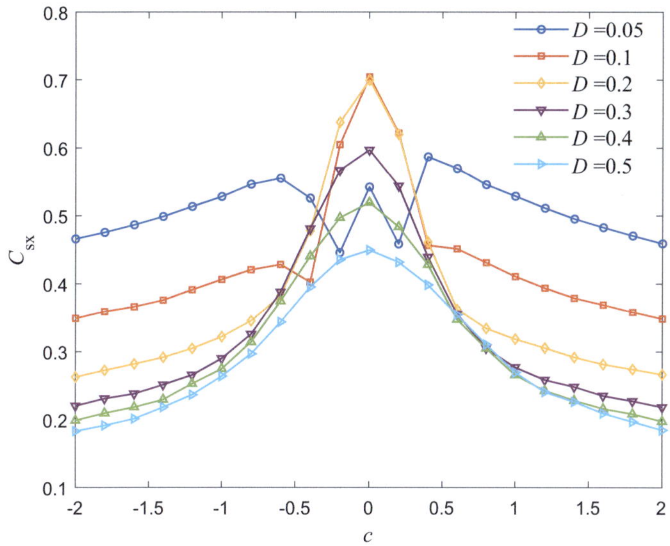

Fig. 4.7 The cross-correlation coefficient C_{sx} is plotted against the asymmetric parameter c under different values of noise intensity D. The simulation parameters are $a = 1$, $b = 1$ and $T = 20$

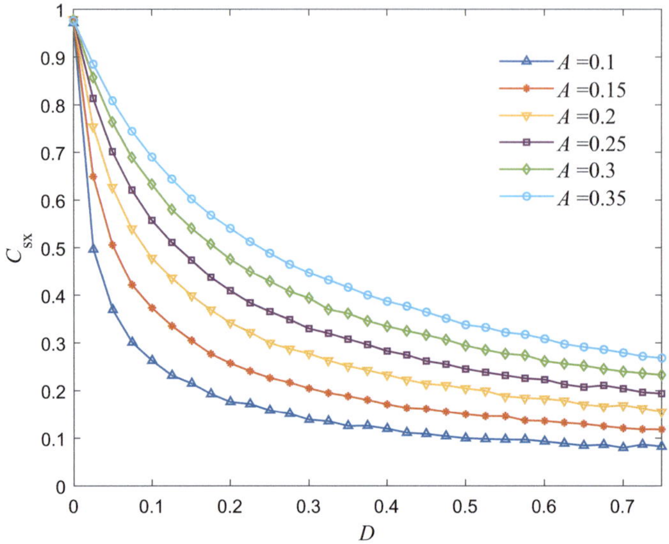

Fig. 4.8 The cross-correlation coefficient C_{sx} is shown as a function of the noise intensity D in the monostable bistable system. The simulation parameters are $a = 1$, $b = 1$ and $T = 20$

When $a > 0$ and $b > 0$, the potential function takes on a monostable shape, with a single equilibrium point at $x = 0$. In Fig. 4.8, we present the cross-correlation coefficient C_{sx} as the noise intensity D increases. It is clear that the cross-correlation coefficient C_{sx} is a decreasing function of the noise intensity D. Specifically, this indicates that there is no aperiodic stochastic resonance present in the monostable system.

We are also interested in exploring aperiodic stochastic resonance in a nonlinear system with a tristable potential function

$$\frac{dx}{dt} = -ax^5 - bx^3 - cx + s(t) + N(t). \tag{4.8}$$

When $a > 0$, $b < 0$ and $c > 0$, the potential function is

$$V(x) = \frac{c}{2}x^2 + \frac{b}{4}x^4 + \frac{a}{6}x^6. \tag{4.9}$$

It features a triple-well configuration with three stable equilibria and two unstable equilibria, as shown in Fig. A.7 of Appendix A.3.

As shown in Fig. 4.9, the cross-correlation coefficient C_{sx} versus the noise intensity D still exhibits aperiodic stochastic resonance. However, the maximum value of C_{sx} is smaller compared to that in the bistable system. This reduction can be attributed to both the parameter selection in this figure and the inherent characteristics of the triple-well shape of the potential function.

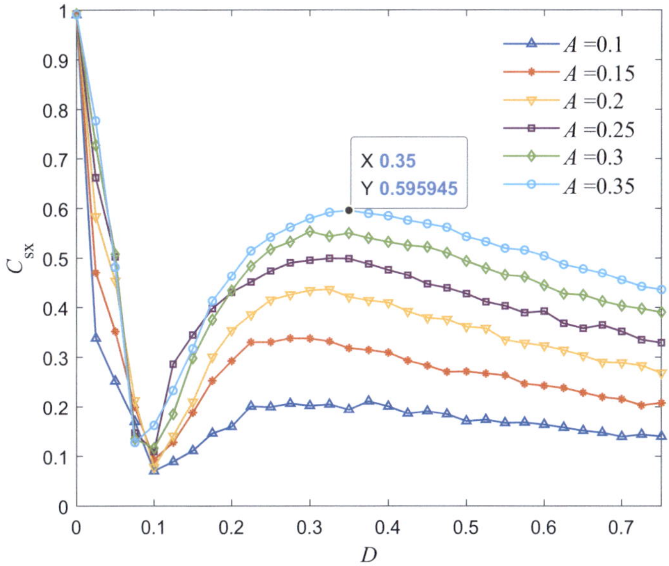

Fig. 4.9 The cross-correlation coefficient C_{sx} versus the noise intensity D in the tri-stable system. The simulation parameters are $a = 1, b = -4, c = 3, T = 20$

To further illustrate the possible reason for the weak aperiodic stochastic resonance in the output of a tristable system, we present Fig. 4.10. Even with the strongest aperiodic stochastic resonance output, as shown in Fig. 4.10c, the synchronization between the output and input signals remains relatively poor. This lack of synchronization is the direct reason for the low value of the cross-correlation coefficient between the system output and the input signal.

The phase diagrams corresponding to the time series in Fig. 4.10 are presented in Fig. 4.11. It is evident that, since $x = 0$ is a stable fixed equilibrium of the system, there is a high probability that the phase diagram of the system response is concentrated near $x = 0$. This observation further underscores the characteristics of the tristable system. Thus, pairing an aperiodic binary system with a tristable system may not be an optimal choice.

While this chapter focuses exclusively on the effects of white Gaussian noise on aperiodic stochastic resonance, it is important to note that a substantial body of work has explored other types of noise, including additive correlated noise [16], multiplicative noise [17], Gaussian colored noise [18], Lévy noise [19, 20], and dichotomous noise [21], among others. Additionally, beyond the cross-correlation coefficient, which quantifies aperiodic stochastic resonance, other indicators such as response amplitude gain and bit error rate can also be employed to measure this phenomenon. We encourage readers to explore these open questions further.

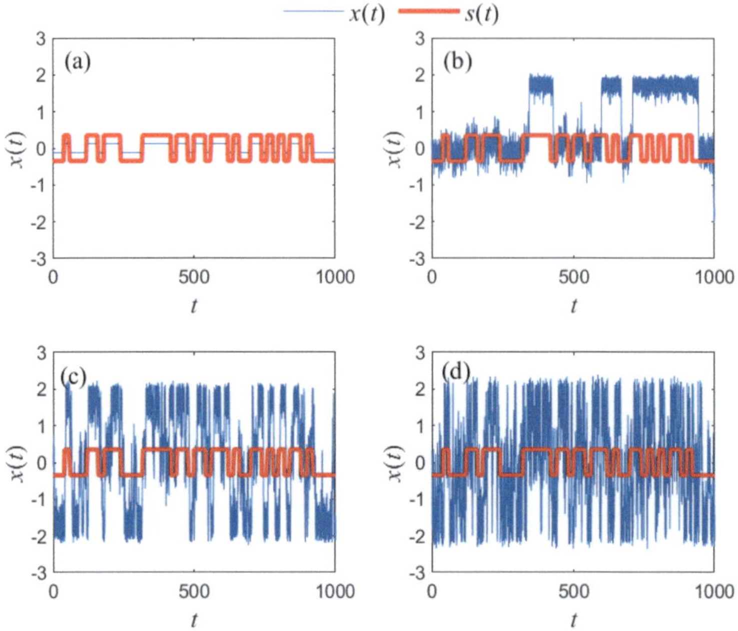

Fig. 4.10 The time series of the output corresponds to different noise intensity D, **a** $D = 0$, **b** $D = 0.1$, **c** $D = 0.35$, **d** $D = 0.75$. The simulation parameters are $a = 1$, $b = -4$, $c = 3$, $A = 0.35$ and $T = 20$

4.2 Aperiodic Stochastic Resonance in Stochastic Fractional Differential System

Aperiodic stochastic resonance is observed in a wide variety of nonlinear systems beyond the typical examples, such as fractional-order systems [22], excitable systems [3], rat SA1 cutaneous mechanoreceptors [23], electrochemical cells [24], and hysteretic populations of cardiac neurons [25]. Notably, aperiodic resonance is also prevalent in fractional-order differential systems. These systems involve equations that incorporate fractional-order calculus operations [26]. To enhance readers' understanding of aperiodic resonance in fractional-order differential systems, we delve into this topic in detail. Fractional-order differential systems are often referred to as the "systems of the $21st$ century" [27], and they find extensive applications in dynamic modeling [28–31], industrial process control [32–34], signal processing [35–37], heat conduction [38], electrochemistry [39], anomalous diffusion [40–42], and image processing [43–45]. The presence of numerous aperiodic signals across various fields highlights the importance of studying aperiodic stochastic resonance in fractional-order differential systems, which holds both theoretical significance [46, 47] and practical engineering application value [48].

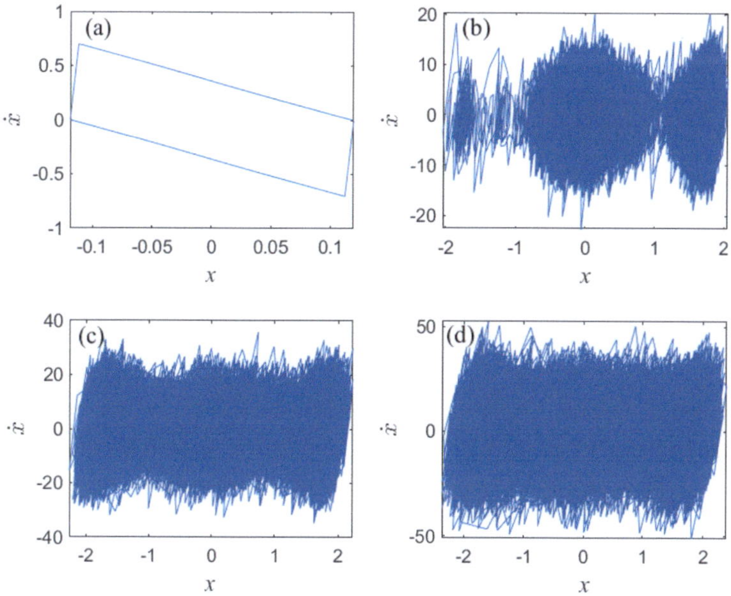

Fig. 4.11 The phase diagrams corresponds to the time series in Fig. 4.10 sequentially

We study the aperiodic stochastic resonance in a fractional-order bistable system [49]

$$\frac{d^{\alpha}x}{dt^{\alpha}} = ax - bx^3 + s(t) + N(t), \alpha \in (0, 2], \tag{4.10}$$

where α is the order of the fractional derivative. Several definitions of the fractional derivative exist, including the Grünwald-Letnikov definition, the Riemann-Liouville definition, and the Caputo definition [26, 50]. The Grünwald-Letnikov definition of fractional derivatives is given by

$$\frac{d^{\alpha}x(t)}{dt^{\alpha}}|_{t=k\Delta t} = \lim_{\Delta t \to 0} \frac{1}{\Delta t^{\alpha}} \sum_{j=0}^{k} (-1)^j \binom{\alpha}{j} f(k\Delta t - j\Delta t), \tag{4.11}$$

where the binomial coefficient is expressed in the form of the Gamma function

$$\binom{\alpha}{j} = \frac{\Gamma(\alpha+1)}{\Gamma(j+1)\Gamma(\alpha-j+1)}. \tag{4.12}$$

Letting

$$w_j^{\alpha} = (-1)^j \binom{\alpha}{j}, \tag{4.13}$$

then, according to the reference [26], we have

$$w_0^\alpha = 1, \quad w_k^\alpha = \left(1 - \frac{\alpha + 1}{k}\right) w_{k-1}^\alpha, \quad k = 1, \quad 2, \quad \cdots, \quad n. \tag{4.14}$$

If $\alpha = 1$, Eq. (4.12) changes to

$$w_0^1 = 1, \quad w_1^1 = -1, \quad w_k^\alpha = 0, \quad k = 2, \quad \cdots, \quad n. \tag{4.15}$$

At this point, the Grünwald-Letnikov definition of the fractional derivative is transformed to the ordinary derivative definition

$$\frac{dx(t)}{dt} = \lim_{\Delta t \to 0} \frac{x(t) - x(t - \Delta t)}{\Delta t}. \tag{4.16}$$

Substituting Eqs. (4.3), (4.11)–(4.14)–(4.10), we directly discretize the stochastic fractional differential equation to

$$\lim_{\Delta t \to 0} \frac{1}{\Delta t^\alpha} \left[x(k) + \sum_{j=1}^{k-1} w_j^\alpha x(k - j) \right] = f\left[x(k-1)\right] + N(k-1). \tag{4.17}$$

By taking a small value for the time step Δt and removing the limit sign, we obtain

$$x(k) = -\sum_{j=1}^{k-1} w_j^\alpha x(k-j) + \Delta t^\alpha \left\{ f\left[x(k-1)\right] + N(k-1) \right\}. \tag{4.18}$$

With the special case $\alpha = 1$, the expression in Eq. (4.18) turns to Eq. (4.4).

We use noise intensity as the control variable. Figure 4.12 illustrates the dependence of the cross-correlation coefficient C_{sx} on noise intensity D for different values of the fractional order α. Similar to the case of the ordinary derivative (i.e., $\alpha = 1$), the curves for various fractional orders exhibit distinct resonance peaks. As the fractional order $\alpha = 1$ increases, the corresponding peak in the curve occurs at a lower noise intensity, while the peak value slightly increases.

Corresponding to the "strongest" aperiodic stochastic resonance, Fig. 4.13 presents the time series for different cases. For each value of α shown in the figure, aperiodic stochastic resonance is consistently observed. The key difference is that a smaller value of α results in a larger range of oscillation within the potential well.

To further illustrate the influence of the fractional order on the response, Fig. 4.14 presents the phase diagrams corresponding one-to-one with the time series in Fig. 4.13. These diagrams intuitively demonstrate the range and velocity of the response oscillating within the potential well. This reinforces the observation that both the oscillation range and velocity are greater for smaller values of the fractional order α.

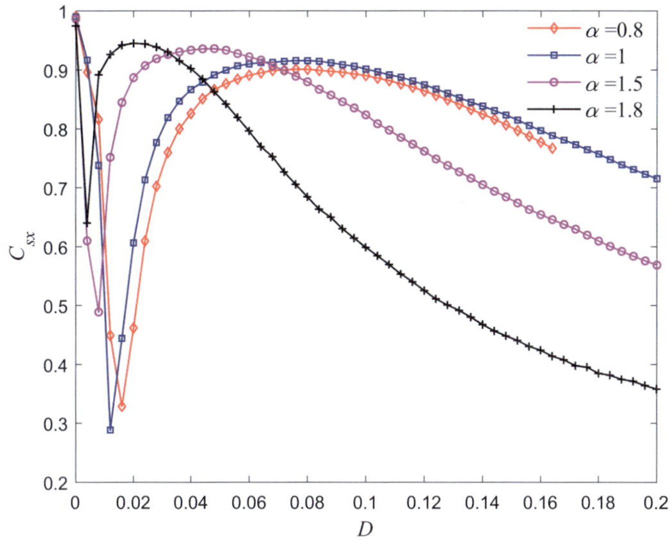

Fig. 4.12 The cross-correlation coefficient C_{sx} versus the noise intensity D under different values of the fractional-order α. The simulation parameters are $a = 1$, $b = 1$, $A = 0.25$ and $T = 200$

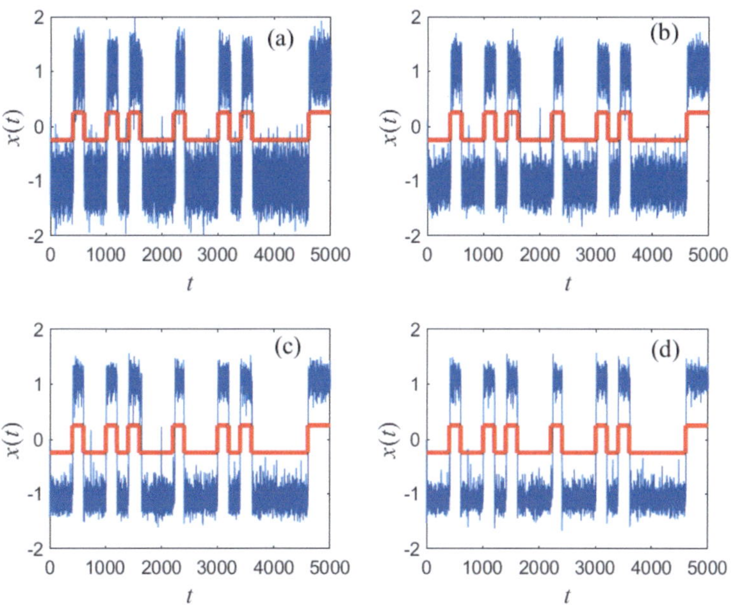

Fig. 4.13 The time series of the output corresponds to the optimal aperiodic stochastic resonance. **a** $\alpha = 0.8$, $D = 0.076$, **b** $\alpha = 1$, $D = 0.072$, **c** $\alpha = 1.5$, $D = 0.044$, **d** $\alpha = 1.8$, $D = 0.02$. The simulation parameters are $a = 1$, $b = 1$, $A = 0.25$ and $T = 200$

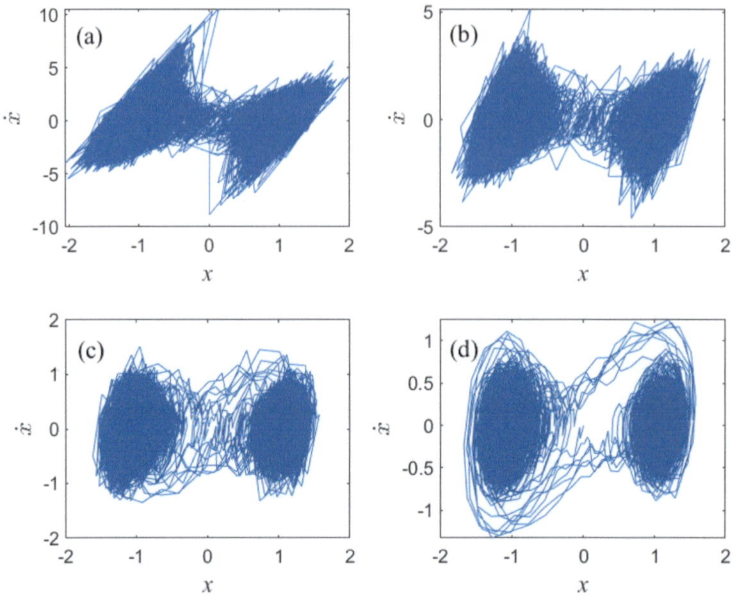

Fig. 4.14 The phase portraits of the output corresponds to the optimal aperiodic stochastic resonance. **a** $\alpha = 0.8$, $D = 0.076$, **b** $\alpha = 1$, $D = 0.072$, **c** $\alpha = 1.5$, $D = 0.044$, **d** $\alpha = 1.8$, $D = 0.02$. The simulation parameters are $a = 1$, $b = 1$, $A = 0.25$ and $T = 200$

Although using the Grünwald-Letnikov definition for numerical calculation is relatively simple, the prerequisite for such calculation is the initial condition $x(0) = 0$. Of course, for some systems, the initial conditions do not affect the steady-state response of the system. However, there are still many systems whose responses are influenced by initial conditions significantly, such as chaotic systems. To lift this restriction, we can use the Caputo definition for discretizing the factional operator [51, 52].

The Caputo definition of the fractional derivative has a good engineering application background, and the predictor-corrector method introduced here is based on the Caputo definition. There are various forms of the predictor-corrector algorithm, and we only introduce the most commonly used as below. The formula for numerically iterating Eq. (4.10) using the predictor-corrector method under any initial condition $x(0)$ is given by

$$\begin{aligned} x(i+1) = x_0 + \tfrac{\Delta t^{\alpha}}{\Gamma(\alpha+2)} \left[a x_p(j) - b {x_p}^3(j) + s(j) + N(j)\right] \\ + \tfrac{\Delta t^{\alpha}}{\Gamma(\alpha+2)} \sum_{j=0}^{i} a_{j,i+1} \left[a x(j) - b x^3(j) + s(j) + N(j)\right] \end{aligned}, \tag{4.19}$$

where $c_{j,i+1}$ is the weight coefficient of the correction term

$$c_{j,i+1} = \begin{cases} i^{\alpha+1} - (i-\alpha)(i+1)^{\alpha}, & j = 0 \\ (i-j+2)^{\alpha+1} + (i-j)^{\alpha+1} - 2(i-j+1)^{\alpha+1}, & 1 \le j \le i \end{cases} \tag{4.20}$$

and $x_p(i+1)$ is the predicted value of $x(i+1)$

$$x_p(i+1) = x_0 + \frac{1}{\Gamma(\alpha)} \sum_{j=0}^{i} d_{j,i+1}[ax(i) - bx^3(i) + s(i) + N(i)], \tag{4.21}$$

and $d_{j,i+1}$ represents the weight coefficient of the predicted term

$$d_{j,i+1} = \frac{\Delta t^{\alpha}}{\alpha} \left[(i-j+1)^{\alpha} - (i-j)^{\alpha}\right]. \tag{4.22}$$

Readers can verify this algorithm by themselves.

4.3 Aperiodic Stochastic Resonance Improved by Vibrational Resonance

There have been several studies on stochastic resonance enhanced by vibrational resonance [53–56]. By incorporating a fast-varying square waveform signal into the excitation in Eq. (4.1), then we have

$$\frac{dx}{dt} = ax - bx^3 + s(t) + F(t) + N(t), \tag{4.23}$$

herein,

$$F(t) = B\mathrm{sgn}[\cos(\Omega t)], \tag{4.24}$$

and we let $\Omega = 2k\pi/T$.

In Fig. 4.15, we observe that for the case $B = 0$, aperiodic stochastic resonance is present. However, for $B = 2$, the peak of the cross-correlation coefficient achieves a significantly larger value compared to that for $B = 0$. As B is further increased, aperiodic stochastic resonance begins to disappear, with the cross-correlation coefficient decreasing as the noise intensity D increases. This indicates that only an appropriately tuned fast-varying signal can induce vibrational resonance and enhance the performance of aperiodic stochastic resonance.

In Fig. 4.15, we primarily examine the effect of the strength of the auxiliary signal on the performance of aperiodic stochastic resonance. In Fig. 4.16, we investigate the impact of the time scale, specifically the period of the auxiliary signal, on aperiodic stochastic resonance. Initially, for the case $k = 5$, no resonance phenomenon

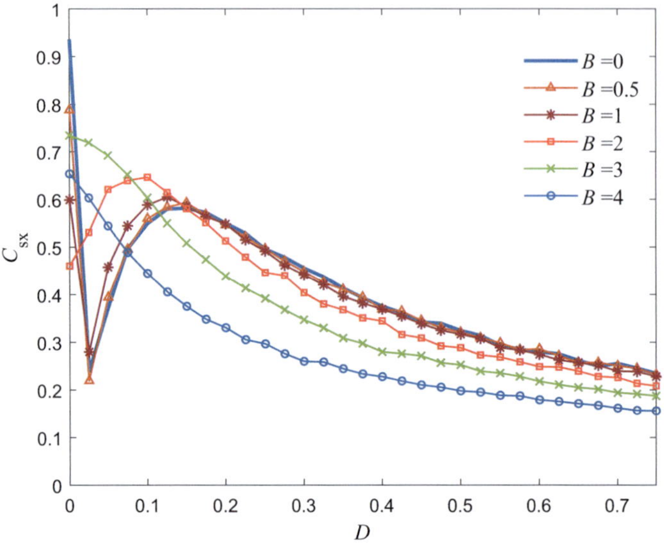

Fig. 4.15 Effect of the fast-varying square waveform on the aperiodic stochastic resonance under different values of B. The simulation parameters are $a = 1$, $b = 1$, $A = 0.2$, $T = 20$ and $k = 20$

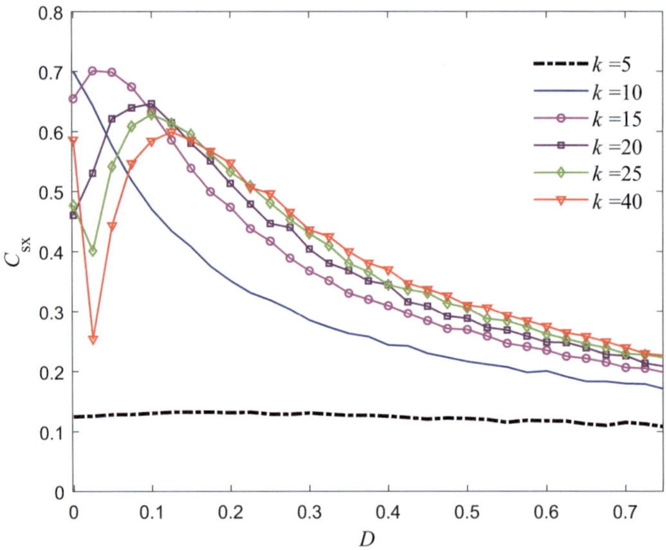

Fig. 4.16 Effect of the fast-varying square waveform on the aperiodic stochastic resonance under different values of k. The simulation parameters are $a = 1$, $b = 1$, $A = 0.2$, $T = 20$ and $B = 2$

is observed, and the original aperiodic stochastic resonance has also disappeared. In this scenario, the auxiliary signal disrupts the synchronization between the system output and the input. As the value of Ω (i.e., k) increases, aperiodic stochastic resonance begins to emerge. However, for $k = 10$, aperiodic stochastic resonance is still absent, and the cross-correlation coefficient decreases with increasing noise intensity D. For the cases $k = 15, 20, 25, 40$, the aperiodic stochastic resonance appears. In summary, there is an optimal value of k that allows aperiodic stochastic resonance to reach its optimal state. The results in Figs. 4.15 and 4.16 clearly illustrate the enhancement of aperiodic stochastic resonance through vibrational resonance.

4.4 Concluding Remarks

In this chapter, we explore aperiodic stochastic resonance in both stochastic ordinary and fractional differential systems, highlighting the beneficial role of noise in facilitating this phenomenon. For a characteristic signal with fixed noise, adaptive aperiodic stochastic resonance can be achieved by selecting appropriate system parameters. To enhance the performance of aperiodic stochastic resonance, we incorporate vibrational resonance, allowing the peak value of the cross-correlation coefficient to reach significant levels.

Our discussions reveal that the bistable system is an ideal nonlinear model for processing aperiodic binary signals. Beyond the nonlinear models, Gaussian white noise, and aperiodic binary signals considered in this chapter, we encourage readers to investigate other nonlinear models, various types of noise, and different aperiodic signals (such as M-ary signals) in the context of aperiodic stochastic resonance. Furthermore, if the characteristic signal takes on a rapidly varying form, aperiodic stochastic resonance can also be realized within a re-scaled theoretical framework. Notably, even when the aperiodic binary signal is obscured by strong noise and its waveform is completely unknown, we can still induce aperiodic resonance and recover the signal [10, 57].

References

1. L. Gammaitoni, P. Hänggi, P. Jung, F. Marchesoni, Stochastic resonance, *Reviews of Modern Physics*, 1998, **70**(1): 223-287.
2. Z. Budrikis, Forty years of stochastic resonance, *Nature Reviews Physics*, **2021**, 3(12): 771-771.
3. J.J. Collins, C.C. Chow, T.T. Imhoff, Aperiodic stochastic resonance in excitable systems, *Physical Review E*, 1995, **52**(4): R3321-R3324.
4. J.J. Collins, C.C. Chow, A.C. Capela, T.T. Imhoff, Aperiodic stochastic resonance, *Physical Review E*, 1996, **54**(5): 5575-5584.
5. K. Park, Y.C. Lai, Z. Liu, A. Nachman, Aperiodic stochastic resonance and phase synchronization, *Physics Letters A*, 2004, **326**(5-6): 391-396.
6. C. Wu, Z. Wang, J. Yang, D. Huang, M.A.F. Sanjuán, Adaptive piecewise re-scaled stochastic resonance excited by the LFM signal, *The European Physical Journal Plus*, 2020, **135**: 130.

7. Z. Shan, Z. Wang, J. Yang, D. Zhou, H. Liu, Extracting non-stationary signal under strong noise background: Time-varying system analysis, *Journal of Vibration and Control*, 2023, **29**(17-18): 4036-4045.
8. C. Yang, J. Yang, D. Zhou, S. Zhang, G. Litak, Adaptive stochastic resonance in bistable system driven by noisy NLFM signal: phenomenon and application., *Philosophical Transactions of the Royal Society A*, 2021, **379**(2192): 20200239.
9. Z. Xu, Z. Wang, J. Yang, M.A.F. Sanjuán, B. Sun, S. Huang, Aperiodic stochastic resonance in a biased monostable system excited by different weak aperiodic pulse signals and strong noise, *The European Physical Journal Plus*, 2023, **138**(5): 386.
10. D. Huang, J. Yang, D. Zhou, M.A.F. Sanjuán, H. Liu, Recovering an unknown signal completely submerged in strong noise by a new stochastic resonance method, *Communications in Nonlinear Science and Numerical Simulation*, 2019, **66**: 156-166.
11. C. Wu, J. Yang, M.A.F. Sanjuán, H. Liu, Stochastic resonance induced by an unknown linear frequency modulated signal in a strong noise background, *Chaos*, 2020, **30**(4): 043128.
12. B. He, Y. Huang, D. Wang, B. Yan, D. Dong, A parameter-adaptive stochastic resonance based on whale optimization algorithm for weak signal detection for rotating machinery, *Measurement* 2019, **136**: 658-667.
13. X. Zhang, Q. Miao, Z. Liu, Z. He, An adaptive stochastic resonance method based on grey wolf optimizer algorithm and its application to machinery fault diagnosis, *ISA Transactions*, 2017, **71**: 206-214.
14. D. Huang, J. Yang, J. Zhang, H. Liu, An improved adaptive stochastic resonance method for improving the efficiency of bearing faults diagnosis, *Proceedings of the Institution of Mechanical Engineers, Part C: Journal of Mechanical Engineering Science*, 2018, **232**(13): 2352-2368.
15. J. Yang, S. Rajasekar, M. A. F. Sanjuán, Vibrational resonance: A review, *Physics Reports*, 2024, **1067**: 1-62.
16. A. Capurro, K. Pakdaman, T. Nomura, S. Sato, Aperiodic stochastic resonance with correlated noise, *Physical Review E*, 1998, **58**(4): 4820.
17. B. Xu, J. Li, J. Zheng, Parameter-induced aperiodic stochastic resonance in the presence of multiplicative noise and additive noise, *Physica A*, 2004, **343**: 156-166.
18. Y. Kang, R. Liu, X. Mao, Aperiodic stochastic resonance in neural information processing with Gaussian colored noise, *Cognitive Neurodynamics*, 2021, **15**: 517-532.
19. L. Zeng, R. Bao, B. Xu, Effects of Lévy noise in aperiodic stochastic resonance, *Journal of Physics A: Mathematical and Theoretical*, 2007, **40**(26): 7175-7185.
20. L. Zeng, B. Xu, Effects of asymmetric Lévy noise in parameter-induced aperiodic stochastic resonance, *Physica A*, 2010, **389**(22): 5128-5136.
21. Y. Jin, Z. Ma, S. Xiao, Coherence and stochastic resonance in a periodic potential driven by multiplicative dichotomous and additive white noise, *Chaos, Solitons & Fractals*, 2017, **103**: 470-475.
22. C. Wu, S. Lv, J. Long, J. Yang, M.A.F. Sanjuán Self-similarity and adaptive aperiodic stochastic resonance in a fractional-order system, *Nonlinear Dynamics*, 2018, **91**: 1697-1711.
23. J.J. Collins, T.T. Imhoff, P. Grigg, Noise-enhanced information transmission in rat SA1 cutaneous mechanoreceptors via aperiodic stochastic resonance, *Journal of Neurophysiology*, 1996, **76**(1): 642-645.
24. I. Tiwari, R. Phogat, P. Parmananda, J.L. Ocampo-Espindola, M. Rivera, Intrinsic periodic and aperiodic stochastic resonance in an electrochemical cell, *Physical Review E*, 2016, **94**(2): 022210.
25. G.C. Kember, G.A. Fenton, K. Collier, J.A. Armour, Aperiodic stochastic resonance in a hysteretic population of cardiac neurons, *Physical Review E*, 2000, **61**(2): 1816-1824.
26. C.A. Monje, Y. Chen, B.M. Vinagre, D. Xue, V. Feliu-Batlle, *Fractional-order Systems and Controls: Fundamentals and Applications* (Springer Science & Business Media, 2010)
27. M.D. Ortigueira, An introduction to the fractional continuous-time linear systems: the 21st century systems, *IEEE Circuits and Systems Magazine*, 2008, **8**(3): 19-26.

28. A. Bonfanti, J.L. Kaplan, G. Charras, A. Kabla, Fractional viscoelastic models for power-law materials, *Soft Matter*, 2020, **16**(26): 6002-6020.
29. M. Sasso, G. Palmieri, D. Amodio, Application of fractional derivative models in linear viscoelastic problems, *Mechanics of Time-Dependent Materials*, 2011, **15**: 367-387.
30. K. Adolfsson, M. Enelund, P. Olsson, On the fractional order model of viscoelasticity, *Mechanics of Time-dependent Materials*, 2005, **9**: 15-34.
31. H. Schiessel, R. Metzler, A. Blumen, T. Nonnenmacher, Generalized viscoelastic models: their fractional equations with solutions, *Journal of Physics A: Mathematical and General*, 1995, **28**(23): 6567-6584.
32. P. Shah, S. Agashe, Review of fractional PID controller, *Mechatronics*, 2016, **38**: 29-41.
33. R. Caponetto, *Fractional Order Systems: Modeling and Control Applications*, (World Scientific, Singapore, 2010)
34. J. Sabatier, P. Lanusse, P. Melchior, A. Oustaloup, Fractional order differentiation and robust control design, *Intelligent Systems, Control and Automation: Science and Engineering*, 2015, **77**: 13-18.
35. S. Das, I. Pan, *Fractional Order Signal Processing: Introductory Concepts and Applications* (Springer Science & Business Media, 2011)
36. R. Magin, M.D. Ortigueira, I. Podlubny, J. Trujillo, On the fractional signals and systems, *Signal Processing*, 2011, **91**(3) 350-371.
37. S.M. Shah, R. Samar, N.M. Khan, M.A.Z. Raja, Fractional-order adaptive signal processing strategies for active noise control systems, *Nonlinear Dynamics*, 2016, **85**: 1363-1376.
38. Y. Povstenko, Fractional heat conduction equation and associated thermal stress, *Journal of Thermal Stresses*, 2004, **28**(1): 83-102.
39. K.B. Oldham, Fractional differential equations in electrochemistry, *Advances in Engineering Software*, 2010, **41**(1): 9-12.
40. L.R. Evangelista, E.K. Lenzi, *Fractional Diffusion Equations and Anomalous Diffusion* (Cambridge University Press, 2018)
41. R. Metzler, W.G. Glöckle, T.F. Nonnenmacher, Fractional model equation for anomalous diffusion, *Physica A*, 1994, **211**(1): 13-24.
42. W. Chen, H. Sun, X. Zhang, D. Korošak, Anomalous diffusion modeling by fractal and fractional derivatives, *Computers & Mathematics with Applications*, 2010, **59**(5): 1754-1758.
43. L. Chen, H. Yin, T. Huang, L. Yuan, S. Zheng, L. Yin, Chaos in fractional-order discrete neural networks with application to image encryption, *Neural Networks*, 2020, **125**: 174-184.
44. Y. Xu, H. Wang, Y. Li, B. Pei, Image encryption based on synchronization of fractional chaotic systems, *Communications in Nonlinear Science and Numerical Simulation*, 2014, **19**(10): 3735-3744.
45. F. Yang, J. Mou, J. Liu, C. Ma, H. Yan, 2020, Characteristic analysis of the fractional-order hyperchaotic complex system and its image encryption application, *Signal Processing*, **169**: 107373.
46. J. Yang, Q. Ma, C. Wu, H. Liu, Aperiodic vibrational resonance in the fractional-order bistable system, *Acta Physica Sinica*, 2018, **67**(5): 054501.
47. C. Wu, S. Lv, J. Long, J. Yang, M.A.F. Sanjuán, Self-similarity and adaptive aperiodic stochastic resonance in a fractional-order system, *Nonlinear Dynamics*, 2018, **91**: 1697-1711.
48. C. Wu, J. Yang, D. Huang, H. Liu, E. Hu, Weak signal enhancement by fractional-order system resonance and its application in bearing fault diagnosis, *Measurement Science and Technology*, 2019, **30**(3): 035004.
49. L. Lin, H. Wang, W. Lv, Stochastic resonance system with linear random frequency fluctuation for aperiodic LFM signal, *Nonlinear Dynamics*, 2017, **88**: 1361-1371.
50. I. Petráš, *Fractional-Order Nonlinear Systems: Modeling, Analysis and Simulation* (Springer Science & Business Media, 2011)
51. K. Diethelm, N.J. Ford, A.D. Freed, A predictor-corrector approach for the numerical solution of fractional differential equations, *Nonlinear Dynamics*, 2002, **29**: 3-22.
52. V. Daftardar-Gejji, Y. Sukale, S. Bhalekar, A new predictorCcorrector method for fractional differential equations, *Applied Mathematics and Computation*, 2014, **244**: 158-182.

53. J. Zhao, J. Yang, J. Zhang, C. Wu, D. Huang, Improving the stochastic resonance in a bistable system with the bounded noise excitation, *Journal of Statistical Physics*, 2018, **173**: 1688-1697.
54. H. Liu, J. Yang, M.A.F. Sanjuán, Improvement in the stochastic resonance in the Duffing oscillator subjected to a Poisson white noise excitation, *The European Physical Journal Plus*, 2021, **136**(5): 528.
55. V.N. Chizhevsky, G. Giacomelli, Improvement of signal-to-noise ratio in a bistable optical system: Comparison between vibrational and stochastic resonance, *Physical Review A*, 2005, **71**(1): 011801.
56. H. Liu, J. Yang, H. Liu, S. Shi, Different fast excitations on the improvement of stochastic resonance in bounded noise excited system, *International Journal of Modern Physics B*, 2020, **34**(26): 2050238.
57. C. Wu, C. Wu, Recovery and enhancement of unknown aperiodic binary signal by adaptive aperiodic stochastic resonance, *Pramana*, 2021, **95**: 36.

Chapter 5
Aperiodic Stochastic Resonance Caused by Frequency-Modulated Signal and Noise

In this chapter, we introduce frequency-modulated signals in the excitation process to induce aperiodic stochastic resonance. We primarily focus on several types of signals: linear frequency-modulated signals, linear frequency-modulated signals with unknown parameters, ultra-high frequency linear frequency-modulated signals, and nonlinear frequency-modulated signals. To achieve aperiodic stochastic resonance with these characterized signals, we present a theoretical framework that incorporates piecewise or real-time re-scaling, fractional Fourier transformation, fractional signal-to-noise ratio, and a parametric asymmetric bistable system. Additionally, we explore the use of signal decomposition methods to address the associated challenges. Under the excitation of frequency-modulated signals, both aperiodic stochastic resonance and inverse periodic stochastic resonance may be observed.

5.1 Linear Frequency-Modulated Signal as the Characteristic Signal

Scholars are increasingly interested in aperiodic stochastic resonance induced by linear frequency-modulated signals. Lin et al. introduced the fractional Fourier transformation to address linear frequency-modulated signals, demonstrating the occurrence of aperiodic stochastic resonance [1, 2]. However, their results primarily focus on scenarios with relatively low instantaneous frequency, without discussing how to manage high instantaneous frequencies. We address this gap using the piecewise re-scaled method [3]. Furthermore, we enhanced this re-scaled method into a real-time re-scaled approach, which we applied to bearing fault diagnosis under time-varying speed conditions [4]. This subchapter emphasizes aperiodic stochastic resonance caused by linear frequency-modulated signals, a topic that has been studied relatively recently and poses greater challenges, yet is widely applicable in engineering fields.

J. Yang and M. A. F. Sanjuan, *Aperiodic Resonances—Theory and Applications*,
https://doi.org/10.1007/978-981-96-6819-9_5

5.1.1 Aperiodic Stochastic Resonance Realized by the Piecewise Re-scaled Method

We continue to use Eq. (4.1) to study aperiodic stochastic resonance, replacing the signal $s(t)$ with a linear frequency-modulated signal as described by Eq. (1.2). The characteristic of the instantaneous frequency of the linear frequency-modulated signal, which changes constantly over time, can result in significant variations in instantaneous frequency. This subsection employs the piecewise re-scaled method to achieve aperiodic stochastic resonance. First, we divide the linear frequency-modulated signal into several segments of equal length. Next, we apply different scale coefficients to each segment of the signal. Each segment is then processed separately in the re-scaled system to obtain the corresponding output. Finally, we concatenate the outputs of each segment in their original order to produce the complete final response.

The system employed for processing each signal segment remains the general re-scaled equation

$$\frac{dx}{dt} = a_1\beta x - b_1\beta x^3 + \beta s(t) + \sqrt{\beta}N(t). \tag{5.1}$$

However, each segment of the linear frequency-modulated signal uses different scale coefficient constant β_0, and it is still defined in Eq. (2.11), while we use the end time of the ith signal segment as t_{ei}.

In Fig. 5.1, we illustrate the processing effect of the linear frequency-modulated signal using aperiodic stochastic resonance achieved through the piecewise re-scaled method. Notably, the optimal noise intensity corresponding to aperiodic stochastic resonance varies with different values of β_0. Specifically, as the scale coefficient constant increases, the required optimal noise intensity decreases. Furthermore, a comparison of the various curves reveals that when the scale coefficient constant is set to 800, the optimal noise intensity corresponds to the highest peak in the cross-correlation coefficient.

For the case of $\beta_0 = 800$, Fig. 5.2 shows the output time series corresponding to the aperiodic stochastic resonance observed in Fig. 5.1, specifically at the noise intensity that maximizes the cross-correlation coefficient. The time-domain waveform indicates that the output is nearly synchronized with the input signal. However, it also shows that as time progresses, the synchronization between the output and input diminishes. This decline in synchronization is primarily due to the fact that each signal segment utilizes the same noise intensity to induce aperiodic stochastic resonance. This results in an averaged approximation across the multiple segments, which may not yield the optimal aperiodic stochastic resonance for each individual segment in practice.

To achieve optimal aperiodic stochastic resonance, it is essential to calculate the optimal noise intensity for each signal segment, thereby obtaining the best resonance response for the entire signal. Specifically, we first plot the $C_{sx} - D$ curve for each signal segment to identify the optimal noise intensity for each segment.

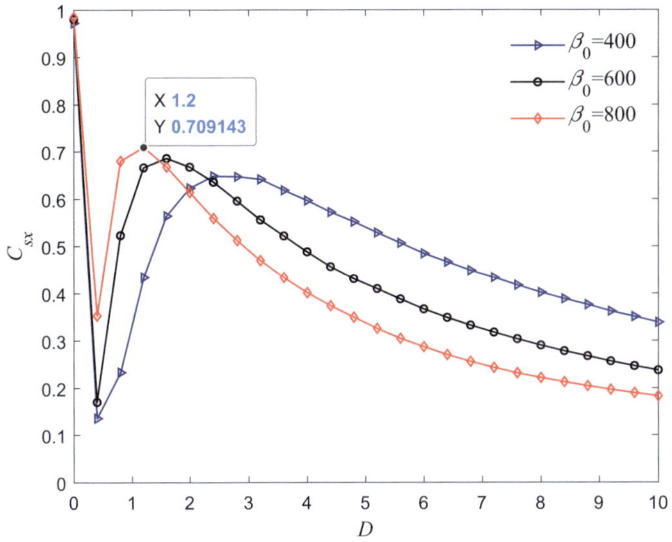

Fig. 5.1 The cross-correlation coefficient C_{sx} is calculated between the entire output and input as a function of the noise intensity D. The simulation parameters used are $a_1 = 1$, $b_1 = 1$, $A = 0.2$, $\gamma = 1$, $f_0 = 1$, and $\phi = 0$). The signal is divided evenly into 5 segments for processing

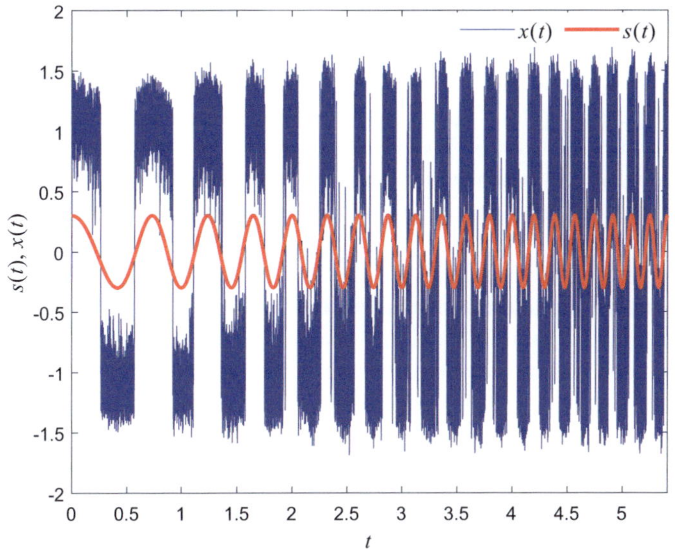

Fig. 5.2 The time series of the output corresponding to a fixed noise which makes the entire output and the input archives the maximal cross-correlation coefficient. The simulation parameters are $a_1 = 1, b_1 = 1, A = 0.2, \gamma = 1, f_0 = 1, \phi = 0, D = 1.2, \beta_0 = 800$

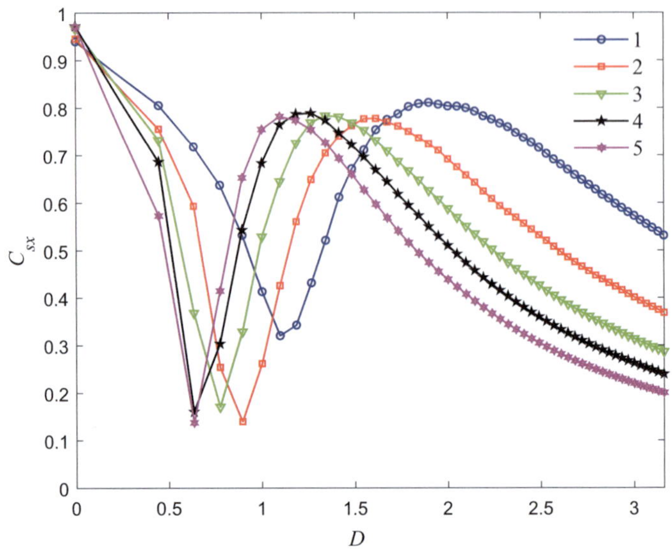

Fig. 5.3 The cross-correlation coefficient C_{sx} is calculated between each segment of the output and the input as a function of the noise intensity D. The simulation parameters used are $a_1 = 1, b_1 = 1$, $A = 0.2$, $\gamma = 1$, $f_0 = 1$, and $\psi = 0$. The signal is evenly divided into 5 segments for processing, and these segments are numbered sequentially from 1 to 5

Subsequently, we apply the optimal aperiodic stochastic resonance to each segment using its respective optimal noise intensity to obtain the optimal resonance output. Finally, we concatenate the optimal resonance outputs of each signal segment in their original order to produce the final system output. Following these steps, the $C_{sx} - D$ curves for the different signal segments are illustrated in Fig. 5.3. It is evident that the optimal noise intensity varies among the segments. For segments 1–5, the corresponding optimal noise intensities are $D = 1.8$, 1.3, 0.9,0.8, and 0.6, respectively.

Following the steps outlined in the previous paragraph, Fig. 5.4 displays the entire output constructed from each optimal resonance output in sequence. Comparing Fig. 5.4 with Fig. 5.2, the time-domain waveform indicates that the aperiodic stochastic resonance output presented in Fig. 5.4 is significantly improved. This improvement means that the signal waveform is better preserved, and synchronization between the output and input is maintained while amplifying the signal amplitude. Therefore, the method of determining the optimal noise intensity for each segment enhances the output of the aperiodic stochastic resonance.

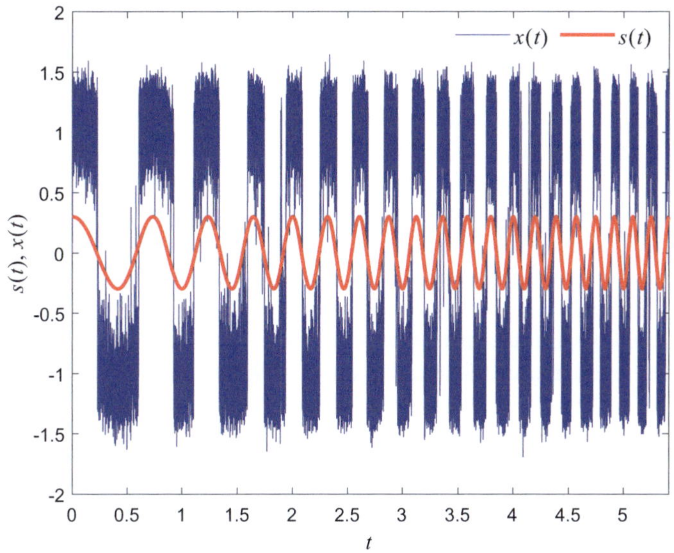

Fig. 5.4 The time series of the output corresponding to a fixed noise intensity, which allows each segment of the output and input to achieve the maximum cross-correlation coefficient, is presented here. The simulation parameters used are $a_1 = 1$, $b_1 = 1$, $A = 0.2$, $\gamma = 1$, $f_0 = 1$, and $\phi = 0$. For the calculation, the noise intensities for segments 1 to 5 are set to $D = 1.8$, 1.3, 0.9, 0.8, and 0.6, respectively

5.1.2 Aperiodic Stochastic Resonance Realized by the Real-Time Re-scaled Method

For frequency signals with rapidly changing frequencies, we utilize the real-time re-scaled theoretical framework to induce optimal aperiodic stochastic resonance. The equation changes to

$$\frac{dx}{dt} = a(t)x - b(t)x^3 + s(t) + N(t). \tag{5.2}$$

Here, $a(t)$ and $b(t)$ are both large parameters that vary over time, and the instantaneous frequency $f(t)$ of the signal $s(t)$ is much greater than 1. We continue to apply the scale transformations in Eqs. (3.9) and (3.10). Consequently, the real-time re-scaled aperiodic stochastic resonance can be achieved through the following equation

$$\frac{dx}{dt} = a_1\beta(t)x - b_1\beta(t)x^3 + \beta(t)s(t) + \sqrt{\beta(t)}N(t). \tag{5.3}$$

We can use the point tracing method to draw the functional relationship between the metric of the aperiodic stochastic resonance and the control variable, and then find the optimal value of the control variable. It is a commonly used method. In addition, we have another method, consisting in using an optimization algorithm to obtain the optimal control variable corresponding to the "strongest" aperiodic stochastic resonance, without the need to first plot the relevant function curve. Using an optimization algorithm to adjust relevant parameters for adapting to noisy signals and achieving the "strongest" resonance response is known as "adaptive stochastic resonance." Within the real-time re-scaled theoretical framework, we select the optimal control parameters through an optimization algorithm, which is referred to as *adaptive real-time re-scaled aperiodic stochastic resonance*. For detailed steps of this method, readers can refer to our previous work [4].

We continue to use the linear frequency-modulated signal defined in Eq. (1.2). For this simulation, we set the amplitude $A = 0.4$, and the instantaneous frequency is defined as $f(t) = 20t + 50$. To simulate a strong noise background, we utilize the AWGN function in MATLAB to add noise to the signal $s(t)$, resulting in a signal-to-noise ratio of -25 dB. The time series and time-frequency spectrum of the input noisy signal are presented in Fig. 5.5a, b. Next, we process the noisy signal using a novel time-frequency mode decomposition method to obtain a filtered signal [5].

By varying the system parameters a_1 and b_1, we calculate the cross-correlation coefficient C_{sx} and the system output $x(t)$, as shown in Fig. 5.6. The figure displays

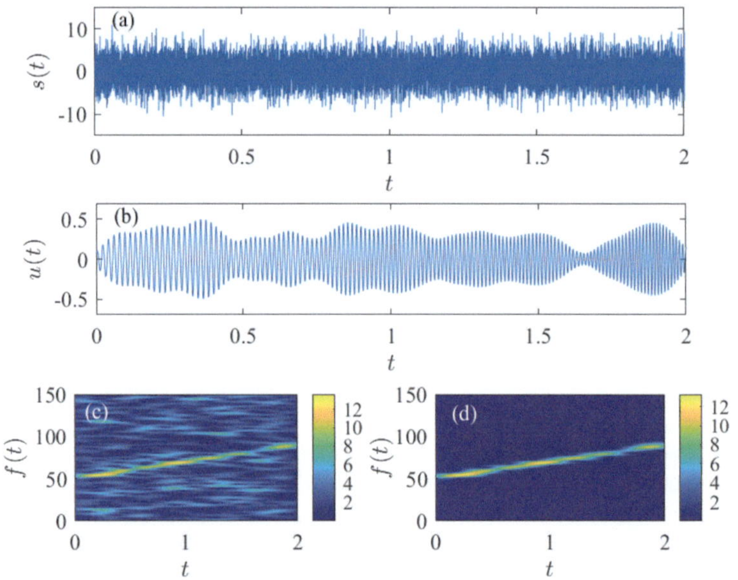

Fig. 5.5 The time series and time-frequency spectrum of the input linear frequency-modulated signal. **a** Time series of the original input noisy signal. **b** Time-frequency spectrum of the original input noisy signal. **c** Time series of $u(t)$, the noisy signal processed by the time-frequency mode decomposition method. **d** Time-frequency spectrum of $u(t)$, the noisy signal processed by the time-frequency mode decomposition method

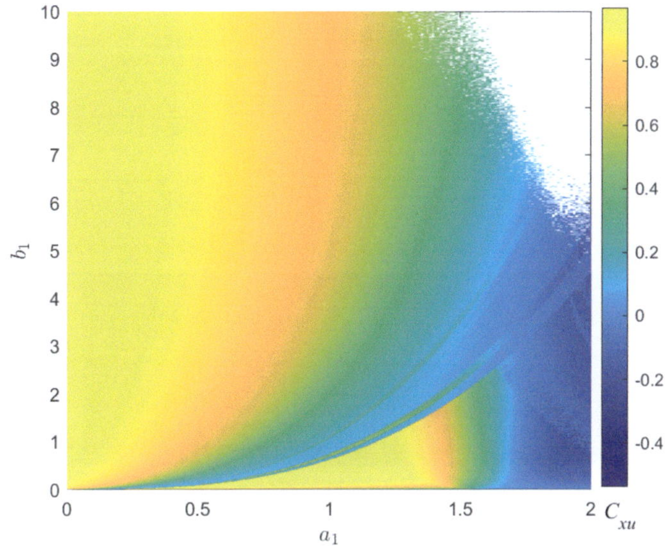

Fig. 5.6 The cross-correlation coefficient C_{sx} calculated between the processed signal $u(t)$ and the output $x(t)$. The scale coefficient constant is $\beta_0 = 100$

two distinct resonance regions, with the resonance region in the upper left exhibiting a particularly large range. Additionally, instances of negative cross-correlation coefficients are observed, primarily due to the significant phase difference between the input signal $u(t)$ and the system response $x(t)$. This figure demonstrates the feasibility of achieving aperiodic stochastic resonance in a noisy linear frequency-modulated signal through the adjustment of system parameters.

First, we select a set of arbitrary values for a_1 and b_1 within the upper left resonance region shown in Fig. 5.6, specifically $a_1 = 0.2$ and $b_1 = 4$. Under these conditions, when aperiodic stochastic resonance occurs, the system response exhibits low noise, stable amplitude, and good signal continuity performance. In contrast, as illustrated in Fig. 5.5b, the amplitude of the signal varies significantly over time. Figure 5.7 clearly demonstrates the advantages of the signal processing method, highlighting the effectiveness of achieving optimal aperiodic stochastic resonance.

To further validate the method, we select another set of arbitrary values for a_1 and b_1 within the upper left resonance region of Fig. 5.6, specifically $a_1 = 0.4$ and $b_1 = 8$. The corresponding time series of the output is presented in Fig. 5.8. While the time series in this figure effectively reflects the signal characteristics, the amplitude is slightly smaller compared to the signal amplitude in Fig. 5.7. This observation reinforces the feasibility of selecting parameters within the resonance region for effective signal processing, as demonstrated by the time-domain and frequency-domain characteristics of the signal shown in Fig. 5.8.

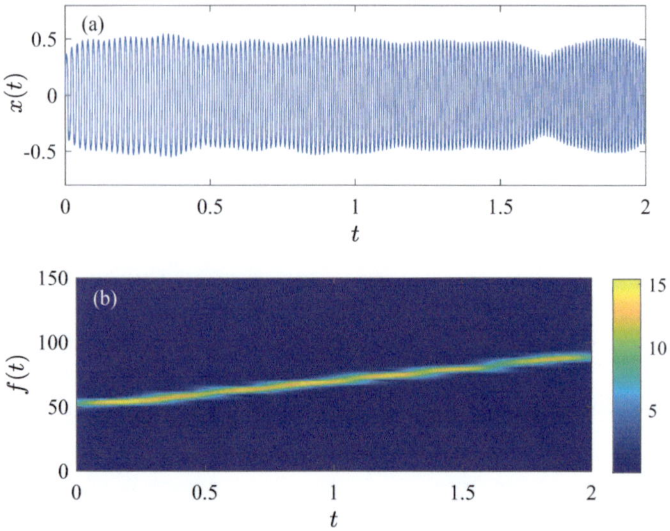

Fig. 5.7 The time series corresponding to the resonance region on the upper left side of Fig. 5.6. **a** The time series of $x(t)$. **b** The time frequency spectrum of $x(t)$. The simulation parameters are $a_1 = 0.2, b_1 = 4, \beta_0 = 100$

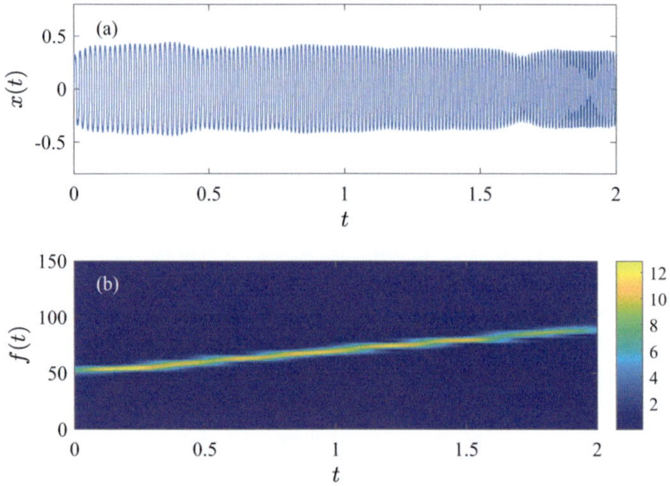

Fig. 5.8 The time series corresponding to the resonance region on the upper left side of Fig. 5.6. **a** The time series of $x(t)$. **b** The time frequency spectrum of $x(t)$. The simulation parameters are $a_1 = 0.4, b_1 = 8, \beta_0 = 100$

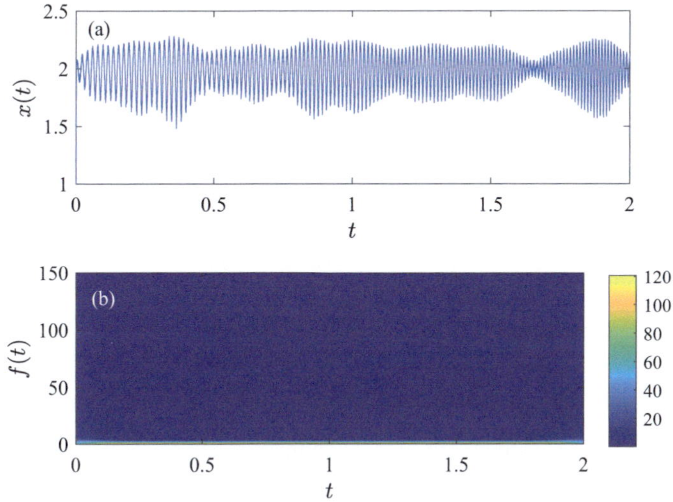

Fig. 5.9 The system output $x(t)$ corresponding to the resonance region on the lower right side of Fig. 5.6. **a** The time series of $x(t)$. **b** The time frequency spectrum of $x(t)$. The simulation parameters are $a_1 = 0.8$, $b_1 = 0.2$, $\beta_0 = 100$

Finally, we choose a set of arbitrary values of a_1 and b_1 in the lower right resonance region in Fig. 5.6, specifically $a_1 = 0.8$ and $b_1 = 0.2$. The corresponding time series and time-frequency spectrum of the output $x(t)$ is presented in Fig. 5.9. For the time series of the output $x(t)$ in Figs. 5.7 and 5.8, the motion of the system response transits between two potential well synchronized with the input signal $u(t)$.

In Fig. 5.9, the motion of the system is confined to oscillating solely within the right potential well. Consequently, the stationarity of the signal amplitude deteriorates, as seen in Fig. 5.9a. Additionally, in Fig. 5.9b, the energy along the instantaneous frequency curve appears to be significantly reduced. However, this is somewhat misleading, as the numerical values indicated in the color bar of the figure also reflect this issue. The system's response, oscillating within a potential well, results in a relatively large constant value within the system response. This leads to a high value appearing at the 0 frequency in the time-frequency spectrum, which can create the illusion of decreased energy at the time-frequency spectrum. Importantly, by comparing Figs. 5.7, 5.8, and 5.9, we observe that the system response performs better when the output $x(t)$ can transition irregularly between the two potential wells. Therefore, when selecting parameters, it is crucial to prioritize those situated within the upper left resonance region.

5.2 Unknown Linear Frequency-Modulated Signal as the Characteristic Signal

Due to the presence of noise in the signal, we lack knowledge of any parameters related to the characteristic signal, making it impossible to calculate quantitative indices such as the cross-correlation coefficient. To measure aperiodic stochastic resonance using this index, prior knowledge of the characteristic signal is required. To address this challenge, we consider an unknown linear frequency-modulated characteristic signal that is submerged in strong noise and present two methods for achieving aperiodic stochastic resonance to identify the signal's parameters. One method involves defining a new index called the piecewise mean value indicator to realize aperiodic stochastic resonance and recover the unknown signal against a strong noise background, aided by the fitting curve of the optimal output [6]. We will not discuss this method in detail here; interested readers can refer to our original work. The second method utilizes the fractional signal-to-noise ratio, a novel index within the fractional Fourier transformation, to realize aperiodic stochastic resonance with the support of an optimization algorithm [7]. This subsection primarily focuses on discussing the latter method in detail.

We continue our investigation of Eq. (5.1), this time with the linear frequency-modulated signal subjected to noise with an intensity of 0.2. The time series of the noisy signal for three different amplitudes A are presented in Fig. 5.10. In these cases, we observe that the signals are completely submerged by noise, making it impossible to extract any meaningful information from the time series.

To mitigate the influence of noise and to account for the characteristics of the instantaneous frequency, we introduce the fractional Fourier transform for preliminary signal processing [9–11]. The fractional Fourier transform is an effective time-frequency analysis tool that can be conceptualized as a transformation that rotates the signal coordinates counterclockwise around the origin in the time-frequency plane. When the signal coordinates are rotated by $\pi/2$, the fractional Fourier transform simplifies to the traditional Fourier transform. Thus, the fractional Fourier transform serves as an extension of the traditional Fourier transform. The definition of the fractional Fourier transform is as follows

$$X_\alpha(u) = F^\alpha[x(t)] = \int_{-\infty}^{\infty} x(t) K_\theta(t, u) \mathrm{d}t, \tag{5.4}$$

where α is the fractional-order and $F^\alpha[\cdot]$ is the fractional Fourier transform operator, θ is the rotation angle and $\theta = \alpha\pi/2$. The specific kernel function is defined as

$$K_\theta(t, u) = \begin{cases} \sqrt{\frac{1 - j\cot\theta}{2\pi}} e^{j(t^2\cot\theta/2 - ut\csc\theta + u^2\cot\theta/2)}, & \theta \neq n\pi \\ \delta(t - u), & \theta = 2n\pi \\ \delta(t + u), & \theta = (2n \pm 1)\pi \end{cases} \tag{5.5}$$

where j is the imaginary unit and $j^2 = -1$.

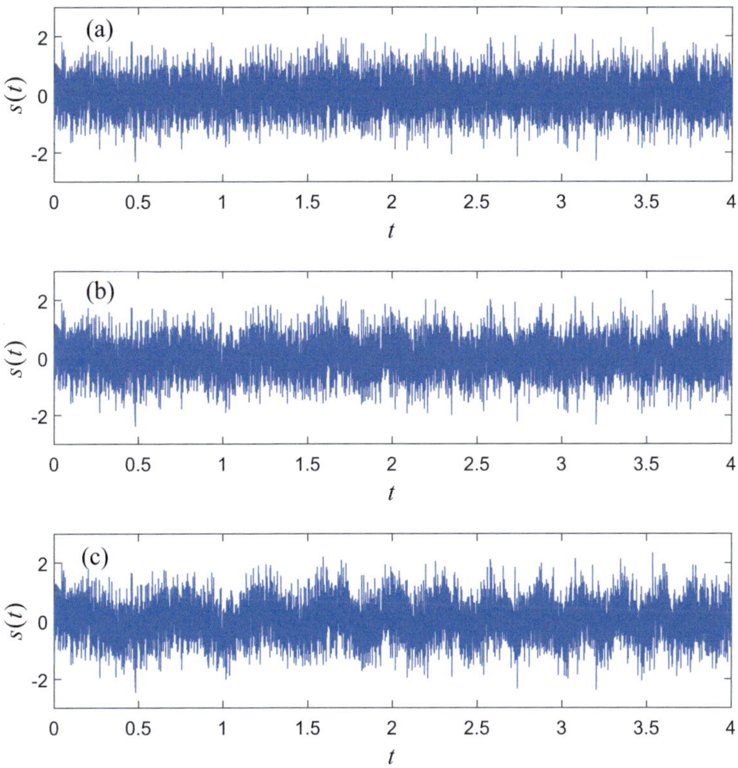

Fig. 5.10 The figure presents a group of noisy linear frequency-modulated signals with varying amplitudes. Specifically, we have **a** $A = 0.05$, **b** $A = 0.15$, **c** $A = 0.25$. The signal parameters for all cases are set to $\gamma = 1$, $f_0 = 1$, $\phi = 0$, and the noise intensity is $D = 0.2$

The fractional Fourier transform offers numerous advantages in signal processing. By utilizing integration along a straight line in the time-frequency plane, it effectively eliminates interference caused by cross terms. Most importantly, the fractional Fourier transform can convert linear frequency-modulated signals into pulses within the fractional Fourier domain, which concentrates the energy of the linear frequency-modulated signal while preventing noise energy from being similarly concentrated in any fractional Fourier domain. This unique characteristic enables the fractional Fourier transform to extract linear frequency-modulated signals from the original noisy signal, significantly reducing other interferences in the process.

The signal-to-noise ratio (SNR) is a widely used metric to describe the relationship between signal energy and noise energy at a given frequency. In the context of stochastic resonance induced by periodic signal excitation, SNR serves as the primary indicator for characterizing the proportion of signal energy at a fixed frequency within the overall output time series. However, because the instantaneous frequency of a linear frequency-modulated signal varies continuously over time, its time-frequency representation displays a frequency band with a specific bandwidth that describes the signal energy in the frequency domain. This continuous variation complicates the application of traditional SNR methods to accurately assess the energy-to-noise ratio within a specific frequency band for linear frequency-modulated signals. To address this challenge, we introduce a new metric termed the *fractional signal-to-noise ratio (FSNR)*, which is defined as follows

$$\mathrm{FSNR} = 10 \log \frac{|X(u_\alpha)|^2}{(\sum_{u=1}^{N} |X(u)|^2) - |X(u_\alpha)|^2}, \tag{5.6}$$

where $|X(u_\alpha)|^2$ and $(\sum_{u=1}^{N} |X(u)|^2) - |X(u_\alpha)|^2$ are the energy of the linear frequency-modulated signal and the energy of the noise in the fractional domain corresponding to the fractional-order α.

The fractional Fourier transform is applied with the transformation order α as a variable, resulting in the fractional Fourier spectrum depicted in Fig. 5.11a. This graph allows us to identify the optimal fractional order corresponding to the fractional Fourier transform that maximizes the concentration of signal energy. By selecting this optimal fractional Fourier transform, defined by the optimal fractional order α, we obtain the rotated signal shown in Fig. 5.11b. In this figure, it is evident that the energy concentrates at $u = 1.9955$ when the fractional order $\alpha = 0.999493$.

By utilizing the fractional signal-to-noise ratio $FSNR$ as the objective function and the noise intensity D as the control variable, Fig. 5.12 illustrates the relationship between $FSNR$ and D. The curves presented in the figure exhibit resonance characteristics similar to those observed in traditional stochastic resonance. As the noise intensity D increases, each curve displays a distinct peak, indicating that noise plays a crucial role in maximizing the enhancement of signal energy at this peak. However, it is important to note that either excessive or insufficient noise can hinder the achievement of this optimal effect.

In Fig. 5.13, we present a comparison of the original input linear frequency-modulated signal, the optimal output, and the fitted curve of the output, all corresponding to the peaks of the $FSNR - D$ curve. This processing method successfully recovers the noisy linear frequency-modulated signal while also amplifying it. Notably, we did not have prior knowledge of the characteristics of the signal, nor did we calculate any quantified index functions based on known signal properties. Instead, we achieved aperiodic stochastic resonance using the unknown linear frequency-modulated signal, demonstrating the effectiveness of our approach in handling signals obscured by noise.

In Fig. 5.14, we illustrate the amplitude spectra of the response $x(t)$ corresponding to the optimal output shown in Fig. 5.13. The subplots reveal that the energy is now concentrated on three discrete lines, indicating a significant noise reduction

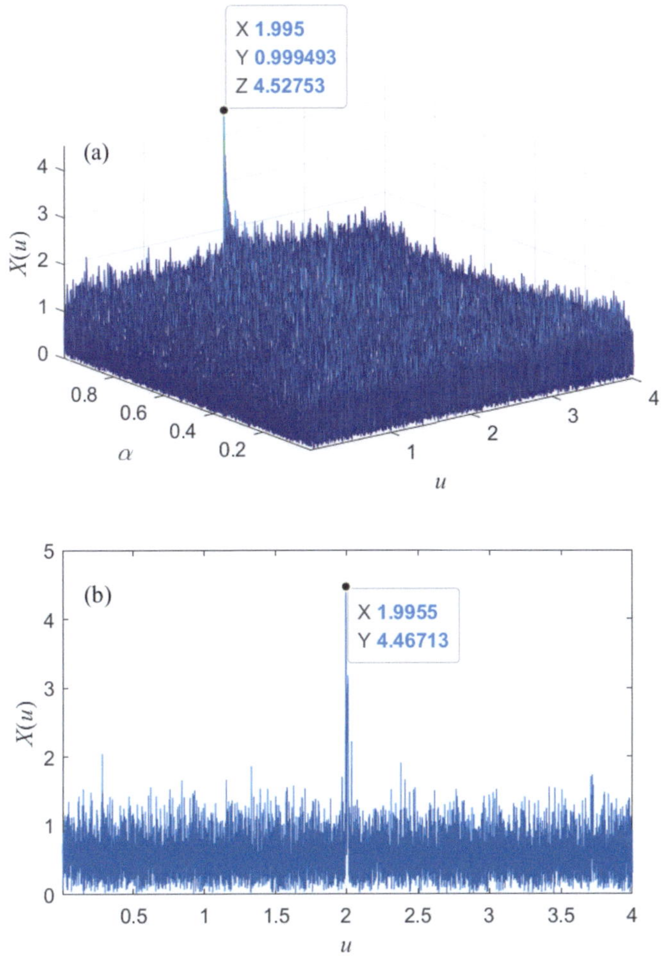

Fig. 5.11 The amplitude spectra of the noisy signal in the fractional Fourier domain. **a** The amplitude spectra depends on the fractional-order α. **b** The amplitude spectra under the optimal fractional-order α. The simulation parameters are $A = 0.25$, $\gamma = 1$, $f_0 = 1$, $\phi = 0$, and $D = 0.2$

effect achieved by our method. This stands in contrast to the amplitude spectrum in Fig. 5.11b, where the interference from noise was clearly evident. The comparison highlights the effectiveness of our approach in extracting the desired signal while minimizing the impact of noise.

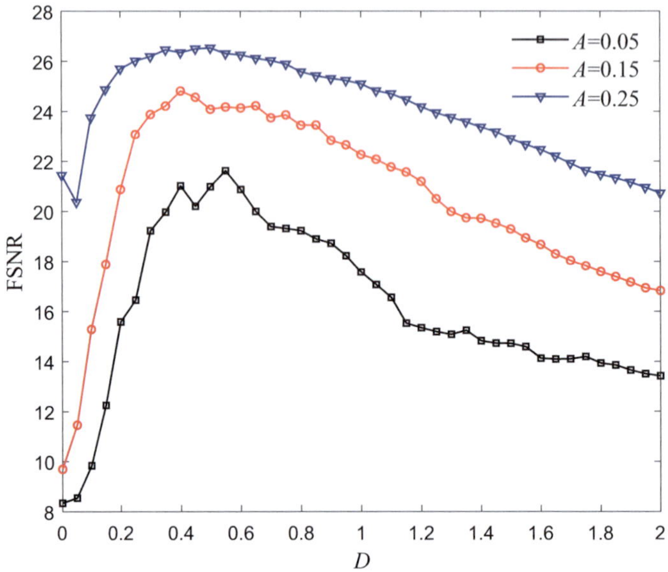

Fig. 5.12 The fractional signal-to-noise ratio versus the noise intensity for different amplitudes of A. The simulation parameters are $\gamma = 1$, $f_0 = 1$, $\phi = 0$, $a_1 = 1$, $b_1 = 1$, and $\beta_0 = 60$

5.3 Ultra-High Frequency Linear Frequency-Modulated Signal as the Characteristic Signal

Ultra-high frequency modulation signals are widely used across various fields, including photonics technology [12], radar technology [13, 14], and laser applications [15, 16]. Processing these signals, particularly when they are contaminated by noise, poses significant challenges. This section introduces the aperiodic stochastic resonance method as a solution to these issues, offering a novel approach for processing ultra-high frequency signals.

To clarify the meaning of the signal parameters, we express the ultra-high frequency linear frequency-modulated signal as

$$s(t) = A \sin\left(2\pi f_0 + \pi \frac{W}{T} t^2\right), \tag{5.7}$$

where the instantaneous frequency is $f = f_0 + \frac{W}{T} t$, with f_0 representing the starting frequency, W is the bandwidth, T is the time width of the linear frequency-modulated signal. In line with mainstream millimeter-wave radar applications, the parameters selected in Eq. (5.7) are $A = 0.4$, $f_0 = 77\,\text{GHz}$, $W = 0.9\,\text{GHz}$, and $T = 0.03\,\mu\text{s}$.

In Fig. 5.15, we present the noisy ultra-high frequency linear frequency-modulated signal, which is generated by adding noise with a signal-to-noise ratio of $-28\,\text{dB}$ to

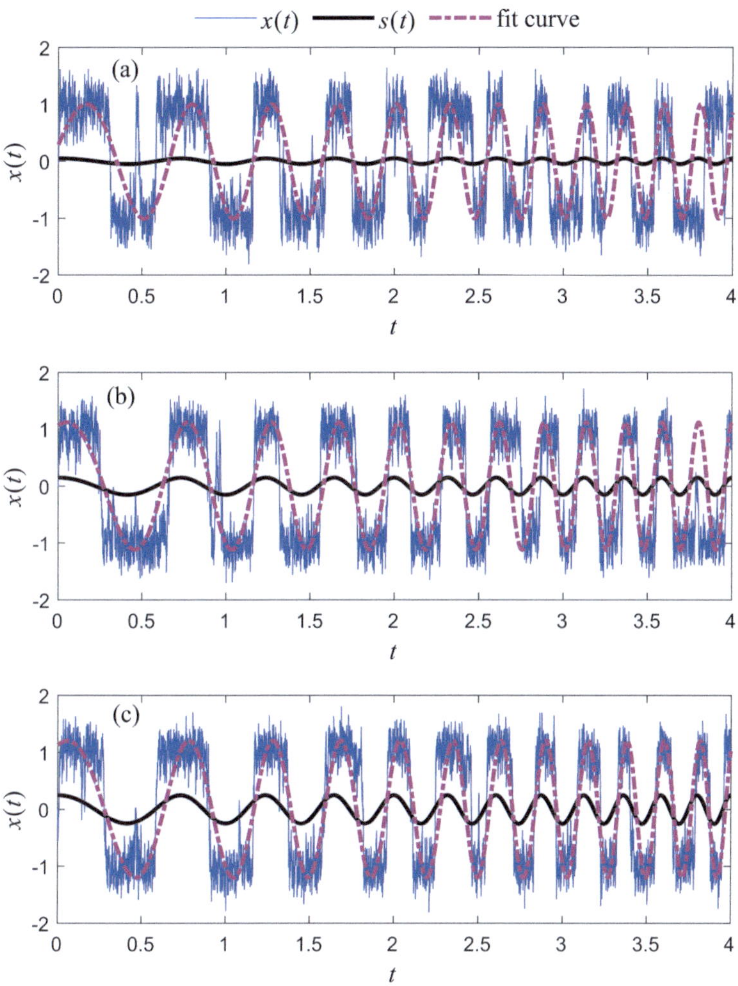

Fig. 5.13 The optimal response corresponding to the aperiodic stochastic resonance. **a** $A = 0.05$, $D = 0.55$, **b** $A = 0.15$, $D = 0.4$, **c** $A = 0.25$, $D = 0.45$. Other simulation parameters are $\gamma = 1$, $f_0 = 1$, $\phi = 0$, $a_1 = 1$, $b_1 = 1$, and $\beta_0 = 60$

the pure signal $s(t)$. As shown in Fig. 5.15a, the time series indicates that the signal is significantly obscured by the noise. Additionally, in Fig. 5.15b, the frequency curve in the time-frequency spectrum is also distorted by noise, making it difficult to discern clearly.

To extract the ultra-high frequency linear frequency-modulated signal from the noise by the aperiodic stochastic resonance method, we investigate the following equation

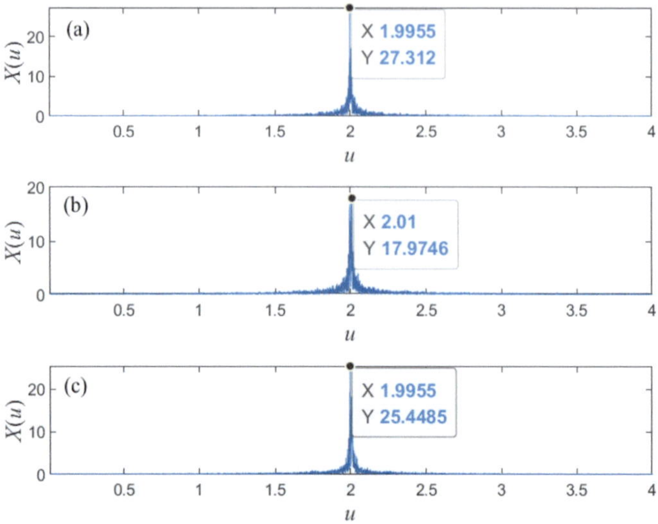

Fig. 5.14 The optimal amplitude spectra of the noisy signal after being processed by aperiodic stochastic resonance. **a** Amplitude spectra as a function of the fractional order α **b** Amplitude spectra at the optimal fractional order α. The spectra correspond to different amplitudes: **a** $A = 0.05$, **b** $A = 0.15$, **c** $A = 0.25$. Other simulation parameters include $\gamma = 1$, $f_0 = 1$, $\phi = 0$, $D = 0.2$, $a_1 = 1$, $b_1 = 1$, and $\beta_0 = 60$

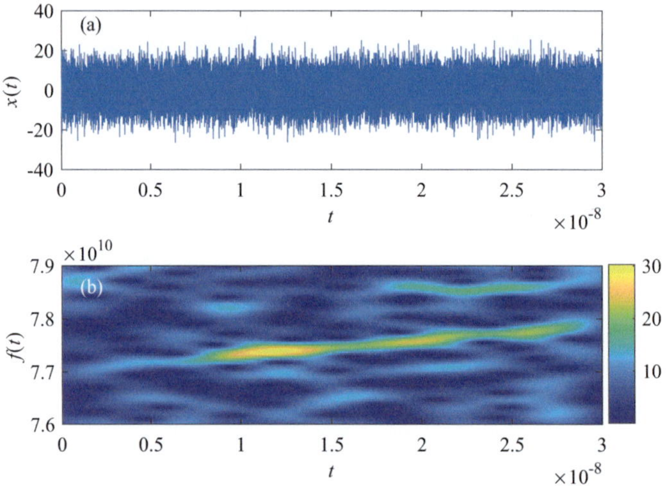

Fig. 5.15 The time series and time-frequency spectrum of the input noisy ultra-high frequency linear frequency-modulated signal. **a** Time series of the original input noisy signal. **b** Time-frequency spectrum of the original input noisy signal

$$\frac{dx(t)}{dt} = a(t)x(t) - r(t)x^2(t) - b(t)x^3(t) + s(t) + N(t). \tag{5.8}$$

When the parameters $a(t)$, $b(t)$ and $r(t)$ degenerate to three constants, the potential function of the system is

$$V(x) = -\frac{1}{2}ax^2 + \frac{1}{3}rx^3 + \frac{1}{4}bx^4. \tag{5.9}$$

The potential function in Eq. (5.9) is configured in an asymmetric bistable form, as illustrated in Fig. A.4 of Appendix A.1.3. The tilt of the potential function is determined by the sign of the asymmetric parameter r. When $r = 0$, the system transitions to a traditional symmetric bistable configuration. To better align the frequency characteristics of the signal with the system parameters, we employ time-varying parameters $a(t)$, $b(t)$, and $r(t)$. Numerous studies have investigated stochastic resonance in asymmetric bistable systems [17–21]. These systems offer advantages in processing certain specialized signals, which is why we have selected the asymmetric bistable model for our analysis.

We use the real-time re-scaled method to simulate the aperiodic stochastic resonance in the following equation

$$\frac{dx(t)}{dt} = a_1\beta(t)x(t) - r_1\beta(t)x^2(t) - b_1\beta(t)x^3(t) + \beta(t)s(t) + \sqrt{\beta(t)}N(t), \tag{5.10}$$

where $\beta(t) = \beta_0 f(t)$. For the noisy input signal, we utilize the concept of adaptive stochastic resonance, which involves adjusting the system parameters to achieve aperiodic stochastic resonance, as depicted in Fig. 5.16. This figure illustrates three distinct aperiodic stochastic resonance regions and two inverse aperiodic stochastic resonance regions. The term *inverse aperiodic stochastic resonance* refers to areas where the absolute values of the cross-correlation coefficient approach 1. In these regions, the phase of the system output is nearly completely opposite to that of the input signal, allowing the output to not only reduce noise but also have a high potential for amplification. In Fig. 5.16, we label three aperiodic stochastic resonance regions as **A**, **B** and **D**, and two inverse aperiodic stochastic resonance regions as **C** and **E**. The figure demonstrates that real-time changes in system parameters can lead to complex resonance patterns in the system's response.

We analyze the system response characteristics of the five resonance regions labelled in Fig. 5.16 and present the system response in the form of a phase diagram in Fig. 5.17. The phase diagram in Fig. 5.17a displays periodic motion transitioning between two potential wells, corresponding to the resonance region **A** in Fig. 5.16. In contrast, Fig. 5.17b, d, which correspond to the resonance regions **B** and **D**, respectively, show the phase diagram exhibiting periodic oscillatory motion confined to a single potential well. Specifically, the output oscillates within the right potential well in Fig. 5.17b, while it oscillates in the left potential well in Fig. 5.17d. For the outputs corresponding to the inverse aperiodic stochastic resonance regions **C** and **E**, the motion of the output oscillates in the right and left potential wells, respectively,

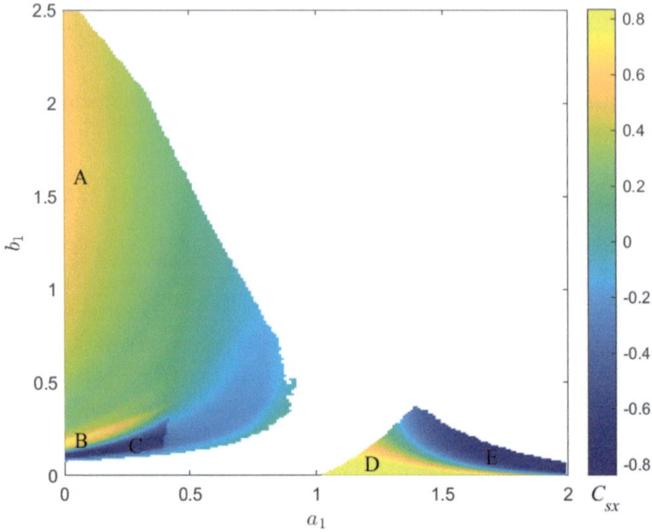

Fig. 5.16 The cross-correlation coefficient C_{sx} calculated between the pure signal $s(t)$ and the output $x(t)$. The simulation parameters are $r_1 = -0.5$, and $\beta_0 = 20$

while the response displays an evident chaotic state. The phase diagrams in Fig. 5.17 further illustrate the complex response of the nonlinear system under the excitation of the ultra-high frequency modulation signal.

In Fig. 5.18, which corresponds to the phase diagram in Fig. 5.17a, we present the output in both the time domain and the time-frequency domain. From Fig. 5.18, we observe pronounced denoising and amplification effects on the noisy signal achieved through the aperiodic stochastic resonance method.

In this section, we present a preliminary study of the asymmetric bistable system excited by a noisy ultra-high frequency linear frequency-modulated signal, revealing intriguing and complex dynamical phenomena. However, there is limited research on the dynamics of nonlinear systems driven by ultra-high frequency modulation signals. This topic warrants further investigation.

5.4 Nonlinear Frequency-Modulated Signal as the Characteristic Signal

In this section, we continue to use Eq. (5.1), but we replace the signal with the NLFM signal defined in Eq. (1.6) [22]. First, we apply the piecewise re-scaled method to divide the noisy signal into p subsegment signals in time. Next, we perform adaptive

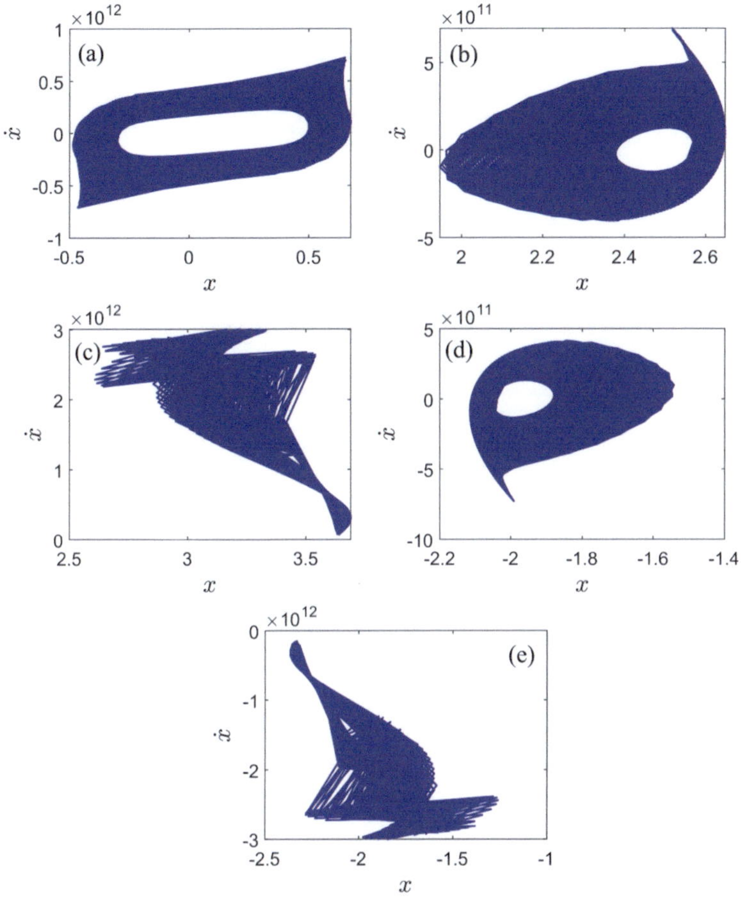

Fig. 5.17 The phase diagrams corresponding to different resonance regions. **a**–**e** Corresponding to the regions **A-E** respectively, and **a** $a_1 = 0.001$, $b_1 = 1.761$, **b** $a_1 = 0.121$, $b_1 = 0.221$, **c** $a_1 = 0.201$, $b_1 = 0.141$, **d** $a_1 = 1.301$, $b_1 = 0.081$, **e** $a_1 = 1.861$, $b_1 = 0.081$. Other simulation parameters are $r_1 = -0.5$, and $\beta_0 = 20$

empirical mode decomposition (EMD) on the segmented subsegment signal $s_i(t)$ to obtain intrinsic mode function (IMF) components that contain different frequencies. The decomposition process of the subsegment signal can be described as

$$s_i(t) = \sum_{k=1}^{K} c_{i,k}(t) + r_{ei}(t). \tag{5.11}$$

Herein, $c_{i,k}(t)$ represents the kth IMF component generated after EMD decomposition of the ith subsegment signal, K is the number of IMFs generated per decomposition, and $r_{ei}(t)$ represents the residual generated after EMD processing of the ith

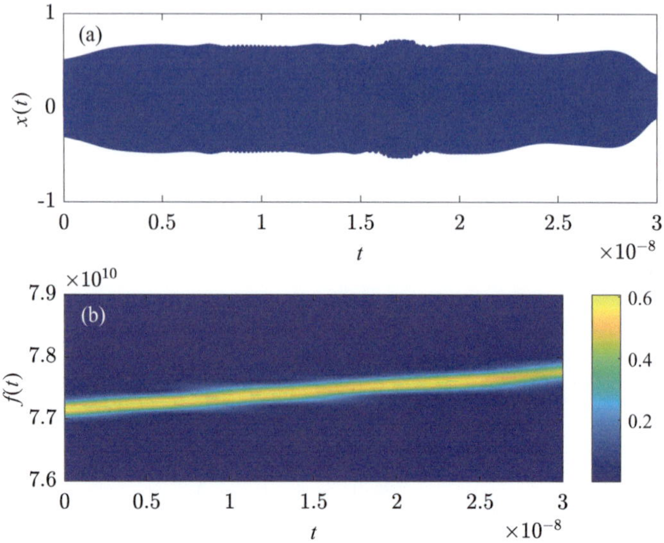

Fig. 5.18 The optimal of the system output after the noisy signal processed by the aperiodic stochastic resonance. **a** The time series of the output. **b**The time frequency spectrum of the output. This figure corresponds to the phase diagram in Fig. 5.17a

subsegment signal. It is worth noting that the characteristics of subsegment signals can often be represented by some key frequency components. Therefore, the cross-correlation coefficient between the ith subsegment signal and its kth IMF component is applied here to choose the optimal IMF component of the subsegment signal $s_i(t)$. Certainly, we choose kth IMF component which makes the cross-correlation coefficient have the maximal value. By repeating the above choosing process for each subsegment signal, the set of optimal IMF components for the entire NLFM signal can be obtained. Take each optimal IMF component as the input of the nonlinear system in sequence. Then, by matching different system parameters, the optimal aperiodic stochastic resonance corresponding to each subsegment signal can be obtained. Finally, the optimal system response induced by the NLFM signal is reconstructed by integrating the optimal stochastic resonance output of the subsegment signals. The process of piecewise re-scaled stochastic resonance has been mentioned multiple times before and will not be described in detail here.

To quantify the optimal stochastic resonance response, the spectral amplification factor in Eq. (1.18) is improved, and the new spectral amplification factor is defined as

$$\eta=\frac{\int_{f_{\min}}^{f_{\max}} X_i\left(f\right)\Big/\int_{f_m-2\delta}^{f_m+2\delta} X_i\left(f\right)}{\int_{f_{\min}}^{f_{\max}} S_i\left(f\right)\Big/\int_{f_m-2\delta}^{f_m+2\delta} S_i\left(f\right)} \qquad i=1,\ldots,p, \tag{5.12}$$

where f_{max} and f_{min} are the maximum and minimum instantaneous frequencies of the ith sub segment signal, and $f_m = (f_{max} + f_{min})/2$ is the average of the instantaneous frequencies f_{max} and f_{min}. 4δ is the bandwidth centered around f_m, ensuring that the actual instantaneous frequency of the ith subsegment is within the frequency band centered around f_m. The value of δ is related to the sampling frequency and number of sampling points of the input signal. Here, 4δ is greater than twice the bandwidth f_{max} and f_{min} of the subsegment signal. $S_i(f)$ and $X_i(f)$ respectively represent the amplitude spectra of the ith input signal and the corresponding response output. The improved spectral amplification factor comprehensively reflects the changes in the concentration of noise energy into signal energy before and after stochastic resonance, as well as the degree of signal energy amplification.

We input the optimal intrinsic mode function (IMF) component from all subsegment signals in chronological order, induce aperiodic stochastic resonance, and reconstruct the output of the entire NLFM signal. Subsequently, we plot the short-time Fourier transform (STFT) time-frequency spectrum of the system response. Finally, we quantify the enhancement effect of the weak NLFM signal using the spectral amplification factor.

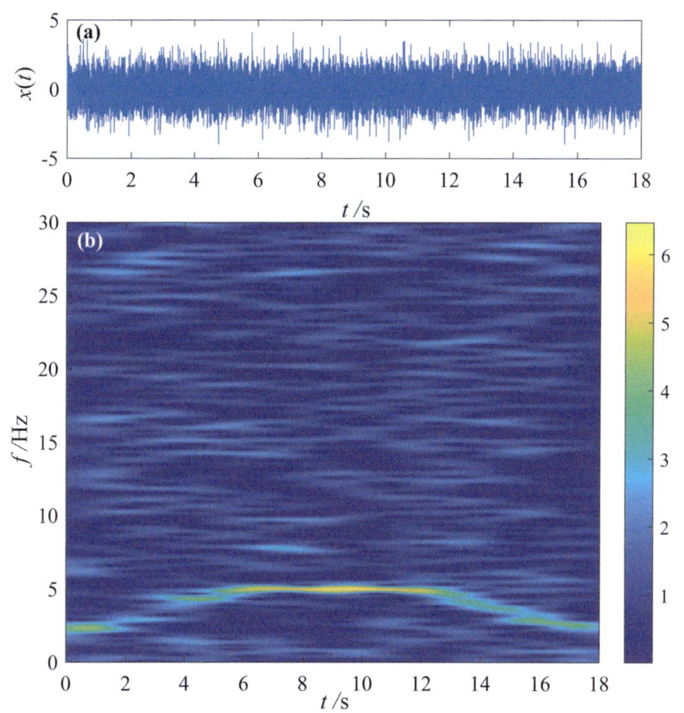

Fig. 5.19 The input NLFM signal. **a** The time domain waveform, **b** the STFT spectrum. The simulation parameters are $A = 0.2$, $\gamma = 0.5$, $f_0 = 2$, $t_1 = 6$, $t_2 = 12$, $= 0$, $D = 0.5$ and $p = 9$

Here, the sampling frequency is $f_s = 1000$ Hz, and the number of sampling points is $n = 18000$. To simulate a strong noise background, Gaussian white noise with a strength of $D = 0.5$ was added to the NLFM signal. Figure 5.19a, b show the time-domain waveform and STFT time-frequency spectrum of the NLFM signal with noise, respectively. As a classic time-frequency analysis method, STFT time-frequency spectrum can reflect the time-frequency characteristics of signals and better demonstrate the trend of instantaneous frequency variation of NLFM signals over time. The color bar on the right side of the STFT spectrum represents the power spectral density of the input signal. As shown in Fig. 5.19b, the NLFM signal is overwhelmed by strong background noise and has poor time-frequency clustering. It should be noted that in the signal frequency band, noise energy and signal energy are at the same amplitude level. Meanwhile, the noise outside the signal frequency band is still very strong. Therefore, it is impossible to accurately obtain the complete instantaneous frequency curve of NLFM signals.

To overcome the above two shortcomings, the method described in this section is adopted. Corresponding to Figs. 5.19, 5.20 shows the time-domain waveform and STFT spectrum of the enhanced NLFM signal reconstructed from the optimal response of each subsegment signal. Compared with Fig. 5.19, the reconstructed

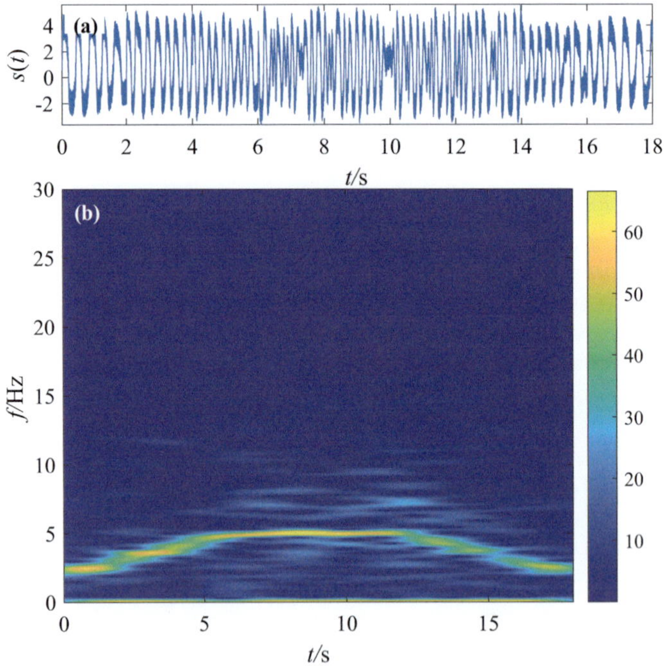

Fig. 5.20 The enhancement effect of the NLFM signal processed by the piecewise re-scaled aperiodic stochastic resonance method proposed in this subchapter. **a** The time domain waveform, **b** the STFT spectrum. The simulation parameters are corresponding to Fig. 5.19

NLFM signal has higher time-frequency aggregation. As shown in Fig. 5.20, the periodic oscillation characteristics of the NLFM signal have been restored, and the signal energy has also been significantly improved. The enhancement effect of weak NLFM signals was numerically evaluated using spectral amplification factor, and the spectral amplification gain of the NLFM signal shown in Fig. 5.20 is calculated to be 2.6. This indicator represents the degree of amplification of signal frequency band energy on the one hand, and the degree of utilization of noise on the other hand. The results indicate that the stochastic resonance method can effectively enhance weak NLFM signals.

In addition to the NLFM signal used in this section, aperiodic stochastic resonance can also be induced by NLFM signals in various forms. We provided several examples and discussions in our previous work [23]. In next section, we focus on the application of this method for extracting fault feature information from rolling bearings under strong noise conditions and time-varying operating circumstances, as discussed in Chap. 12.

5.5 Concluding Remarks

This chapter explores aperiodic resonance induced by various frequency-modulated signals. In particular, aperiodic stochastic resonance induced by linear frequency-modulated signals is realized through both piecewise re-scaled and real-time re-scaled theoretical frameworks. For unknown linear frequency-modulated signals, we employ a new fractional signal-to-noise ratio in the fractional Fourier domain to achieve aperiodic stochastic resonance and recover the signal. When the instantaneous frequency of the frequency-modulated signal is ultra-high, the response exhibits different resonance behaviors, including aperiodic stochastic resonance and inverse stochastic resonance. For NLFM signals submerged in strong noise, we utilize the empirical mode decomposition (EMD) method in conjunction with the piecewise re-scaled aperiodic stochastic resonance method to extract signal features from the noisy background. Various frequency-modulated signals are widely applied in engineering and scientific fields, and research on aperiodic stochastic resonance caused by these signals is still in its infancy. Given its inherent complexity and the resulting nonlinear system responses, this topic warrants further in-depth investigation.

References

1. L. Lin, H. Wang, W. Lv, S. Zhong, A novel parameter-induced stochastic resonance phenomena in fractional Fourier domain, *Mechanical Systems and Signal Processing*, 2016, **76**: 771-779.
2. L. Lin, H. Wang, W. Lv, Stochastic resonance system with linear random frequency fluctuation for aperiodic LFM signal, *Nonlinear Dynamics*, 2017, **88**: 1361-1371.
3. C. Wu, Z. Wang, J. Yang, D. Huang, M.A.F. Sanjuán, Adaptive piecewise re-scaled stochastic resonance excited by the LFM signal, *The European Physical Journal Plus*, 2020, **135**: 130.

4. Z. Shan, Z. Wang, J. Yang, D. Zhou, H. Liu, Extracting non-stationary signal under strong noise background: Time-varying system analysis, *Journal of Vibration and Control*, 2023, **29**(17-18): 4036-4045.
5. Z. Shan, Z. Wang, J. Yang, Q. Ma, T. Gong, Novel time-frequency mode decomposition and information fusion for bearing fault diagnosis under varying-speed condition, *IEEE Transactions on Instrumentation and Measurement*, 2023, **72**: 3511210.
6. D. Huang, J. Yang, D. Zhou, M.A.F. Sanjuán, H. Liu, Recovering an unknown signal completely submerged in strong noise by a new stochastic resonance method, *Communications in Nonlinear Science and Numerical Simulation*, 2019, **66**: 156-166.
7. C. Wu, J. Yang, M.A.F. Sanjuán, H. Liu, Stochastic resonance induced by an unknown linear frequency modulated signal in a strong noise background, *Chaos*, 2020, **30**(4): 043128.
8. D.H. Bailey, P.N. Swarztrauber, The fractional Fourier transform and applications, *SIAM Review*, 1991, **33**(3): 389-404.
9. H.M. Ozaktas, M.A. Kutay, D. Mendlovic, Introduction to the fractional Fourier transform and its applications, *Advances in Imaging and Electron Physics*, 1999, **106**: 239-291.
10. A. Bultheel, H.E.M. Sulbaran, Computation of the fractional Fourier transform, *Applied and Computational Harmonic Analysis*, 2004, **16**(3): 182-202.
11. E. Sejdić, I. Djurović, L. Stanković, Fractional Fourier transform as a signal processing tool: An overview of recent developments, *Signal Processing*, 2011, **91**(6): 1351-1369.
12. S. Wang, Z. Lu, N. Idrees, S. Zheng, X. Jin, X. Zhang, X. Yu, Photonic generation and dechirping of broadband THz linear-frequency-modulated signals, *IEEE Photonics Technology Letters*, 2019, **31**(11): 881-884.
13. S. Xu, W. Zou, G. Yang, J. Chen, Ultra-high range resolution demonstration of a photonics-based microwave radar using a high-repetition-rate mode-locked fiber laser, *Chinese Optics Letters*, 2018, **16**(6): 062801.
14. R. Appelman, Z. Zalevsky, D. Mendlovic, G. Shabtay, Hybrid optical-RF system for generating an improved linear frequency modulated pulses for radar applications, *In Record of the IEEE 2000 International Radar Conference*, Cat. No. 00CH37037, 775-780, IEEE, 2000, May.
15. G. Wu, J. Zheng, T. Pu, J. Li, X. Zhang, S. Chen, J. Zhao, Y. Zhang, X. Chen, Simple multiband linearly frequency-modulated signal generation with a multiplying bandwidth based on a gain-switching laser, *Applied Optics*, 2023, **62**(7): 1822-1828.
16. J. Yang, T. Yang, Z. Wang, D. Jia, C. Ge, A novel method of measuring instantaneous frequency of an ultrafast frequency modulated continuous-wave laser, *Sensors*, 2020, **20**(14): 3834.
17. Y. Li, Q. Zhu, Y. Xu, R. Tian, Enhanced fault diagnosis via stochastic resonance in a piecewise asymmetric bistable system, *Chaos*, 2024, **34**(1): 011101.
18. G. Zhang, Y. Shu, T. Zhang, Adaptive modulation stochastic resonance of two-dimensional asymmetric bistable system and its application in bearing fault diagnosis, *IEEE Access*, 2021, **9**: 104722-104733.
19. P. Xia, H. Xu, M. Lei, S. Zhang, An improved underdamped asymmetric bistable stochastic resonance method and its application for spindle bearing fault diagnosis, *IEEE Access*, 2020, **8**: 46824-46836.
20. B. Zhou, W. Xu, Stochastic resonance in an asymmetric bistable system driven by multiplicative colored noise and additive white noise, *Chaos, Solitons & Fractals*, 2008, **38**(4): 1146-1151.
21. J.H. Li, Effect of asymmetry on stochastic resonance and stochastic resonance induced by multiplicative noise and by mean-field coupling, *Physical Review E*, 2002, **66**(3): 031104.
22. C. Yang, J. Yang, D. Zhou, S. Zhang, G. Litak, Adaptive stochastic resonance in bistable system driven by noisy NLFM signal: phenomenon and application, *Philosophical Transactions of the Royal Society A*, 2021, **379**(2192): 20200239.
23. J. Yang, S. Zhang, M.A.F. Sanjuán, H. Liu, Time-frequency analysis of a new aperiodic resonance, *Communications in Nonlinear Science and Numerical Simulation*, 2020, **85**: 105258.

Chapter 6
Aperiodic Vibrational Resonance Caused by Aperiodic Binary Signal and Fast-Varying Auxiliary Signal

This chapter examines aperiodic vibrational resonance induced by an aperiodic binary signal across various nonlinear models. We analyze this phenomenon in both ordinary differential systems with diverse potential functions and in fractional differential systems. Aperiodic vibrational resonance exhibits distinct properties within overdamped symmetric bistable systems, underdamped bistable systems, asymmetric bistable systems, monostable systems, and fractional differential systems. Additionally, the influence of noise on aperiodic vibrational resonance is discussed in detail.

6.1 Aperiodic Vibrational Resonance in Ordinary Differential Systems

Vibrational resonance was first proposed by Landa and McClintock in 2000 [1], and it has since attracted increasing attention [2]. In recent years, although extensive studies have focused on vibrational resonance induced by periodic excitations [3–7], aperiodic vibrational resonance caused by various aperiodic signals remains an active research area and a significant component of vibrational resonance studies [8–16]. As mentioned in previous chapters, a wide range of aperiodic signals are commonly employed across different fields. Most of these aperiodic signals can induce aperiodic vibrational resonance under appropriate conditions.

J. Yang and M. A. F. Sanjuan, *Aperiodic Resonances—Theory and Applications*,
https://doi.org/10.1007/978-981-96-6819-9_6

6.1.1 Aperiodic Vibrational Resonance in Overdamped Symmetric Bistable System

The aperiodic vibrational resonance will be studied by using the following ordinary differential equation

$$\frac{dx(t)}{dt} = ax(t) - bx^3(t) + s(t) + F(t), \tag{6.1}$$

where $a > 0$, $b > 0$ are the system parameters. Here, $s(t)$ represents the aperiodic binary signal that was defined in Eq. (1.1), while $F(t)$ is an auxiliary fast-varying signal that facilitates the occurrence of aperiodic vibrational resonance. In this analysis, we choose the auxiliary fast-varying signal as defined in Eq. (4.24).

Aperiodic vibrational resonance is typically quantified using the cross-correlation coefficient, as defined in Eq. (1.17). Figure 6.1 shows the cross-correlation coefficient C_{sx} plotted against the amplitude of the auxiliary signal. Notably, the cross-correlation coefficient reaches a peak at $B = 1.5$, suggesting that aperiodic vibrational resonance may occur at this amplitude. While the cross-correlation coefficient effectively indicates the similarity between the system response and the input signal time series, it does not directly measure whether the input signal has been amplified through the system. Therefore, analyzing the time series is necessary to further observe the extent of input signal amplification.

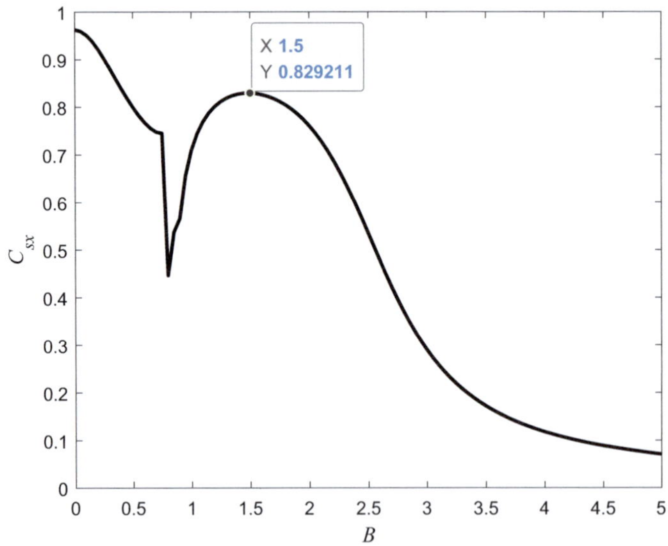

Fig. 6.1 The cross-correlation coefficient C_{sx} versus the signal amplitude B. The simulation parameters are $a = 1$, $b = 1$, $A = 0.3$, $T = 20$, $k = 10$

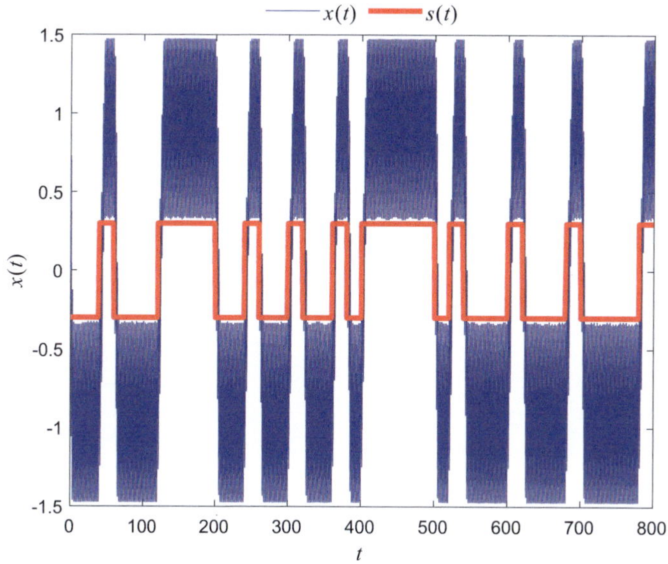

Fig. 6.2 The time series corresponding to the input binary signal (thick line) and the output of the aperiodic vibrational resonance (thin line). The simulation parameters are $a = 1$, $b = 1$, $A = 0.3$, $T = 20$, $B = 1.5$, $k = 10$

To further illustrate the amplification of the input signal by aperiodic vibrational resonance at the peak value of C_{sx}? shown in Fig. 6.1, the time series of the output compared to the input is presented in Fig. 6.2. It is evident that the input signal is significantly amplified, providing an intuitive demonstration of aperiodic vibrational resonance.

In both Figs. 6.1 and 6.2, the aperiodic binary signal varies slowly, meaning that the minimum pulse width T is large. If the aperiodic binary signal instead varies quickly, i.e., T is small, the re-scaled method introduced in Chap. 2 must be applied to induce aperiodic vibrational resonance. Using the re-scaled theoretical framework in Sect. 2.3 and corresponding to the original equation, Eq. (6.1), aperiodic vibrational resonance should then be achieved with the following equation

$$\frac{dx(t)}{dt} = a_1\beta x(t) - b_1\beta x^3(t) + \beta s(t) + \beta F(t), \tag{6.2}$$

where a_1 and a_2 are small parameters and β is still the scale coefficient, as explained in Sect. 2.3.

According to the re-scaled theoretical framework, when $T = 0.2$ and $\beta = 100$, the dynamical behavior of Eq. (6.2) is equivalent to that of Eq. (6.1) corresponding to $T = 20$. Figure 6.3 confirms this assertion. Both in Figs. 6.1 and 6.3a, the cross-correlation coefficient reaches a peak at $B = 1.5$. The peak values in these curves are nearly identical, with negligible error between them, underscoring the effectiveness

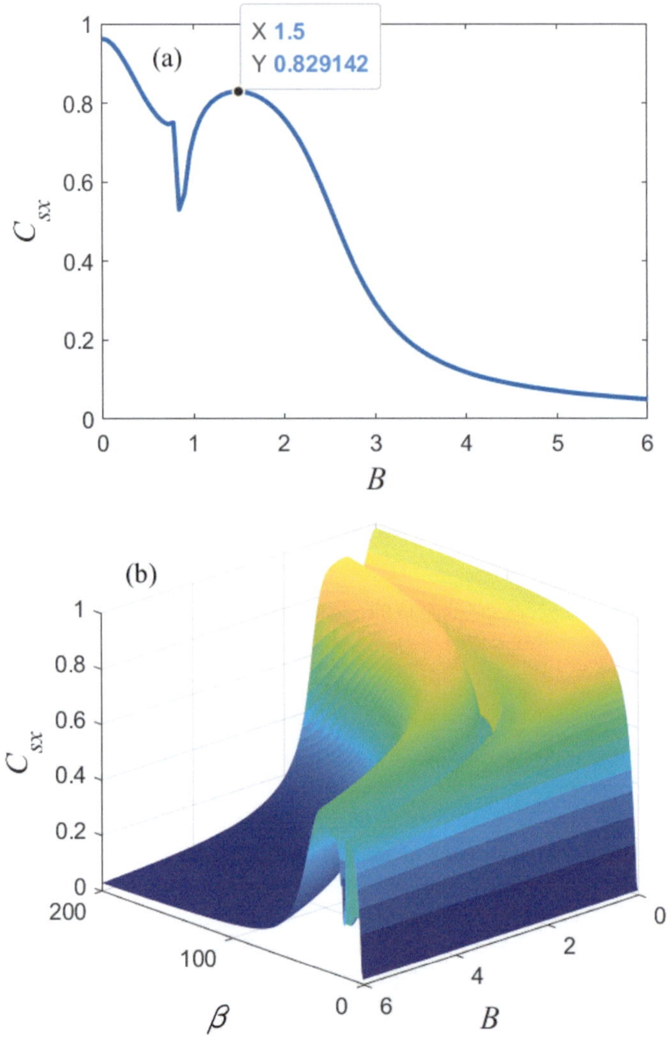

Fig. 6.3 The cross-correlation coefficient C_{sx} versus the signal amplitude B and the scale coefficient β. **a** The curve of C_{sx} under the condition $\beta = 100$. **b** The three-dimensional curve of the cross-correlation coefficient. The simulation parameters are $a = 1$, $b = 1$, $A = 0.3$, $T = 0.2$, $k = 10$

of the re-scaled method here. In Fig. 6.3b, we also observe that increasing the scaling coefficient does not always yield better results; rather, there exists a resonance relationship between the scaling coefficient and the cross-correlation coefficient.

In Fig. 6.4, the time series of the input signal and the output corresponding to the "strongest" aperiodic resonance, as shown in Fig. 6.4a, are presented together. Although the time scale in Fig. 6.4 is just one percent of the time scale in Fig. 6.2,

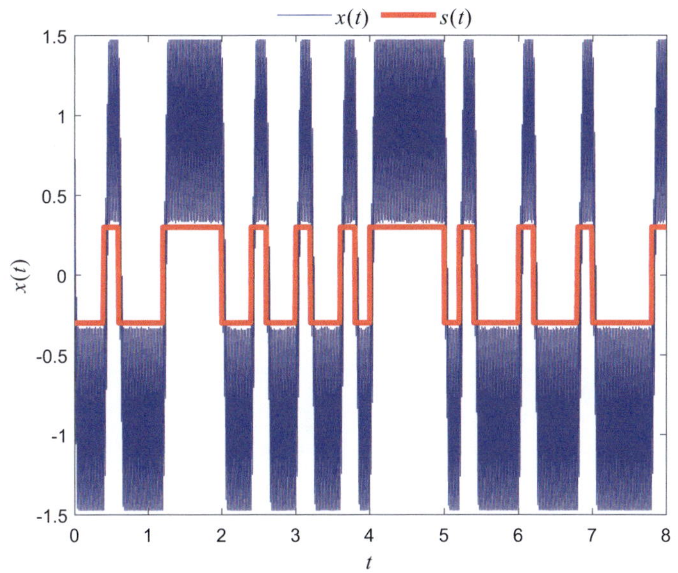

Fig. 6.4 The time series corresponding to the input binary signal (thick line) and the output of the aperiodic vibrational resonance (thin line). The simulation parameters are $a_1 = 1, b_1 = 1, \beta = 100$, $A = 0.3$, $T = 0.2$, $B = 1.5$, $k = 10$

the form of their time series and the resulting resonance effect are identical. Numerically, this satisfies $T \times \beta = 0.2 \times 100 = 20$, where $T = 0.2$ in Fig. 6.2 and $T = 20$ in Fig. 6.4, with 100 as the scaling coefficient. This demonstrates that the scaling coefficient enables the dynamic responses of the two systems to exhibit identical characteristics across different time scales.

To further validate the effectiveness of the re-scaled vibrational resonance method, we chose another set of parameters and present Fig. 6.5 here. In the simulation parameters of Fig. 6.5, the value of T and the scale coefficient β satisfies $T \times \beta = 0.1 \times 200 = 20$. As a result, the curve of the coefficient C_{sx} in this figure is almost the same as that in Figs. 6.1 and 6.3. The "strongest" aperiodic vibrational resonance will be achieved at the same signal amplitude B, as labeled in these three figures.

In Fig. 6.6, the time series of the response corresponding to the resonance peak in Fig. 6.6 is shown in a short time scale. Compared to the time series in Figs. 6.2 and 6.4, their dynamical behaviors are the same in the three different time scales.

This subsection discusses several parameters affecting vibrational resonance. Additionally, other parameters, such as system parameters, the parameters of the aperiodic binary signal (A and T), and the parameter of the fast-varying signal (k in Eq. (6.1)), also have significant impacts on aperiodic vibrational resonance. Readers are encouraged to explore these factors further in their own research.

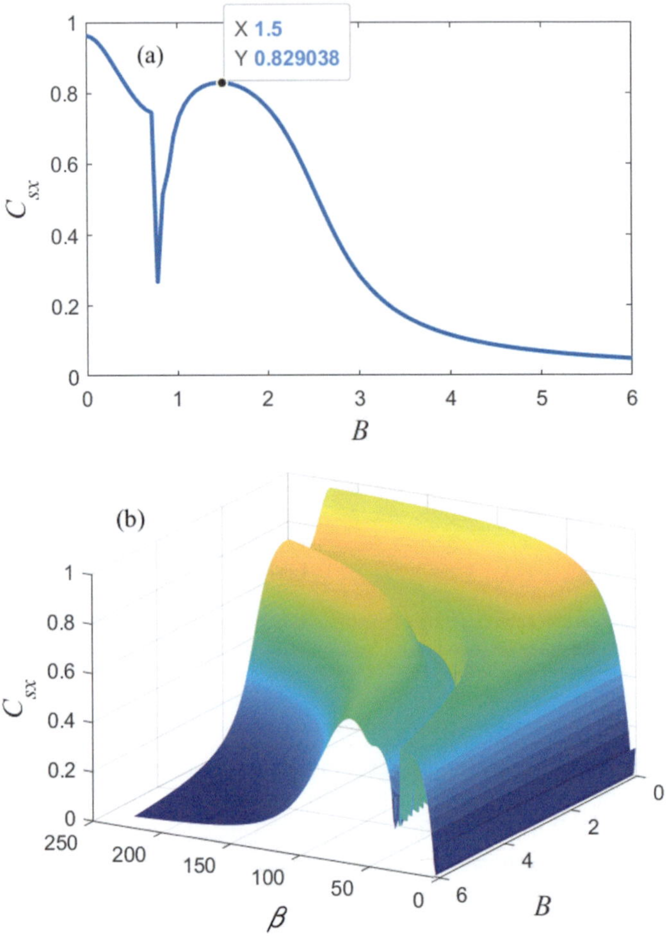

Fig. 6.5 The cross-correlation coefficient C_{sx} versus the signal amplitude B and the scale coefficient β. **a** The curve of C_{sx} under the condition $\beta = 200$. **b** The three-dimensional curve of the cross-correlation coefficient. The simulation parameters are $a = 1$, $b = 1$, $A = 0.3$, $T = 0.1$, $k = 10$

We also considered the case where the amplitude of the binary aperiodic signal is modulated, as discussed in [16]. Specifically, the corresponding equation is

$$\frac{dx(t)}{dt} = ax(t) - bx^3(t) + s(t), \tag{6.3a}$$

$$s(t) = (A + F(t)) \sum_{m=-\infty}^{+\infty} R_m \Gamma(t - mT). \tag{6.3b}$$

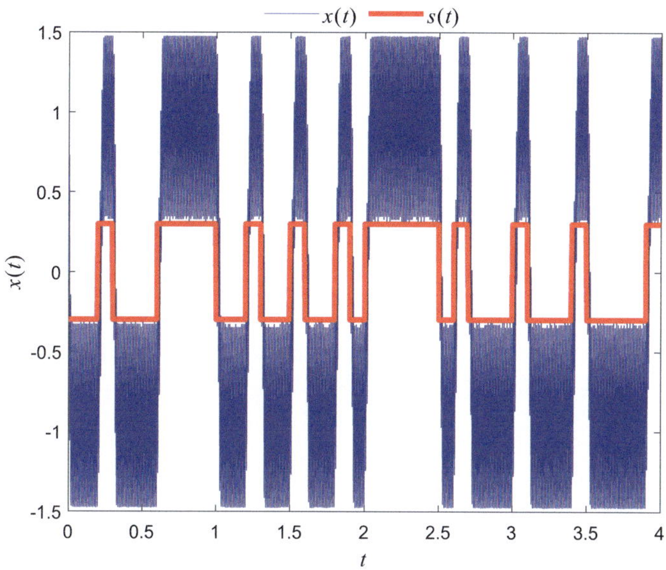

Fig. 6.6 The time series corresponding to the input binary signal (thick line) and the output of the aperiodic vibrational resonance (thin line). The simulation parameters are $a_1 = 1, b_1 = 1, \beta = 200$, $A = 0.3$, $T = 0.1$, $B = 1.5$, $\beta_1 = 10$

When the re-scaled method is used, the aperiodic vibrational resonance should be realized in the equation

$$\frac{dx(t)}{dt} = a_1\beta x(t) - b_1\beta x^3(t) + \beta s(t), \tag{6.4a}$$

$$s(t) = (A + F(t)) \sum_{m=-\infty}^{+\infty} R_m \Gamma(t - mT). \tag{6.4b}$$

In Eqs. (6.3) and (6.4), we continue to use the same expression for the fast-varying auxiliary signal. Specifically, the amplitude of the binary aperiodic signal is modulated by a fast-varying square waveform. The coexistence of both fast and slow varying signals in the excitation creates the potential for aperiodic vibrational resonance to occur.

In Fig. 6.7, we validate the aforementioned prediction. The curve in Fig. 6.7 illustrates the aperiodic vibrational resonance induced by the aperiodic binary signal with amplitude modulation from a fast-varying signal. While the form of the characteristic signal differs, all other parameters remain consistent with those in Fig. 6.3. In the amplitude-modulated case, a clearly observable aperiodic vibrational resonance persists.

Under different values of the amplitude B, the time series of the output from Eq. (6.4) are presented for three cases. In both the small and large amplitude scenarios,

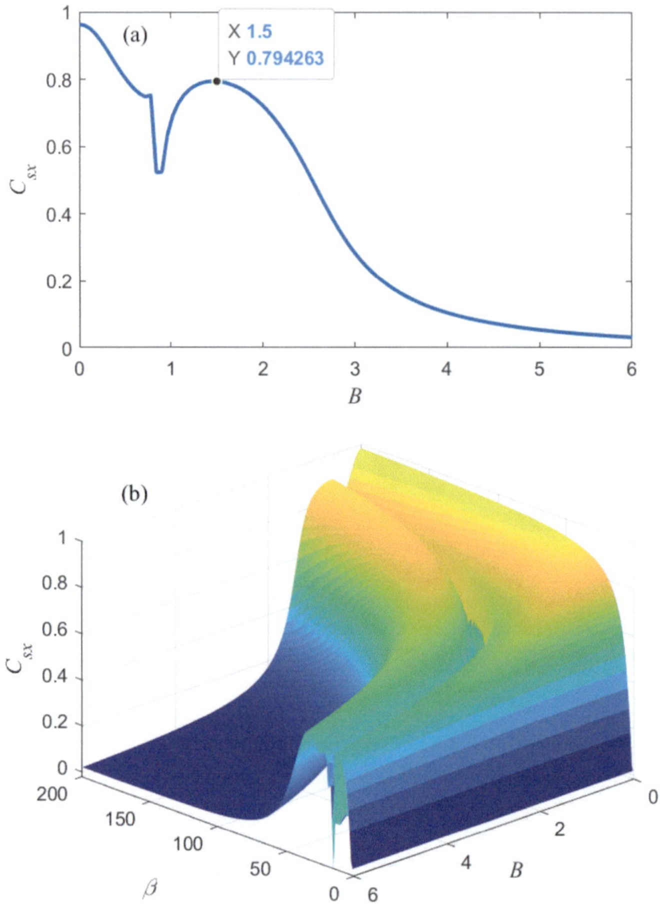

Fig. 6.7 The cross-correlation coefficient C_{sx} versus the signal amplitude B and the scale coefficient β when the amplitude of the binary signal is modulated by a fast varying signal. **a** The curve of C_{sx} under the condition $\beta = 100$. **b** The three-dimensional curve of the cross-correlation coefficient. The simulation parameters are $a = 1$, $b = 1$, $A = 0.3$, $T = 0.2$, $k = 10$

the binary aperiodic signal in the output time series shown in Fig. 6.8a, c is not amplified. However, corresponding to the peak value in Fig. 6.7, the output time series depicted in Fig. 6.8b exhibits aperiodic vibrational resonance. In this case, the output and input change synchronously, and the amplitude of the output signal is relatively large.

When we further decrease the period of the fast-varying signal, as shown in Fig. 6.9, the curve of the cross-correlation coefficient displays a pronounced resonance phenomenon. Both variations in the scale coefficient and the amplitude of the

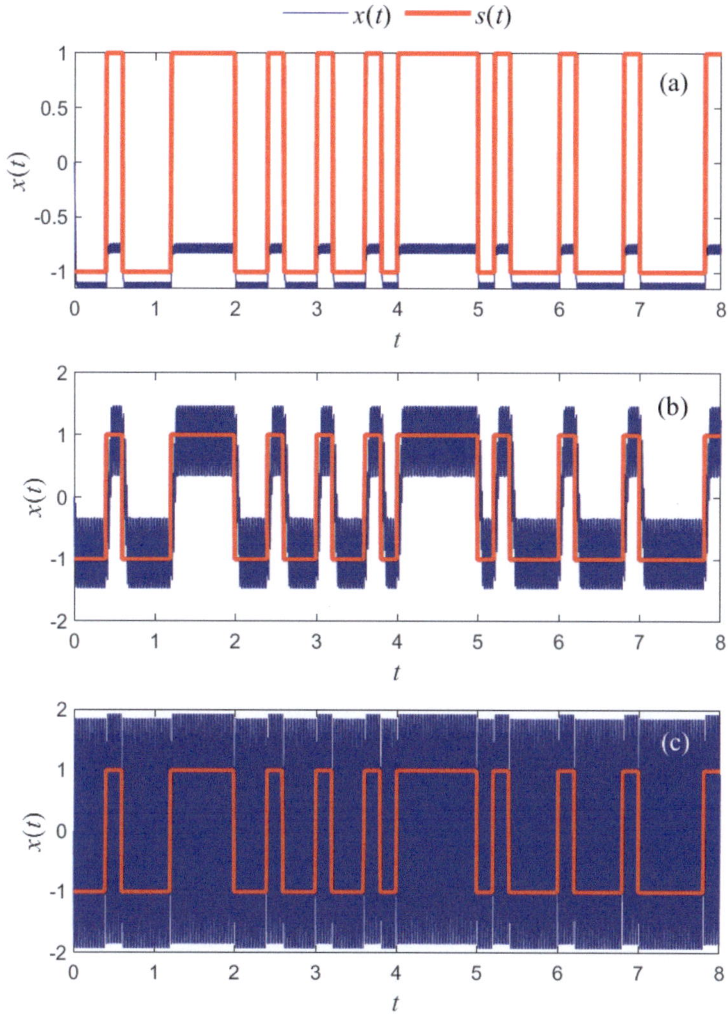

Fig. 6.8 The time series corresponding to the input binary signal (thick line) and the output of the aperiodic vibrational resonance (thin line) when the amplitude of the binary signal is modulated by a fast varying signal. **a** $B = 0.1$, **b** $B = 1.5$, **c** $B = 5$. The simulation parameters are $a_1 = 1$, $b_1 = 1$, $\beta = 100$, $A = 0.3$, $T = 0.2$, $k = 10$

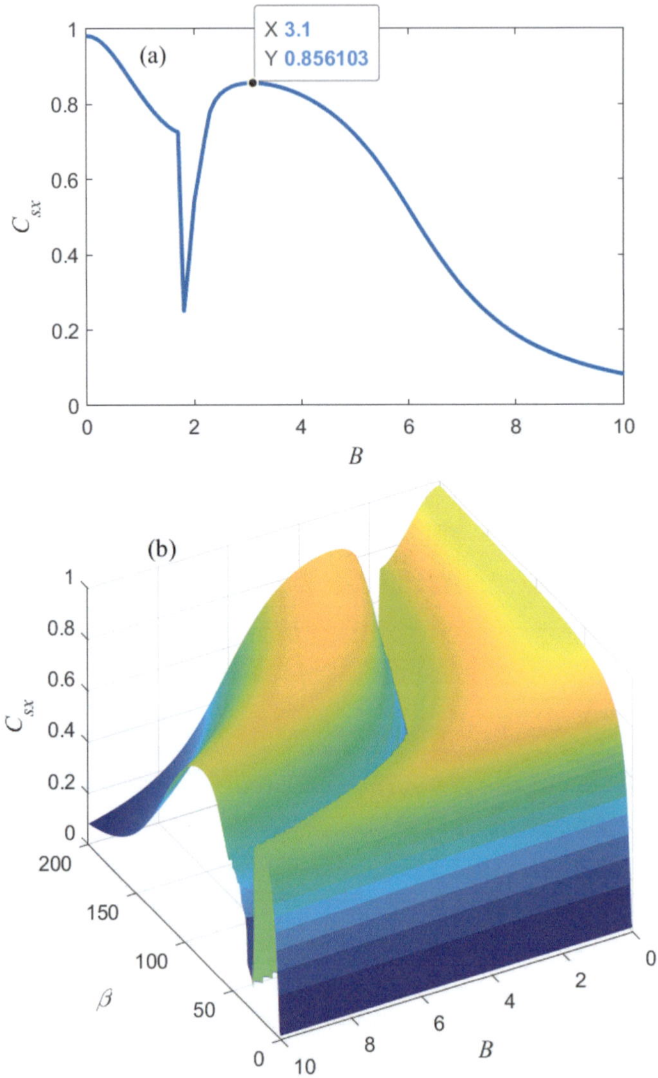

Fig. 6.9 The cross-correlation coefficient C_{sx} versus the signal amplitude B and the scale coefficient β. **a** The curve of C_{sx} under the condition $\beta = 200$. **b** The three-dimensional curve of the cross-correlation coefficient. The simulation parameters are $a = 1$, $b = 1$, $A = 0.3$, $T = 0.2$, $k = 50$

fast-varying signal can induce aperiodic resonance, leading to the maximization of the cross-correlation coefficient value.

In Fig. 6.10, the output time series varies with the value of amplitude B. In Fig. 6.10a, for small value of B, it is not possible to synchronize the system output across two potential wells with the input signal. In Fig. 6.10b, it corresponds the

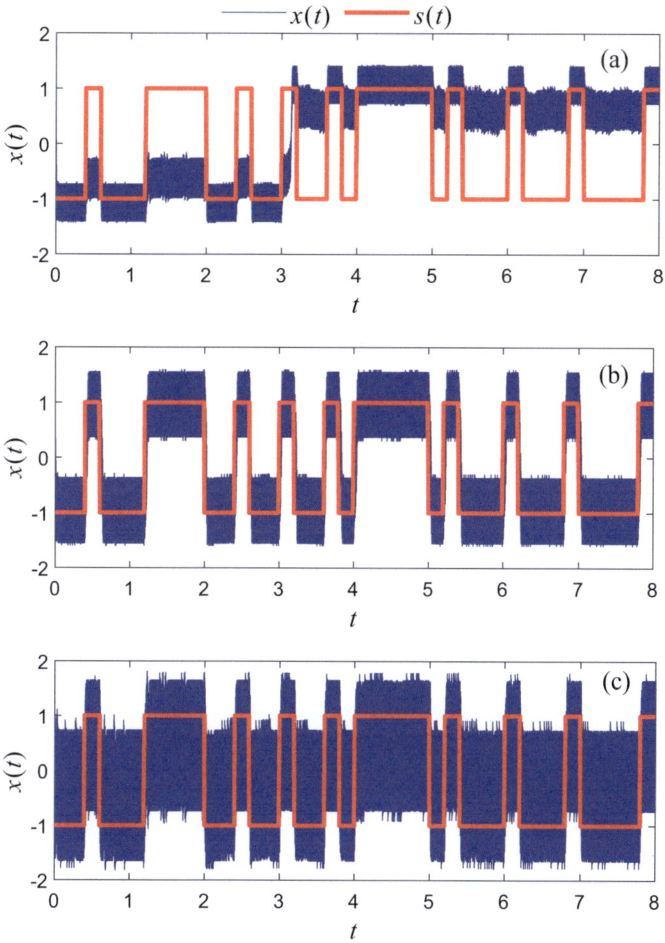

Fig. 6.10 The time series corresponding to the input binary signal (thick line) and the output of the aperiodic vibrational resonance (thin line). **a** $B = 1.8$, **b** $B = 3.1$, **c** $B = 6$. The simulation parameters are $a_1 = 1, b_1 = 1, \beta = 200, A = 0.3, T = 0.2, k = 50$

"strongest" aperiodic vibrational resonance. Time series presents rich forms both in Figs. 6.8 and 6.10

We replace the fast-varying signal of Eq. (6.4) by a fast-varying harmonic signal $F(t) = B\cos(\Omega t)$. Specifically, the frequency Ω is high and the period of the auxiliary signal is much less than T which is the minimal pulse width of the aperiodic binary signal. As shown in Fig. 6.11, it still can induce an almost identical aperiodic vibrational resonance phenomenon in the curve of the cross-correlation coefficient C_{sx}.

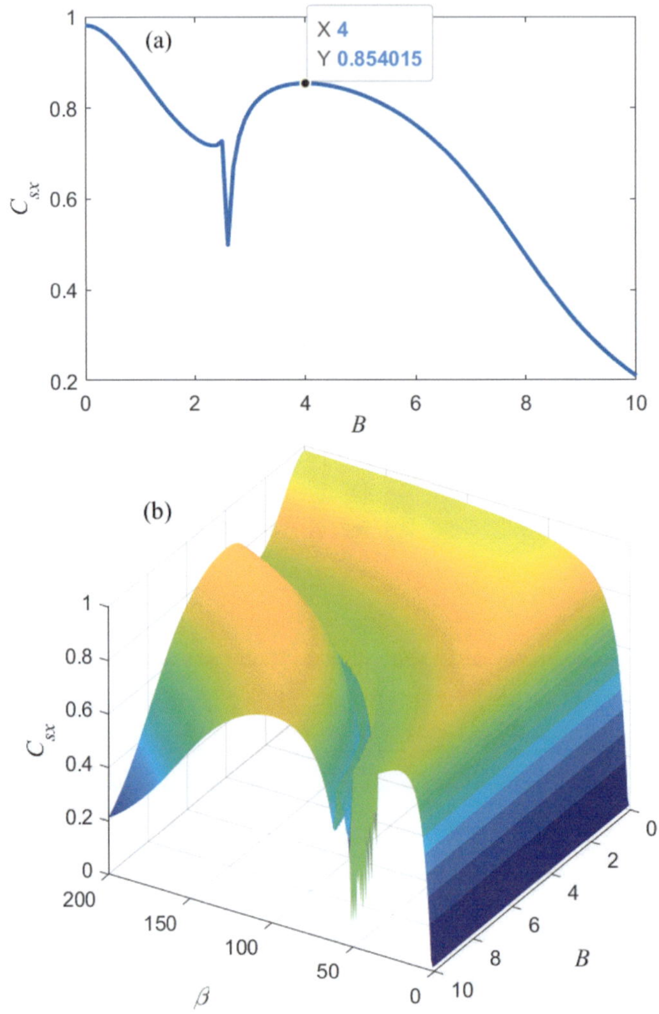

Fig. 6.11 The cross-correlation coefficient C_{sx} versus the signal amplitude B and the scale coefficient β, $F(t) = B\cos(\Omega t)$. **a** The curve of C_{sx} under the condition $\beta = 200$. **b** The three-dimensional curve of the cross-correlation coefficient. The simulation parameters are $a = 1, b = 1$, $A = 0.3, T = 0.2, k = 50$

6.1.2 Aperiodic Vibrational Resonance in Underdamped Symmetric Bistable System

We also consider the aperiodic vibrational resonance in the underdamped bistable system

$$\ddot{x}(t) + \delta\dot{x}(t) - ax(t) + bx^3(t) = s(t) + F(t). \tag{6.5}$$

For the damping parameter δ Fig. 6.12 presents the curves of C_{sx} versus B for four different values of δ. When the damping coefficient δ is relatively small, it is challenging to observe the traditional aperiodic vibrational resonance curve between the cross-correlation coefficient C_{sx} and the auxiliary amplitude B, as illustrated in Fig. 6.12a. In this case, C_{sx}responds sensitively to changes in B, leading to significant fluctuations in the curve. As the damping coefficient δ gradually increases, as shown from Fig. 6.12b–d, the curve transitions toward the traditional aperiodic vibrational resonance curve. The sensitivity of the curve to changes in B diminishes, revealing a clear peak where aperiodic vibrational resonance can occur. Notably, the phenomenon of C_{sx} exhibiting high sensitivity to the auxiliary amplitude B at low damping coefficients is an example of ultra-sensitive resonance, which will be discussed in detail in Chaps. 9 and 10.

Figure 6.13 illustrates the impact of the damping parameter δ on the cross-correlation coefficient C_{sx}. The figure reveals that different values of B also influence the relationship between C_{sx} and δ. Specifically, in Fig. 6.13a, b, when B is relatively small, the cross-correlation coefficient C_{sx} attains a relatively large and nearly constant value, particularly when the damping coefficient δ is high, especially for $\delta > 1$. Conversely, for larger values of B, C_{sx} decreases with increasing damping coefficient when δ is large.

6.1.3 Aperiodic Vibrational Resonance in Overdamped Bistable System with a Fractional Power Nonlinearity

Another symmetric bistable system is represented by a case where the nonlinear term has a fractional exponent. We have previously considered aperiodic system resonance in this context, as shown in Eq. (3.6). The corresponding nonlinear differential equation is given by

$$\dot{x}(t) = ax(t) - bx(t)|x(t)|^{\alpha-1} + s(t) + F(t). \tag{6.6}$$

The characteristic signal and the auxiliary signal used here are the same as those in Eq. (4.24). Furthermore, if the characteristic signal in a fast-varying form, the re-scaled equation to realize aperiodic vibrational resonance becomes

$$\dot{x}(t) = a_1\beta x(t) - b_1\beta x(t)|x(t)|^{\alpha-1} + \beta s(t) + \beta F(t). \tag{6.7}$$

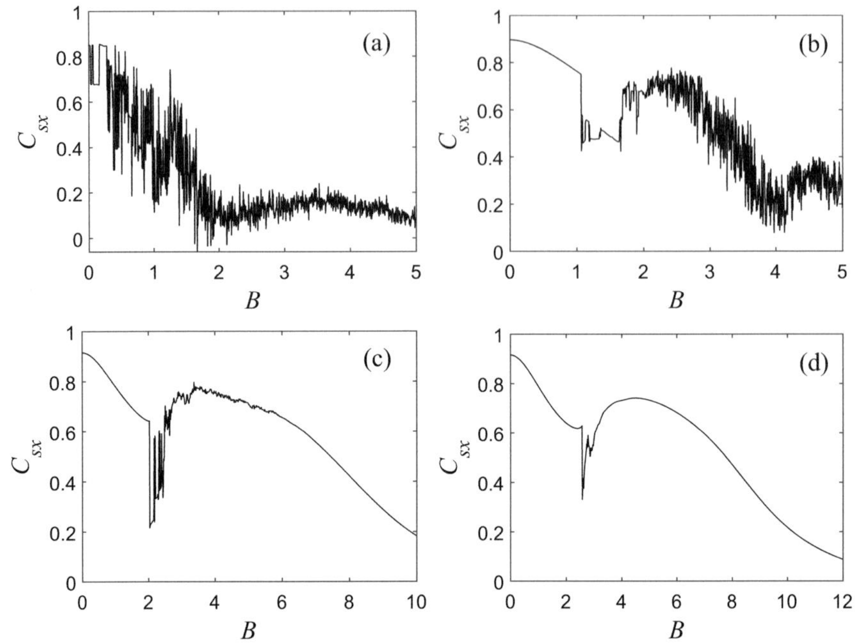

Fig. 6.12 The cross-correlation coefficient C_{sx} versus the signal amplitude B and the damping coefficient δ of the underdamped system in Eq. (6.5). **a** $\delta = 0.2$, **b** $\delta = 0.5$, **c** $\delta = 0.8$, **d** $\delta = 1.2$. The simulation parameters are $a = 1$, $b = 1$, $A = 0.3$, $T = 0.2$, $k = 10$

In Fig. 6.14, the dependence of C_{sx} on the signal amplitude B is given for different values of the fractional exponent α. Interestingly, for different values of α, the cross-correlation coefficient reaches its maximum peak at almost the same value of B. The larger the value of α, the more pronounced the resonance peak. The reasons for this phenomenon are also worth considering.

If α is used as the control variable, a significant aperiodic vibrational resonance phenomenon can still be observed when both the characteristic signal and the auxiliary signal are fixed, as illustrated in Fig. 6.15. Notably, the resonance peak does not occur at $\alpha = 3$, indicating that in unconventional bistable systems, aperiodic vibrational resonance may be more likely to arise when the external excitation signal is predetermined. Additionally, we found that the strength of the external excitation auxiliary signal inversely affects the width of the resonance region: the stronger the external excitation, the narrower the resonance region becomes. For instance, when the amplitude of the external excitation is sufficiently small, as shown by the curve corresponding to $B = 1$ in the figure, the cross-correlation coefficient can achieve larger values across a wide range of α. However, it is crucial to note that this does not necessarily imply that the characteristic signal has been amplified over this broad range. As previously emphasized, a high cross-correlation coefficient is a necessary but not sufficient condition for the amplification of the characteristic signal.

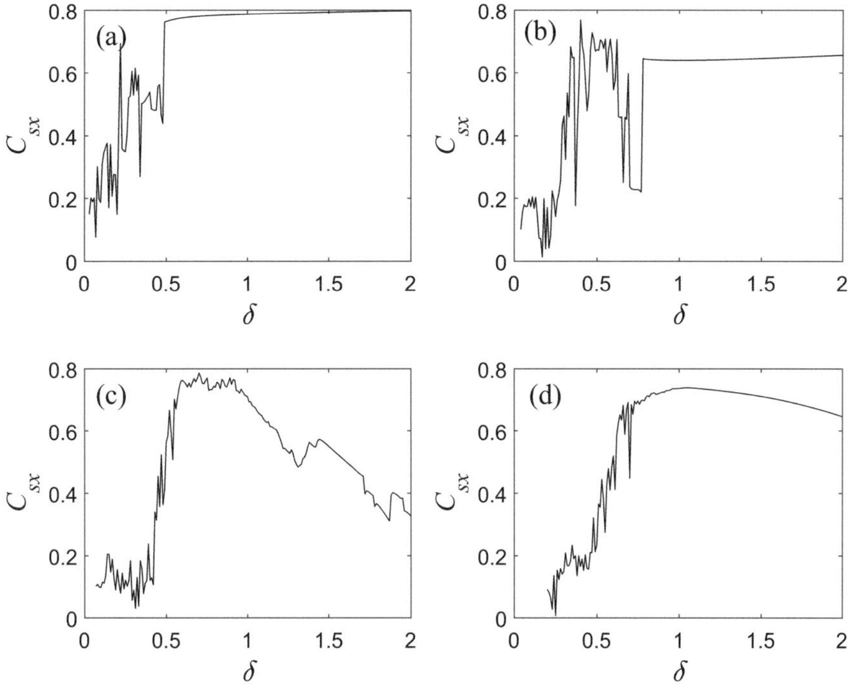

Fig. 6.13 The cross-correlation coefficient C_{sx} versus the damping coefficient δ under different values of the signal amplitude B of the underdamped system in Eq. (6.6). **a** $B = 1$, **b** $B = 2$, **c** $B = 3$, **d** $B = 5$. The simulation parameters are $a = 1$, $b = 1$, $A = 0.3$, $T = 0.2$, $k = 10$

To further investigate the characteristics of aperiodic vibrational resonance in systems with fractional exponents, we present Fig. 6.16. This figure illustrates the relationship between the cross-correlation coefficient and the varying parameters α and β in a two-dimensional plane. When α is held constant, a clear resonance region emerges between the cross-correlation coefficient and the variable scale coefficient β. Interestingly, as the value of α decreases, the resonance region expands. This observation suggests that, for a fixed α, a larger variable scale coefficient does not always yield better results. In scenarios where β is fixed within a variable scale system, three distinct patterns can be observed concerning the cross-correlation coefficient relative to α: (1) a clear resonance relationship when β is small; (2) a sustained high value over a broad range when β is large; and (3) a peak followed by a rapid decline in value when β is very large.

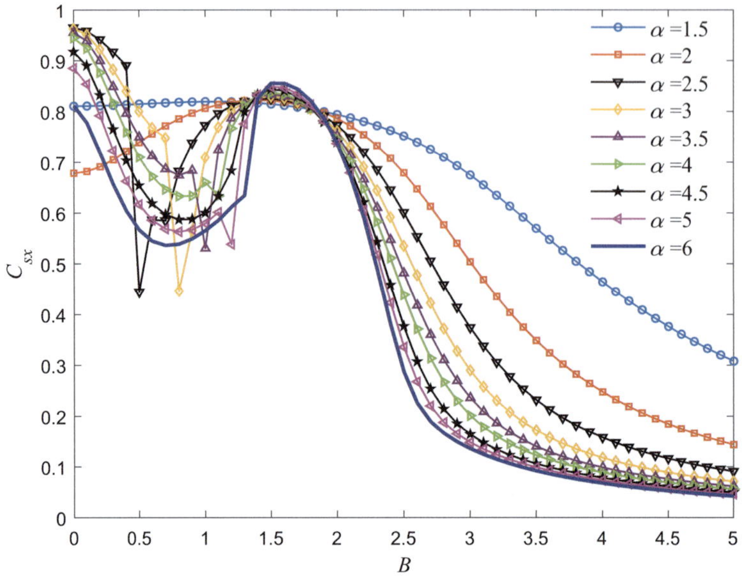

Fig. 6.14 The cross-correlation coefficient C_{sx} versus the signal amplitude B under different values of the fractional exponent α. The simulation parameters are $a = 1, b = 1, A = 0.3, T = 20, k = 10$

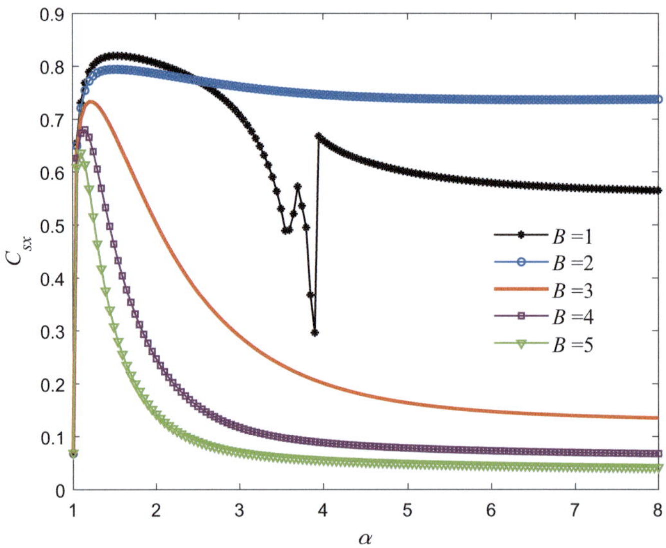

Fig. 6.15 The cross-correlation coefficient C_{sx} versus the fractional exponent α under different values of the signal amplitude B. The simulation parameters are $a = 1$, $b = 1$, $A = 0.3$, $T = 20$, $k = 10$

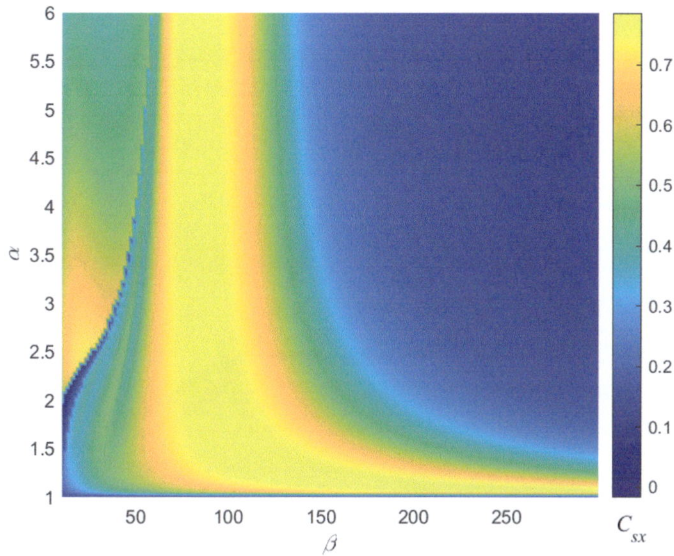

Fig. 6.16 The cross-correlation coefficient C_{sx} versus the fractional exponent α and the scale coefficient β. The simulation parameters are $a = 1$, $b = 1$, $A = 0.3$, $T = 0.2$, $k = 10$

6.1.4 Aperiodic Vibrational Resonance in Asymmetric Bistable System

Vibrational resonance in asymmetric bistable systems reveals distinct resonance patterns compared to symmetric ones. We previously investigated vibrational resonance caused by harmonic characteristic signals in asymmetric bistable systems [17]. In this study, we focus on an asymmetric bistable system excited by an aperiodic binary signal and an auxiliary signal

$$\frac{dx(t)}{dt} = ax(t) - bx^3(t) + r + s(t) + F(t), \tag{6.8}$$

where r is the bias parameter. $F(t)$ is still a fast varying rectangular waveform.

The value of r influences the asymmetry of the system's potential function, affecting the response characteristics and ultimately reflected in the cross-correlation coefficient curve, as shown in Fig. 6.17. Interestingly, when r is relatively small, the curve of the cross-correlation coefficient closely resembles that of a conventional bistable system. As r increases, such as for $r = 0.1$ and $r = 0.2$, a unimodal relationship appears between the cross-correlation coefficient C_{sx} and the signal amplitude B. This characteristic is significant as it facilitates the identification of the B value that induces aperiodic vibrational resonance, particularly benefiting optimization algorithms. When r is large, such as $r = 0.5$ and $r = 1$, the cross-correlation coefficient primarily exhibits a decreasing trend in relation to B.

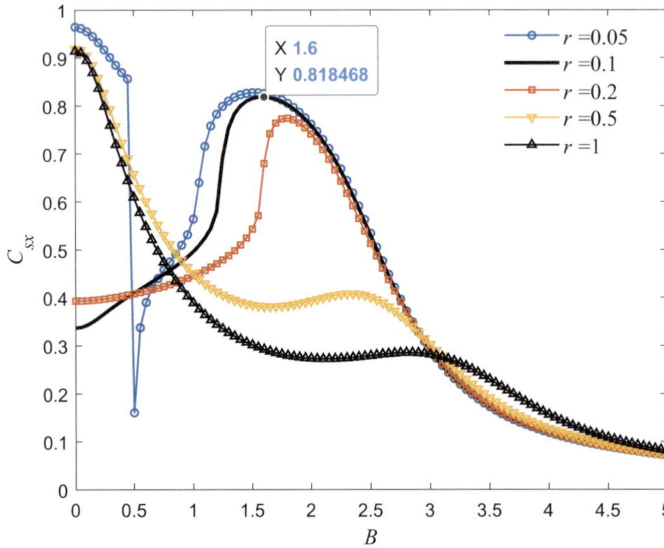

Fig. 6.17 The cross-correlation coefficient C_{sx} versus the signal amplitude B under different values of the bias parameter r. The simulation parameters are $a = 1$, $b = 1$, $A = 0.3$, $T = 20$, $k = 10$

The bias parameter r determines the tilt direction of the potential wells. When r is negative, the tilt direction of the potential well is opposite to that when r is positive. A smaller value of r results in a lesser tilt of the potential well. In Fig. 6.18, the curve of the cross-correlation coefficient relative to the amplitude B is presented for negative values of r. When the tilt of the potential well is small, there is a pronounced peak in the cross-correlation coefficient as a function of B. However, when the tilt is significant, the cross-correlation coefficient predominantly exhibits a decreasing trend with respect to B.

In Fig. 6.19, we present the optimal response corresponding to the "strongest" aperiodic vibrational resonance observed in Figs. 6.17 and 6.18 for the cases $r = 0.1$ and $r = -0.1$, respectively. For the asymmetric bistable system, regardless of the tilt direction of the potential wells, synchronization occurs between the output and input of the system. This synchronization leads to the amplification of the aperiodic binary signal, resulting in pronounced aperiodic vibrational resonance phenomena.

Using the bias parameter r as the control variable, Fig. 6.20 shows the relationship between the cross-correlation coefficient and r for different amplitudes of the auxiliary signal. Clearly, when r approaches 0, the cross-correlation coefficient increases significantly. This suggests that the symmetric bistable system is more effective than the asymmetric bistable system in inducing aperiodic vibrational resonance. However, if the value of B is excessively large, such as $B = 4$, adjusting the value of r will not successfully induce aperiodic vibrational resonance.

There are many types of asymmetric bistable systems, which we have summarized in our review paper [2]. In this section, we consider aperiodic vibrational resonance in a different form of asymmetric bistable system, specifically

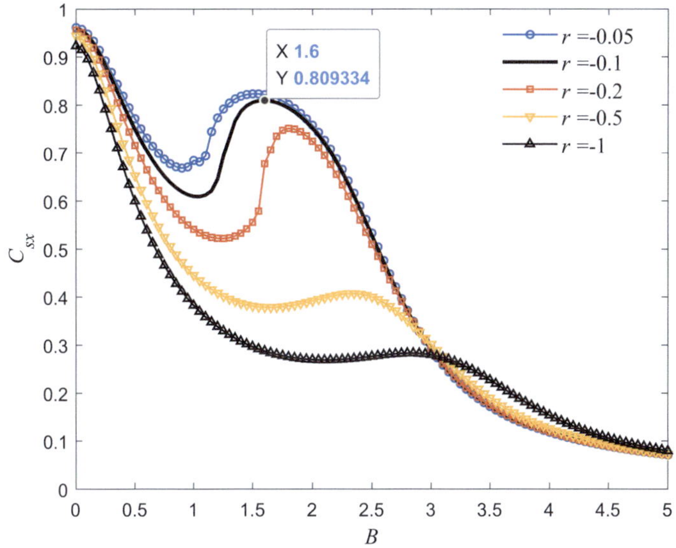

Fig. 6.18 The cross-correlation coefficient C_{sx} versus the signal amplitude B under different values of the bias parameter r. The simulation parameters are $a = 1$, $b = 1$, $A = 0.3$, $T = 20$, $k = 10$

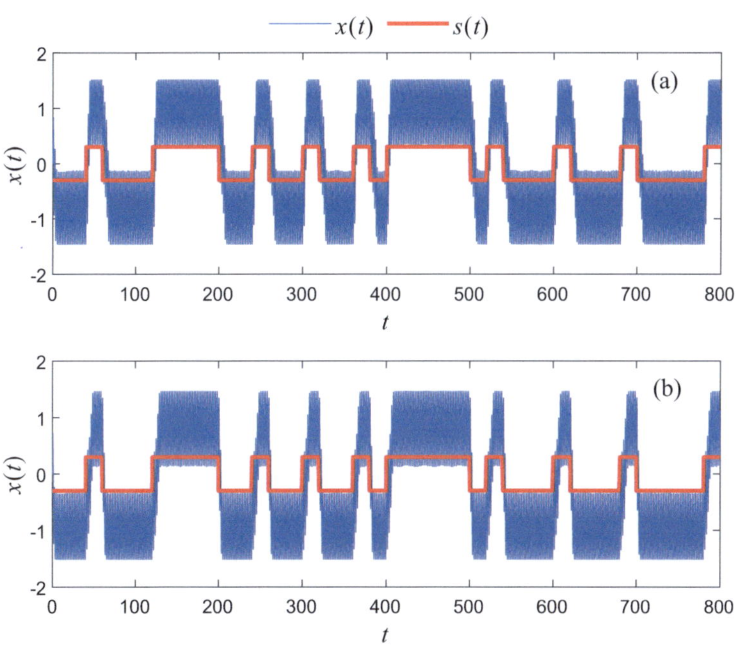

Fig. 6.19 The "strongest" aperiodic vibrational resonance of the asymmetric bistable system. **a** $r = 0.1$, **b** $r = -0.1$. The simulation parameters are $a = 1$, $b = 1$, $A = 0.3$, $T = 20$, $B = 1.6$, $k = 10$

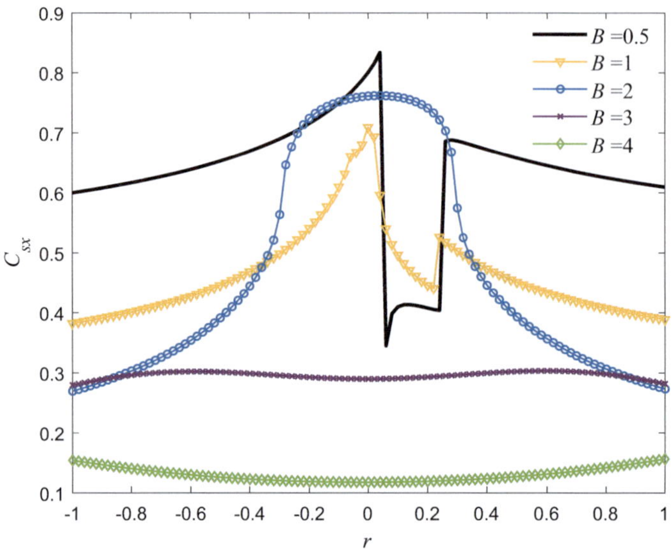

Fig. 6.20 The cross-correlation coefficient C_{sx} versus the bias parameter r under different values of the signal amplitude B. The simulation parameters are $a = 1$, $b = 1$, $A = 0.3$, $T = 20$, $k = 10$

$$\frac{dx(t)}{dt} = ax(t) + rx^2(t) - bx^3(t) + s(t) + F(t), \tag{6.9}$$

where r is the bias parameter, and the excitation signals are the same as those in Eq. (6.8).

For the asymmetric bistable system described by Eq. (6.9), Fig. 6.21 illustrates the dependence of the cross-correlation coefficient on the signal amplitude B. A unimodal curve is still observed, as seen in the case where $r = 0.5$. Although the equations (potential functions) of the asymmetric systems in Eqs. (6.8) and (6.9) are slightly different, both can induce aperiodic vibrational resonance phenomena.

Changing the positive or negative sign of the bias parameter r, i.e., changing the tilt direction of the potential well, Fig. 6.22 shows the cross-correlation coefficient curve under different values of r, indicating that the aperiodic vibrational resonance can still occur in this kind of asymmetric bistable systems. However, similar to the curves in Figs. 6.17 and 6.18, the greater the tilt direction of the potential well, the lower the probability of the aperiodic vibrational resonance occurring.

In Fig. 6.23, we give the time series corresponding to the peaks of the curves in Figs. 6.21 and 6.22, which indicates the appearance of the "strongest" aperiodic vibrational resonance. Similar to Fig. 6.19, the synchronization and amplification of the characteristic signal is presented.

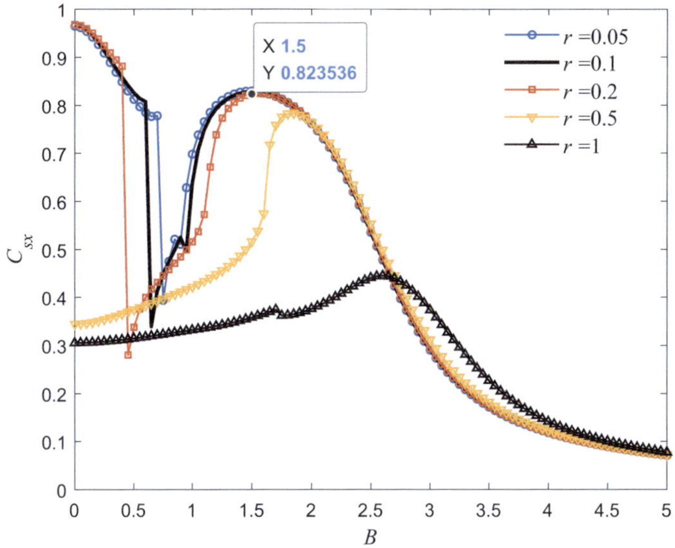

Fig. 6.21 The cross-correlation coefficient C_{sx} versus the signal amplitude B under different values of the bias parameter r. The simulation parameters are $a = 1$, $b = 1$, $A = 0.3$, $T = 20$, $k = 10$

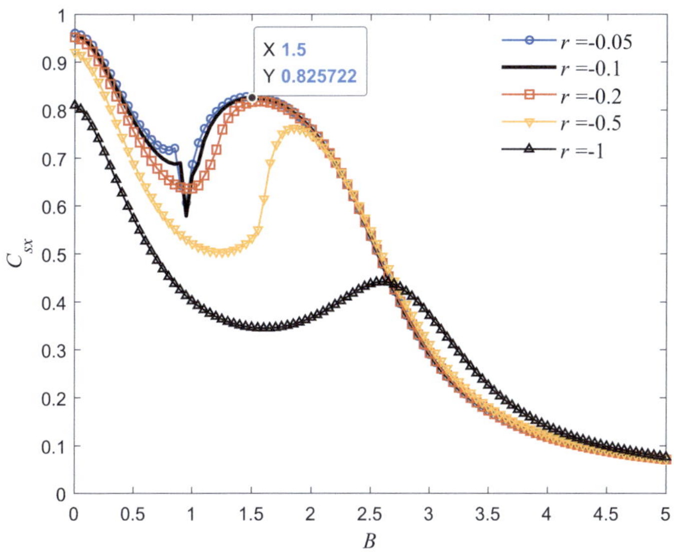

Fig. 6.22 The cross-correlation coefficient C_{sx} versus the signal amplitude B under different values of the bias parameter r. The simulation parameters are $a = 1$, $b = 1$, $A = 0.3$, $T = 20$, $k = 10$

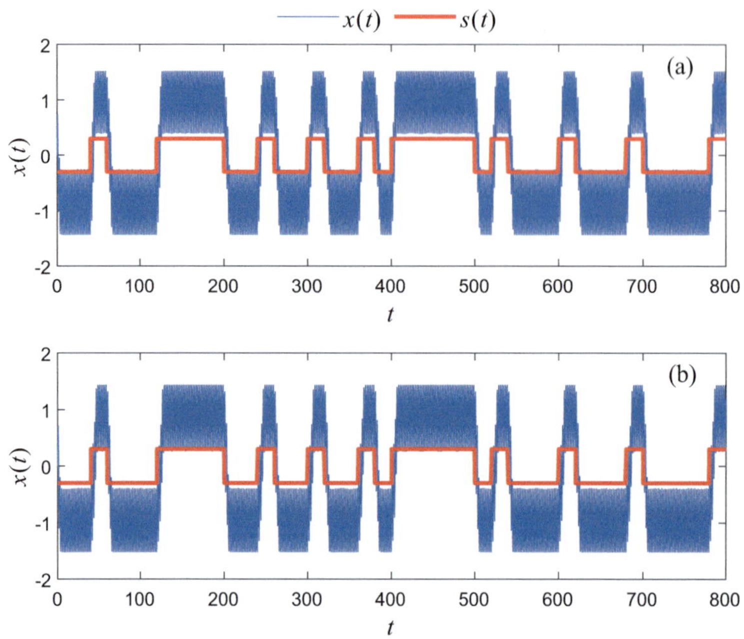

Fig. 6.23 The "strongest" aperiodic vibrational resonance of the asymmetric bistable system. **a** $r = 0.1$, **b** $r = -0.1$. The simulation parameters are $a = 1$, $b = 1$, $A = 0.3$, $T = 20$, $B = 1.5$, $k = 10$

By varying the bias parameter r in Eq. (6.9) for different values of the signal amplitude B, the curves in Fig. 6.24 exhibit distinct shapes. Notably, when $B = 2$, a very wide resonance region is observed. At $B = 0.5$, the curve shows a significant resonance peak, which rapidly decreases as B increases beyond this point. In contrast, for $B = 3$ and $B = 4$, the curves indicate a lack of aperiodic vibrational resonance, with consistently low values for the cross-correlation coefficient.

6.1.5 Aperiodic Vibrational Resonance in Monostable System

To investigate whether aperiodic vibrational resonance occurs in monostable system, we consider two typical types of monostable systems, namely

$$\frac{dx(t)}{dt} = -ax(t) - bx^3(t) + rx^2(t) + s(t) + F(t), \tag{6.10}$$

and

$$\frac{dx(t)}{dt} = -ax(t) - bx^3(t) + r + s(t) + F(t). \tag{6.11}$$

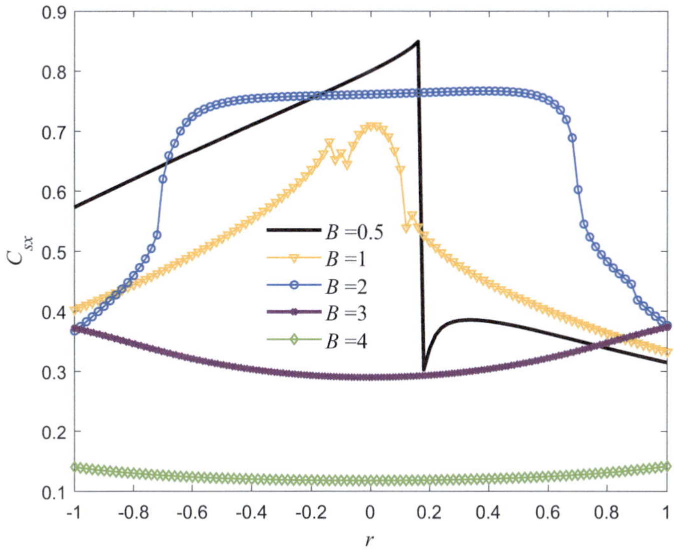

Fig. 6.24 The nonlinear variation of the cross-correlation coefficient C_{sx} versus the bias parameter r under different values of the signal amplitude B. The simulation parameters are $a = 1$, $b = 1$, $A = 0.3$, $T = 20$, $k = 10$

The potential functions of Eqs. (6.10) and (6.11) are $V(x) = \frac{a}{2}x^2 - \frac{r}{3}x^3 + \frac{b}{4}x^4$ and $V(x) = -rx + \frac{a}{2}x^2 + \frac{b}{4}x^4$, respectively. The corresponding potential function curves are given in Fig. A.6.

Clearly, as shown in Fig. 6.25, there is no aperiodic vibrational resonance in the monostable system. For the two different forms of monostable systems described in Eqs. (6.10) and (6.11), no aperiodic vibrational resonance phenomenon occurs. However, it is noteworthy that under periodic excitations, a monostable system can exhibit vibrational resonance, a phenomenon we have summarized in our review paper [2].

Although this subsection discusses aperiodic vibrational resonance in several nonlinear systems, many typical systems remain unexamined. The aperiodic vibrational resonances in these systems also warrant further investigation.

6.2 Aperiodic Vibrational Resonance in Fractional Differential Systems

In this section, we discuss aperiodic vibrational resonance caused by both slow-varying and fast-varying characteristic aperiodic binary signals.

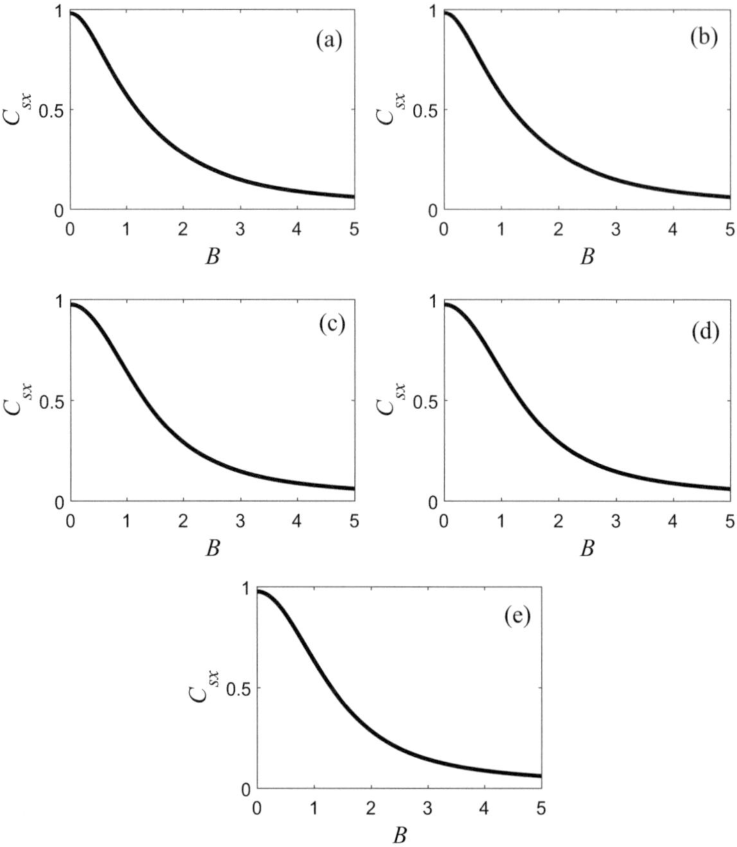

Fig. 6.25 The cross-correlation coefficient C_{sx} versus the signal amplitude B under different values of the bias parameter r. **a** $r = -0.5$, Eq. (6.11), **b** $r = 0.5$, Eq. (6.11), **c** $r = -0.5$, Eq. (6.12), **d** $r = 0.5$, Eq. (6.12). The simulation parameters are $a = 1, b = 1, A = 0.3, T = 20, k = 10$

6.2.1 Aperiodic Vibrational Resonance Realized by the Traditional Method

We modify the ordinary differential equation in Eq. (6.1) to incorporate a fractional-order differential operator. Subsequently, we investigate aperiodic vibrational resonance in the following fractional-order bistable system

$$\frac{d^{\alpha}x(t)}{dt^{\alpha}} = ax(t) - bx^3(t) + s(t) + F(t), \tag{6.12}$$

Initially, we introduce three aperiodic binary signals with different waveforms as the input characteristic signals to investigate aperiodic vibrational resonance. Their

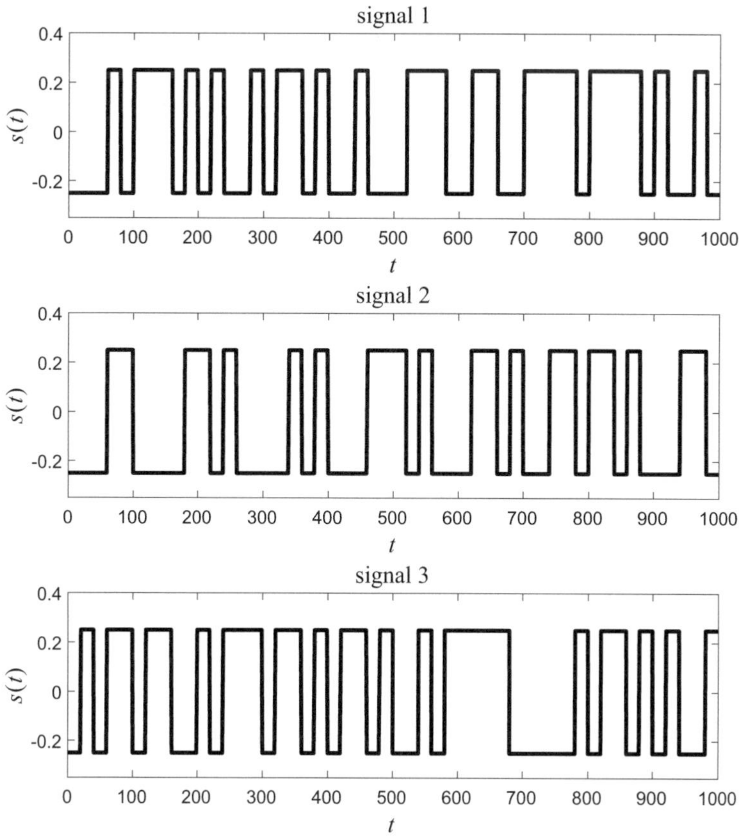

Fig. 6.26 Three waveforms of the aperiodic binary signal. The simulation parameters are $A = 0.25$, $T = 20$

waveforms are illustrated in Fig. 6.26. Although the amplitude and pulse width of the signals are identical, the time series of the characteristic signals differ due to the randomness of R_m in the expression for $s(t)$ in Eq. (1.1). Additionally, since the characteristic signals are presented in a slow-varying form, aperiodic vibrational resonance can be achieved within the traditional theoretical framework.

In Eq. (6.12), we maintain the same expression for $F(t)$ as previously used in this chapter. Under varying values of the fractional order, Fig. 6.27 presents the curves of aperiodic vibrational resonance corresponding to the three signals depicted in Fig. 6.26. Several important phenomena are shown in Fig. 6.27, which we analyze as follows.

Firstly, the relationship between C_{sx} and B is largely unaffected by the waveform of the input signal. Due to the fact that the cross-correlation coefficient is a statistical parameter calculated based on a large amount of data, increasing the amount of data

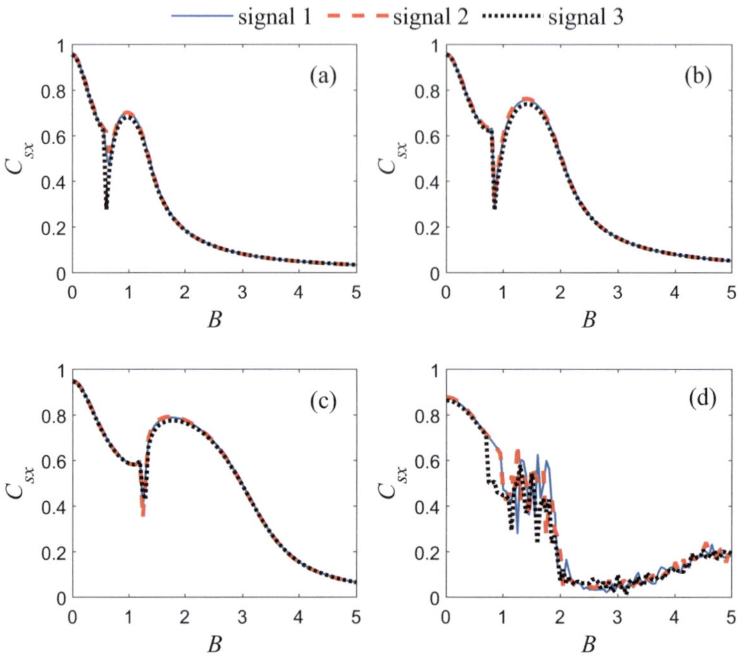

Fig. 6.27 The nonlinear variation of the cross-correlation coefficient C_{sx} versus the signal amplitude B under different values of fractional-order. **a** $\alpha = 0.5$, **b** $\alpha = 0.8$, **c** $\alpha = 1.2$, **d** $\alpha = 1.8$. The simulation parameters are $a = 1$, $b = 1$, $A = 0.25$, $T = 20$, $k = 10$

used in the calculation can further reduce the small errors in the $C_{sx} - B$ curve caused by different characteristic signal waveforms.

Secondly, as the amplitude of the auxiliary signal increases, the value of C_{sx} first decreases and then increases, resulting in a clear "resonance region" where aperiodic vibrational resonance occurs. The "resonance zone" is an index that describes the response characteristics of a linear system under harmonic excitation. Generally, the region on the frequency response curve that is located on both sides of the resonance peak and satisfies a value greater than $1/\sqrt{2}$ times the peak value is defined as the "resonance region" [17]. For aperiodic vibrational resonance, there is no unified concept of "resonance region". For ease of description, following the concept of "resonance region" in linear system under harmonic excitation, this chapter defines the region on the C_{sx} versus B curve that is located on the right side of the valley and on both sides of the resonance peak, and satisfies the condition that C_{sx} is greater than the $1/\sqrt{2}$ times the resonance peak value, as the "resonance region" of aperiodic vibrational resonance.

Thirdly, when the value of B is small, the value of the cross-correlation coefficient C_{sx} is large, indicating a high similarity between the output time series and the input signal, but the characteristic signal part in the output time series has not been amplified. At the resonance peak of each curve, the "similarity" between the output

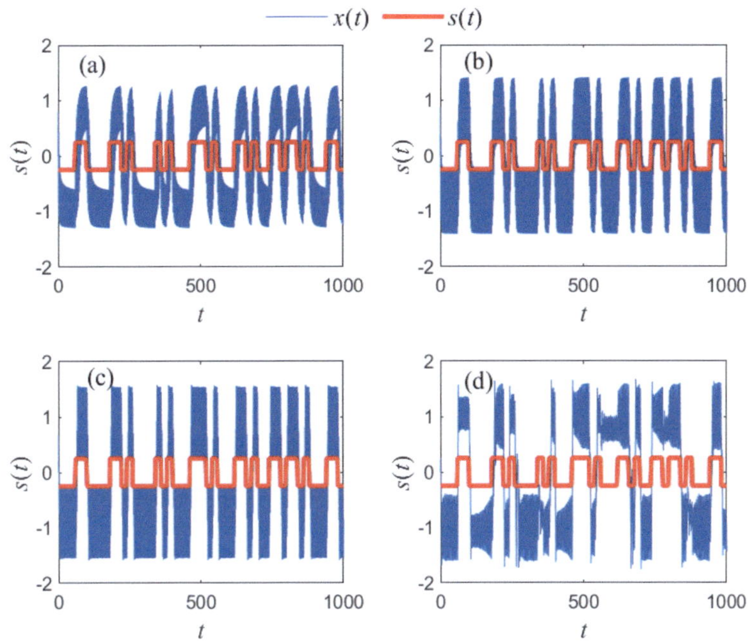

Fig. 6.28 The time series corresponding to the "strongest" aperiodic vibrational resonance in Fig. 6.27. **a** $\alpha = 0.5$, $B = 0.95$, **b** $\alpha = 0.8$, $B = 1.4$, **c** $\alpha = 1.2$, $B = 1.75$, **d** $\alpha = 1.8$, $B = 1.25$. The simulation parameters are $a = 1$, $b = 1$, $A = 0.25$, $T = 20$, $k = 10$

time series and the input characteristic signal is high, and the characteristic signal components contained in the output time series are amplified. This can be found in the subsequent analysis of the time series.

Corresponding to the resonance peak (signal 2 is chosen as the characteristic signal) in Fig. 6.27, in Fig. 6.28, we show the "strongest" output when the aperiodic vibrational resonance occurs for different fractional-order values. Compared to the original input signal $s(t)$, the characteristic component in the output waveform has been greatly enhanced.

6.2.2 Aperiodic Vibrational Resonance Realized by the Re-Scaled Method

When the aperiodic binary signal is in a fast-varying form, aperiodic vibrational resonance should be realized by the re-scaled method.

First, we introduce the scale transformation in Eq. (2.4). Then, Eq. (6.12) changes to

$$\frac{d^{\alpha}z(\tau)}{dt^{\alpha}}=\frac{a}{\beta^{\alpha}}z(\tau)-\frac{b}{\beta^{\alpha}}z^{3}(\tau)+\frac{A}{\beta^{\alpha}}\sum_{m=-\infty}^{+\infty}R_{m}\Gamma(\tau-m\beta T)+\frac{B}{\beta^{\alpha}}\mathrm{sgn}\left[\cos\left(\frac{\Omega\tau}{\beta}\right)\right].\tag{6.13}$$

Letting

$$a_{1}=\frac{a}{\gamma^{\alpha}},\quad b_{1}=\frac{b}{\gamma^{\alpha}},\tag{6.14}$$

and recovering the signal amplitudes to the original levels, we have

$$\frac{d^{\alpha}z(\tau)}{dt^{\alpha}}=a_{1}z(\tau)-b_{1}z^{3}(\tau)+A\sum_{m=-\infty}^{+\infty}R_{m}\Gamma(\tau-m\beta T)+B\mathrm{sgn}\left[\cos\left(\frac{\Omega\tau}{\beta}\right)\right].\tag{6.15}$$

The dynamical properties of Eq. (6.15) on the time scale τ is equivalent to that of Eq. (6.12) on the time scale t. Based on the above analysis process, we can achieve aperiodic vibrational resonance caused by fast-varying aperiodic binary signal in the time scale t in the following equation

$$\frac{d^{\alpha}x(t)}{dt^{\alpha}}=a_{1}\beta^{\alpha}x(t)-b_{1}\beta^{\alpha}x^{3}(t)+A\beta^{\alpha}\sum_{m=-\infty}^{+\infty}R_{m}\Gamma(t-mT)+B\beta^{\alpha}\mathrm{sgn}[\cos(\Omega t)].\tag{6.16}$$

To illustrate the re-scaled method in this section, a set of binary aperiodic signals is presented in Fig. 6.29, with waveforms that are completely similar to the corresponding signals in Fig. 6.26, but with different time scales, i.e., different pulse widths. Compared to the value of the pulse width in Fig. 6.26, the pulse width in Fig. 6.29 is reduced to one percent of the pulse width in Fig. 6.26. It can be considered that the characteristic signal in Fig. 6.26 is a slow-varying signal, while the characteristic signal in Fig. 6.29 is a fast-varying signal. Fast-changing characteristic signals are widely utilized in engineering fields, making the study of aperiodic vibrational resonance caused by these signals both theoretically significant and valuable for practical applications.

For the aperiodic binary signal of the three waveforms in Fig. 6.29, the aperiodic vibrational resonance phenomenon is achieved using the re-scaled method in Fig. 6.30. Obviously, the matching of the signal in Fig. 6.29 with the system parameters used in Fig. 6.30 is achieved based on equation Eq. (6.16). By using a scale transformation of $\beta = 100$, it is equivalent to transforming the pulse width of the characteristic signal $T = 0.2$ into a system equivalent to the pulse width of the characteristic signal $T = 20$. Therefore, the $C_{sx} - B$ curve in Fig. 6.30 is basically the same in shape as the corresponding curve in Fig. 6.27, which also indicates the equivalence between Eqs. (6.16) and (6.12) after the re-scaled transformation. By the re-scaled method can achieve aperiodic vibrational resonance in fractional order nonlinear system under the excitation of fast varying characteristic signal, without being affected by pulse width and signal waveform.

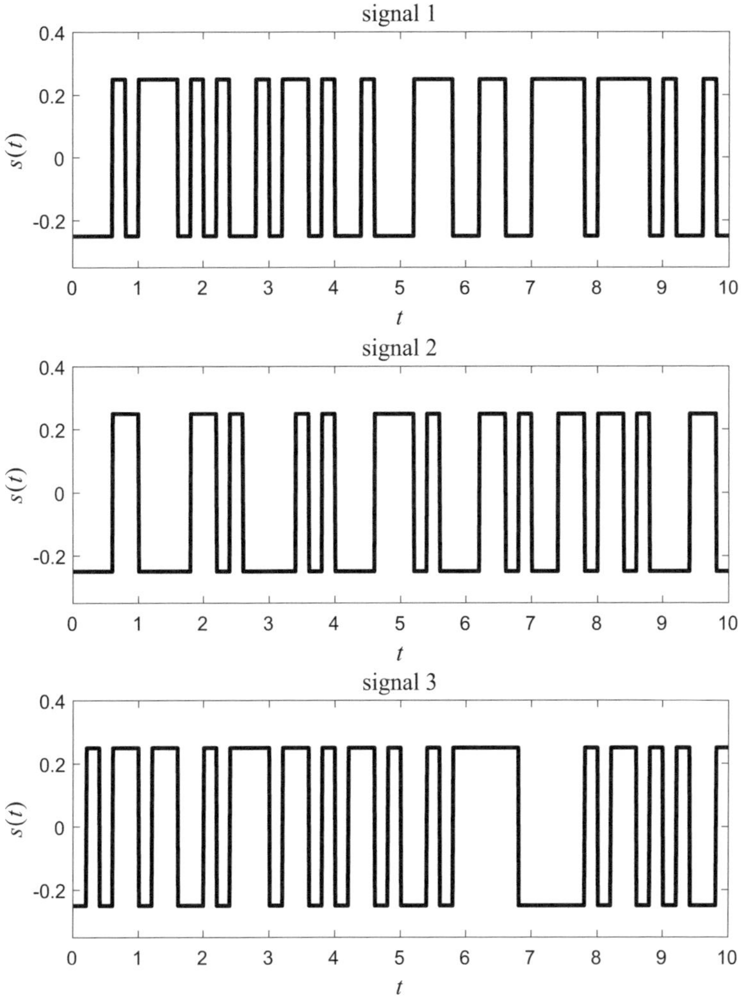

Fig. 6.29 Three waveforms of the aperiodic binary signal. The simulation parameters are $A = 0.25$, $T = 0.2$

Referring to the curves in Fig. 6.30, the output corresponding to the "strongest" vibrational resonance (signal 2 as the characteristic signal) under re-scaled formulation is given in Fig. 6.31. Firstly, the aperiodic output achieves vibrational resonance phenomenon when the binary signal has a relatively small pulse width, and the characteristic signal $s(t)$ is amplified simultaneously. Secondly, by comparing Figs. 6.31 and 6.28, it is found that the output waveform and the amplitude B of the corresponding auxiliary signal are basically the same when the system undergoes aperiodic vibrational resonance. A slight small error can be attributed to numerical calculations. The effectiveness of the re-scaled method in processing aperiodic

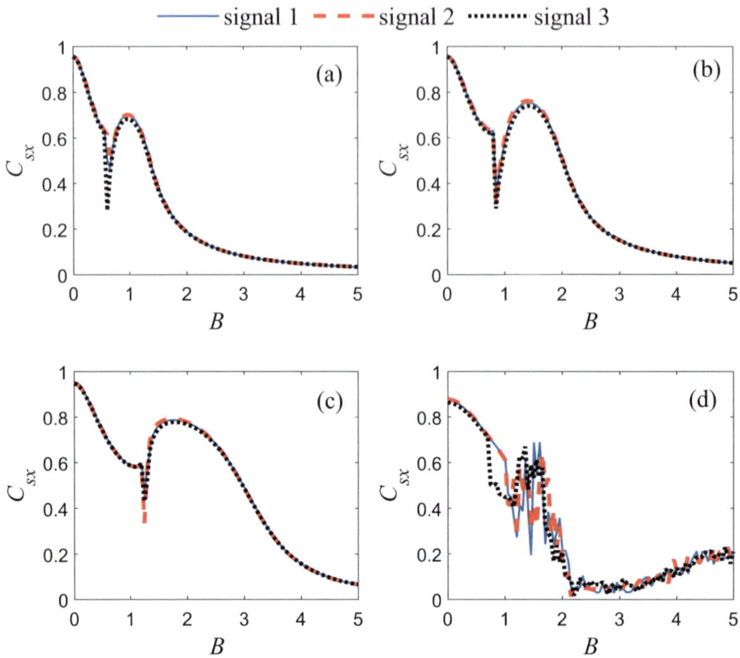

Fig. 6.30 The nonlinear variation of the cross-correlation coefficient C_{sx} versus the signal amplitude B under different values of fractional-order in the re-scaled theoretical framework. **a** $\alpha = 0.5$, **b**) $\alpha = 0.8$, **c** $\alpha = 1.2$, **d** $\alpha = 1.8$. The simulation parameters are $a = 1, b = 1, A = 0.25, T = 0.2$, $k = 10, \beta = 100$

fast-varying characteristic signal and the equivalence between Eqs. (6.16) and (6.12) after re-scaled transformation are once again demonstrated from the time series in Fig. 6.31.

6.3 Effect of Noise on Aperiodic Vibrational Resonance

In this section, we utilize Eq. (4.21) to investigate the effect of noise on aperiodic vibrational resonance. In Fig. 6.32, the curves illustrate the cross-correlation coefficient as a function of noise intensity for different values of noise intensity. Clearly, for the case of $\sigma = 0$, the system's response with respect to the auxiliary signal amplitude B exhibits the traditional aperiodic vibrational resonance curve, with the peak corresponding to this case showing the maximum value. For the cases $\sigma = 0.05$ or 0.1, the curves display a single peak, but the valley of the curve is absent. This change indicates that the amplitude of the fast-varying signal influences the equilibrium points of the equivalent system, resulting in the resonance curve displaying inflection points with peaks and valleys. However, for $\sigma = 0.05$ and 0.11, only one

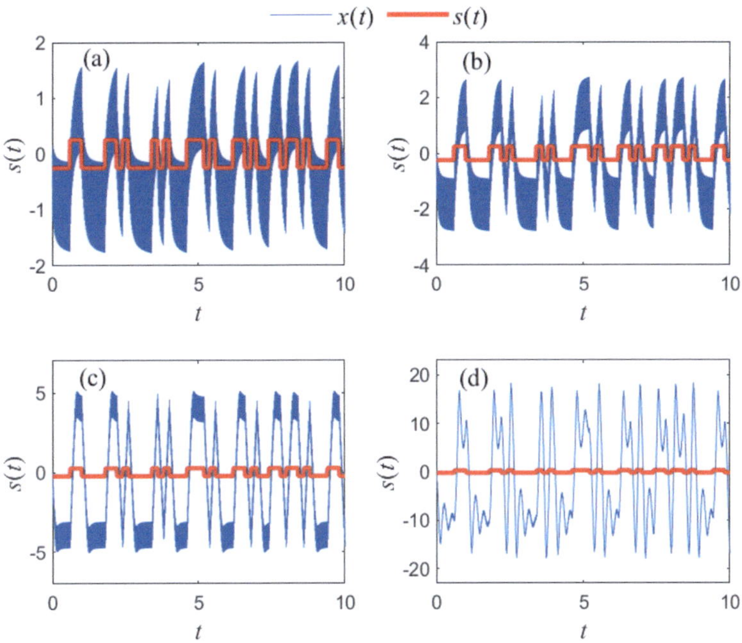

Fig. 6.31 The time series corresponding to the “strongest” aperiodic vibrational resonance in Fig. 6.30 in the re-scaled theoretical framework. **a** $\alpha = 0.5$, $B = 0.95$, **b** $\alpha = 0.8$, $B = 1.4$, **c** $\alpha = 1.2$, $B = 1.75$, **d** $\alpha = 1.8$, $B = 1.65$. The simulation parameters are $a = 1$, $b = 1$, $A = 0.25$, $T = 0.2$, $k = 10$, $\beta = 100$

peak appears on each curve. Consequently, we speculate that the introduction of noise prevents the fast-varying signal from altering the number of equilibrium points in the equivalent system. As the noise intensity increases further, such as $\sigma = 0.25$, 0.5 and 1, the phenomenon of aperiodic vibrational resonance vanishes completely. From Fig. 6.32, it is evident that noise significantly impacts aperiodic vibrational resonance, with different noise intensities leading to varying resonance patterns in the curves.

Consider the aperiodic vibrational resonance induced by the signal parameter k, which indicates the time scale of the auxiliary signal. In Fig. 6.33, for the case $k = 0$, the parameter k induces a clear aperiodic vibrational resonance, characterized by a pronounced peak in the curve. When the noise intensity increases to 0.05, the resonance remains evident, although the peak becomes smaller and the resonance region widens. As the noise intensity increases further, the resonance peak disappears completely. When k exceeds a certain value, the cross-correlation coefficient exhibits minimal variation, approaching a state of saturation.

Throughout Figs. 6.32 and 6.33, the presence of noise disrupts the traditional curve pattern of the aperiodic vibrational resonance. However, this does not mean

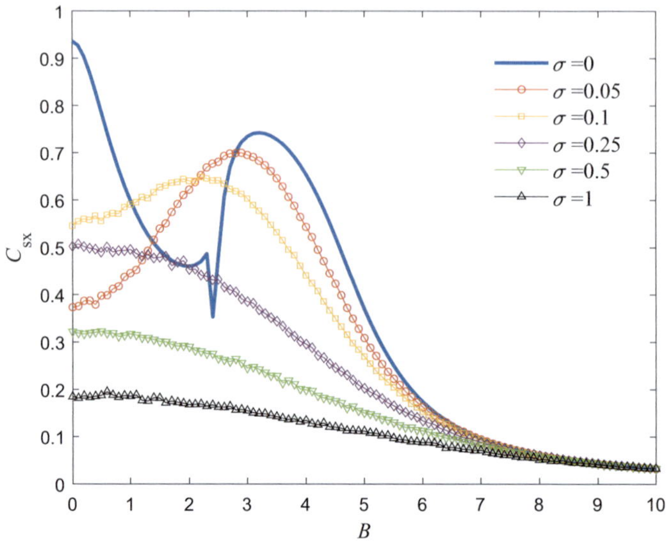

Fig. 6.32 Effect of noise on aperiodic vibrational resonance that induced by the amplitude B of the fast-varying the auxiliary signal. The simulation parameters are $a = 1, b = 1, A = 0.2, T = 20$ and $k = 20$

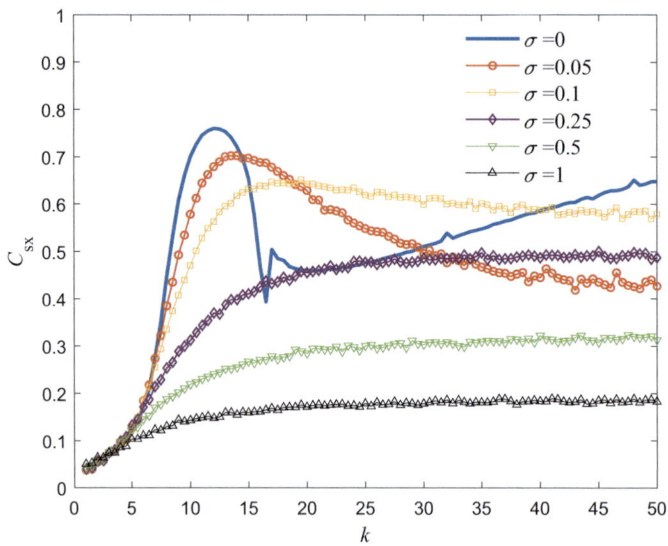

Fig. 6.33 Effect of noise on the aperiodic vibrational resonance induced by the parameter k of the fast-varying auxiliary signal. The simulation parameters are $a = 1$, $b = 1$, $A = 0.2$, $T = 20$ and $B = 2$

that noise cannot play a positive role in the aperiodic resonance response. For example, in Fig. 6.32, when $B = 2$, the cross-correlation coefficient values corresponding to $\sigma = 0.05$ and $\sigma = 0.1$ is much larger than that corresponding to $\sigma = 0$. This indicates that noise contributes to the synchronization between the system's output and the input characteristic signal, exemplifying the beneficial effects of noise. Readers can verify this conclusion by examining the time series of the system output. A similar phenomenon is evident in Fig. 6.33; for instance, when $k = 20$, the advantageous impact of noise on enhancing the cross-correlation coefficient is once again demonstrated.

6.4 Concluding Remarks

Aperiodic vibrational resonance is discussed in detail with respect to characteristic signals in aperiodic binary waveforms. We primarily focus on two classes of nonlinear systems: those described by ordinary differential equations and those governed by fractional differential equations. Additionally, the potential function of the nonlinear system influences the output performance. Our discussion includes symmetric bistable systems (both overdamped and underdamped cases), bistable systems with fractional power nonlinearity, asymmetric bistable systems, and monostable systems. For cases where the characteristic signal exhibits a fast-varying waveform, we utilize a rescaling method to achieve aperiodic vibrational resonance. We also examine the effects of noise on the patterns of aperiodic vibrational resonance, highlighting the positive contributions of noise. Looking ahead, in this research direction, we recommend exploring a broader range of nonlinear models and addressing more practical engineering issues.

References

1. P.S. Landa, P.V.E. McClintock, Vibrational resonance, *Journal of Physics A: Mathematical and General*, 2000, **33**(45): L433-L438.
2. J. Yang, S. Rajasekar, M.A.F. Sanjuán, Vibrational resonance: A review, *Physics Reports*, 2024, **1067**: 1-62
3. J. Yang, *Bifurcation and Resonance in Fractional-order Systems* (Science Press, Beijing, 2017)
4. S. Rajasekar, M.A.F. Sanjuán, *Nonlinear resonances* (Springer International Publishing, Switzerland, 2016)
5. I.I. Blekhman, P.S. Landa, Conjugate resonances and bifurcations in nonlinear systems under biharmonical excitation, *International Journal of Non-Linear Mechanics*, 2004, **39**(3): 421-426.
6. C. Jeevarathinam, S. Rajasekar, M.A.F. Sanjuán, Theory and numerics of vibrational resonance in Duffing oscillators with time-delayed feedback, *Physical Review E*, 2011, **83**(6): 066205.
7. J.H. Yang, H. Zhu, Vibrational resonance in Duffing systems with fractional-order damping, *Chaos*, 2012, **22**(1):013112.
8. V.N. Chizhevsky, G. Giacomelli, Vibrational resonance and the detection of aperiodic binary signals, *Physical Review E*, 2008, **77**(5): 051126.

9. J.H. Yang, Q. Ma, C.J. Wu, H.G. Liu, Aperiodic vibrational resonance in the fractional-order bistable system, 2018, *Acta Physica Sinica*, **67**(5): 054501.
10. P.X. Jia, C.J. Wu, J.H. Yang, M.A.F. Sanjuán, G.X. Liu, Improving the weak aperiodic signal by three kinds of vibrational resonance, *Nonlinear Dynamics*, 2018, **9**: 2699-2713.
11. V.N. Chizhevsky, G. Giacomelli, Vibrational resonance and the detection of aperiodic binary signals, *Physical Review E*, 2008, **77**(5): 051126.
12. T. Gong, J. Yang, M.A.F. Sanjuán, H. Liu, Z. Shan, Vibrational resonance by using a real-time scale transformation method, *Physica Scripta*, 2022, **97**(4): 045207.
13. P. Jia, J. Yang, Y. Leng, The echo chirp signal amplification by the vibrational information fusion method, *International Journal of Modern Physics B*, 2020, **34**(6): 2050041.
14. P. Jia, J. Yang, X. Zhang, M.A.F. Sanjuán, On the LFM signal improvement by piecewise vibrational resonance using a new spectral amplification factor, *IET Signal Processing*, 2019, **13**(1): 65-69.
15. P. Jia, J. Yang, C. Wu, M.A.F. Sanjuán, Amplification of the LFM signal by using piecewise vibrational methods, *Journal of Vibration and Control*, 2019, **25**(1): 141-150.
16. P. Jia, J. Yang, H. Liu, E. Hu, Improving amplitude-modulated signals by re-scaled and twice sampling vibrational resonance methods, *Pramana*, 2018, **91**(38): 1-11.
17. J.H. Yang, M.A.F. Sanjuán, F. Tian, H.F. Yang, Saddle-node bifurcation and vibrational resonance in a fractional system with an asymmetric bistable potential, *International Journal of Bifurcation and Chaos*, 2015, **25**(02): 1550023.
18. B. Balachandran, E.B. Magrab, Vibrations (Cengage Learning, Australia, 2008)

Chapter 7
Aperiodic Vibrational Resonance Caused by Frequency-Modulated Signal and Fast-Varying Auxiliary Signal

This chapter examines aperiodic vibrational resonance induced by frequency-modulated signals in bistable systems. We focus on the distinct characteristics of linear and nonlinear frequency-modulated signals, analyzing each form separately. In particular, for nonlinear frequency-modulated signals, we explore three typical types in detail. To achieve aperiodic vibrational resonance, we employ the piecewise re-scaled method for simpler linear frequency-modulated signals, while utilizing a real-time re-scaled theoretical approach for more complex nonlinear signals, underscoring the advantages of this method.

7.1 Linear Frequency-Modulated Signal as the Characteristic Signal

Consider the bistable system excited by two linear frequency-modulated signal

$$\dot{x} = ax - bx^3 + A\cos\left(\pi\gamma t^2 + 2\pi f t\right) + B\cos\left(k\pi\gamma t^2 + 2k\pi f t\right), \tag{7.1}$$

where $k \gg 1$. In Eq. (7.1), the first signal is referred to as the feature signal, while the second is the auxiliary signal. Both signals are structurally identical linear frequency-modulated signals, differing primarily in their time scales: the time scale of the auxiliary signal is significantly longer than that of the feature signal. This distinction enables a comprehensive analysis of vibrational resonance effects when the two signals interact within the system.

We use the piecewise re-scaled method introduced in Sect. 3.1 to achieve aperiodic vibrational resonance in the nonlinear system of Eq. (7.1). We divide the time series into four segments on average and use the piecewise re-scaled method for the signal processing. According to the piecewise re-scaled theoretical formulation, aperiodic vibrational resonance can be achieved in the system

J. Yang and M. A. F. Sanjuan, *Aperiodic Resonances—Theory and Applications*,
https://doi.org/10.1007/978-981-96-6819-9_7

$$\dot{x} = a_1\beta x - b_1\beta x^3 + \beta A \cos\left(\pi\gamma t^2 + 2\pi f t\right) + \beta B \cos\left(k\pi\gamma t^2 + 2k\pi f t\right), \quad (7.2)$$

where a_1, a_2, β retain the same meanings as in previous chapters. In this system, the scale coefficient β is selected as the starting frequency of each segment, multiplied by 500.

In Fig. 7.1, the spectrum amplification factor and cross-correlation coefficient are used to quantify the occurrence of aperiodic vibrational resonance. Figure 7.1a shows a clear resonance region in the $\eta - B$ curve, while Fig. 7.1b displays a distinct resonance peak in the $C_{sx} - B$ curve. Both curves reach a high value at $B = 0.43$, indicating that these indices align well in measuring the presence of aperiodic vibrational resonance. This consistency highlights the effectiveness of both metrics in capturing resonance behavior within the system.

Fig. 7.2 gives the time series of the input and the output corresponding to the "strongest" aperiodic vibrational resonance. Based on Figs. 7.1 and 7.2, we can conclude that when the "strongest" aperiodic vibrational resonance occurs, the amplitude of the input linear frequency-modulated signal is amplified by 12.71 times.

Both Figs. 7.1 and 7.2 present the aperiodic vibrational resonance. Certainly, the appearance of the "strongest" vibrational resonance depends on the cooperation of the system parameters, the scale coefficient, and the parameters of the signal. In order to find the parameters that induce the "strongest" vibrational resonance, we can use an optimization algorithm to realize the optimal response, which will not be discussed in detail here.

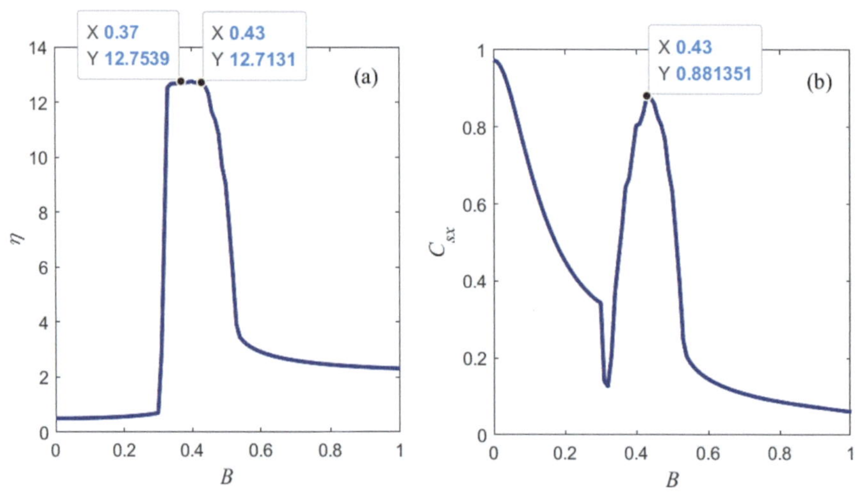

Fig. 7.1 The aperiodic vibrational resonance quantified by two kinds of indices. **a** The aperiodic vibrational resonance is quantified by spectrum amplification factor η. **b** The aperiodic vibrational resonance is quantified by the cross-correlation coefficient C_{sx}. The simulation parameters are $a_1 = 1, b_1 = 1, A = 0.1, k = 10, \gamma = 1, f = 1$

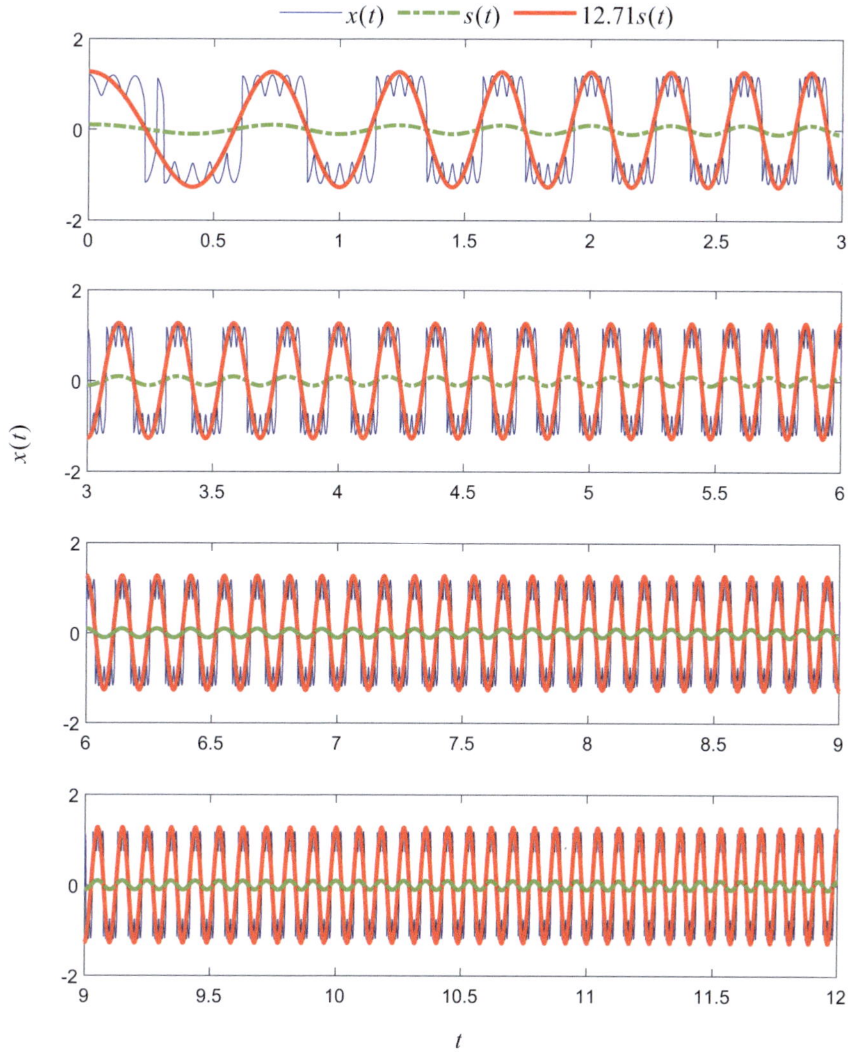

Fig. 7.2 The time series corresponding to the input signal and the output of the aperiodic vibrational resonance. The simulation parameters are $a_1 = 1$, $b_1 = 1$, $A = 0.1$, $B = 0.43$, $k = 10$, $\gamma = 1$, $f = 1$

7.2 Nonlinear Frequency-Modulated Signal as the Characteristic Signal

7.2.1 Nonlinear Frequency-Modulated Signal with Segmented Single Ridge in Time-Frequency Domain

We consider the bistable system with time-varying parameters excited by a characteristic signal $s(t)$ in nonlinear frequency-modulated form described by Eq. (1.16) and a fast-varying auxiliary signal $F(t)$, i.e.,

$$\frac{\mathrm{d}x(t)}{\mathrm{d}t} = a(t)x(t) - b(t)x^3(t) + s(t) + F(t). \tag{7.3}$$

In Eq. (7.3), $a(t) > 0$ and $b(t) > 0$ are time-varying parameters of the system.

We first consider $s(t)$ in the form defined in Eq. (1.6), with $F(t) = B\cos(\Omega(t)t)$, and $\Omega(t) = \Omega_0 f(t)$, where $f(t)$ represents the instantaneous frequency of the nonlinear frequency-modulated signal $s(t)$. Here, Ω_0 is a positive constant, making $F(t)$ vary at a much faster rate than the time scale of $s(t)$. To achieve aperiodic vibrational resonance in Eq. (7.3), we apply the real-time re-scaled method, with detailed derivations found in Sect. 3.2 or Ref. [1]. According to the real-time re-scaled theoretical framework, aperiodic vibrational resonance should be realized in the following equation:

$$\frac{\mathrm{d}x(t)}{\mathrm{d}t} = a_1\beta(t)x(t) - b_1\beta(t)x^3(t) + \beta(t)s(t) + \beta(t)F(t), \tag{7.4}$$

where a_1 and b_1 are small constants. The real-time scaling coefficient, $\beta(t)$, is defined as $\beta(t) = \beta_0 f(t)$, with β_0 representing a constant scale coefficient.

In Fig. 7.3, the curve shows the occurrence of aperiodic vibrational resonance induced by variations in the amplitude of the auxiliary signal. The $\eta - B$ curve reaches a maximum at $B = 0.58$, with a corresponding peak value of $\eta_{max} = 9.37141$. This indicates that, even under the excitation of a nonlinear frequency-modulated signal with complex frequency modulation characteristics, the system response can still exhibit aperiodic vibrational resonance.

In Fig. 7.4b, the output corresponding to the peak value in Fig. 7.3 illustrates the strongest aperiodic vibrational resonance. Compared with the original input signal shown in Fig. 7.4a, the amplitude of the characteristic frequency is significantly amplified. Nevertheless, some interference components also appear in the output, primarily manifested as multiples of the characteristic frequency. This phenomenon presents a worthwhile question for further investigation.

To further investigate the impact of the scale coefficient constant, β_0, on the spectral amplification factor η, Fig. 7.5 illustrates the aperiodic vibrational resonance induced by varying β_0 at different values of B. It is evident that the scale coefficient

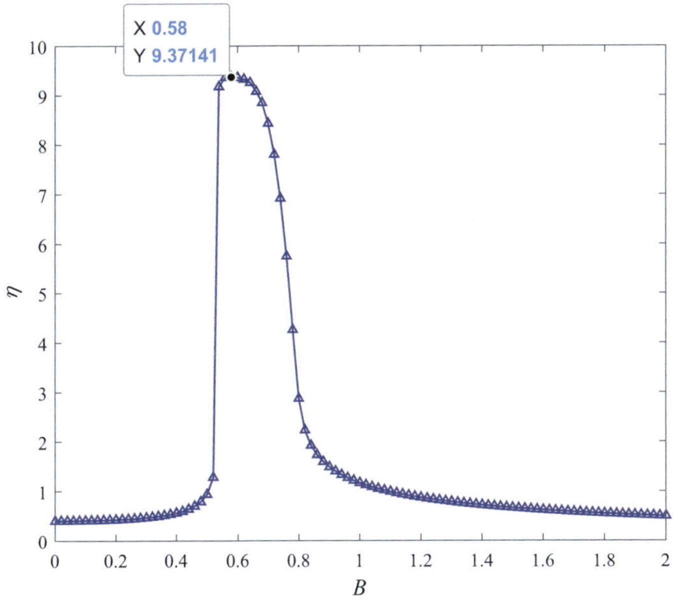

Fig. 7.3 The nonlinear variation of the spectral amplification factor η versus the signal amplitude B. The simulation parameters are $a_1 = 1, b_1 = 1, A = 0.1, \gamma = 10, f_0 = 10, \psi = 0, t_1 = 5, t_2$=15, $\beta_0 = 200$, $\Omega_0 = 20$

constant can also give rise to resonance-like phenomena. In other words, a larger scale coefficient constant does not necessarily lead to better results; rather, an optimal range exists. Furthermore, the curves in Fig. 7.5 show how the amplitude B influences both the position and magnitude of the resonance peak. Specifically, as B increases, both the scale coefficient constant and the peak value of the resonance increase.

7.2.2 *Nonlinear Frequency-Modulated Signal with Multiple Ridges in Time-Frequency Domain*

Now, we consider a more complex nonlinear frequency-modulated signal. Its mathematical expression is

$$\begin{cases} s(t) = \sum\limits_{h=1}^{H} A_h \exp\{-2\delta\pi f_{\mathrm{n}}[t - \upsilon_h(t)]\} \sin\{2\pi f_{\mathrm{n}}[t - \upsilon_h(t)] + \phi\} u[t - \upsilon_h(t)] , \\ \upsilon_h(t) = \sum\limits_{i=1}^{h} T_i . \end{cases} \tag{7.5}$$

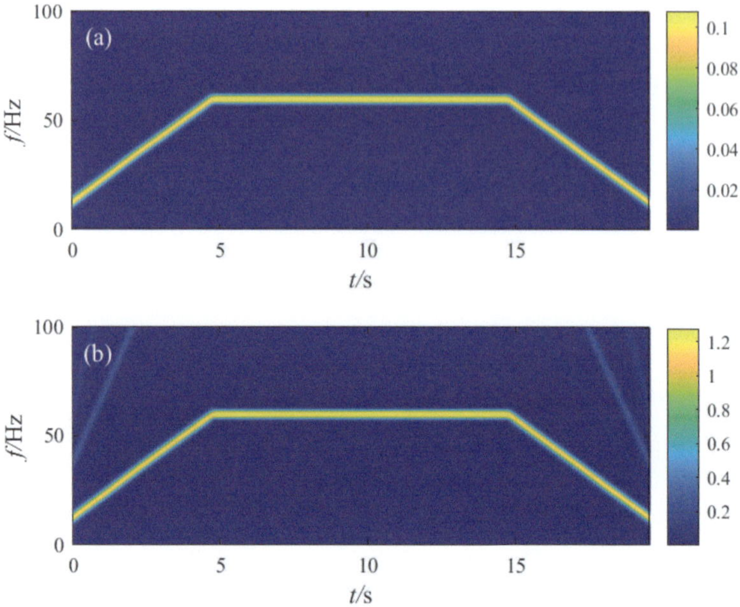

Fig. 7.4 Comparison of the time-frequency spectrum of the input and the output processed by aperiodic vibrational resonance. **a** The time-frequency spectrum of the input signal, **b** the time-frequency spectrum of the output. The simulation parameters are $a_1 = 1$, $b_1 = 1$, $A = 0.1$, $\gamma = 10$, $f_0 = 10$, $\varphi = 0$, $t_1 = 5$, t_2=15, $\beta_0 = 200$, $B = 0.58$, $\Omega_0 = 20$

This kind of nonlinear frequency-modulated signal is used to characterize the vibrational signal of a damaged bearing under speed time-varying condition [2–8]. In Eq. (7.5), H denotes the number of pulses, A_h is the amplitude of the hth pulse, δ is the damping coefficient, f_n represents the bearing's natural frequency, $v_h(t)$ is the cumulative time for the occurrence of h pulses, ϕ is the initial phase, $u(t)$ is the unit step function, and T_i specifies the repetition period of the hth pulse at its time of occurrence. The simulation parameters are set as follows $f_s = 10240$ Hz, δ = 120000, $f_n = 2000$ Hz, $\phi = 0$, $A_h = 0.08$.

In Fig. 7.6, we present the segmented time-domain waveform of the nonlinear frequency-modulated signal, which is composed of a series of non-stationary pulses and exhibits significant nonlinear time-frequency characteristics. Due to the large variation of the instantaneous frequency, in order to display the signal characteristics more clearly, the time series for 0–3 s, 5–5.5 s, and 7–10 s are plotted. The pulse interval of the signal initially decreases, then remains constant, and finally gradually increases, in accordance with the law of instantaneous frequency variation.

We continue to use Eq. (7.4) to achieve aperiodic vibrational resonance. First, we perform Hilbert demodulation and empirical mode decomposition (EMD) filtering as preprocessing steps on the simulated NLFM signal to remove interference and extract characteristic frequency information. Next, we apply the real-time scale trans-

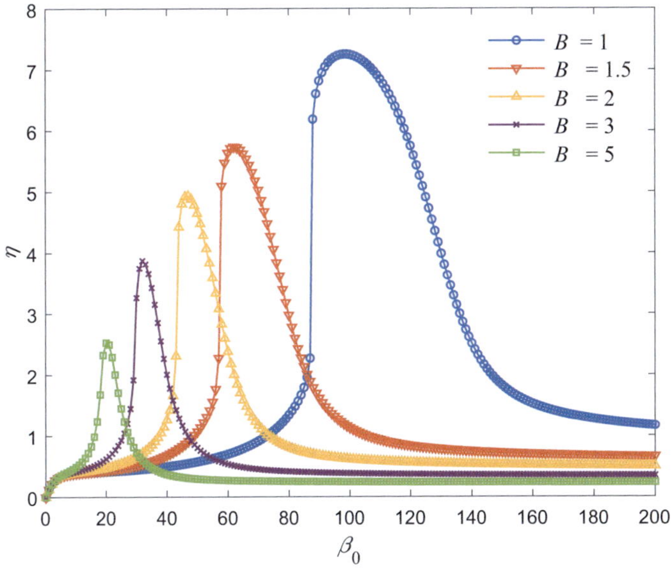

Fig. 7.5 The nonlinear variation of the spectral amplification factor η versus the scale coefficient constant β_0. The simulation parameters are $a_1 = 1$, $b_1 = 1$, $A = 0.1$, $\gamma = 10$, $f_0 = 10$, $\psi = 0$, $t_1 = 5$, t_2=15, $\Omega_0 = 20$

formation method, leveraging aperiodic vibrational resonance technology to enhance the processed signal. Figure 7.7 shows the relationship between the real-time spectral amplification factor η and the auxiliary signal amplitude B. As B gradually increases, η first rises and then decreases. The maximum value occurs at $B = 0.5$, indicating that the optimal aperiodic vibrational resonance response is achieved at this amplitude.

Fig. 7.8 presents the STFT spectrum of the input NLFM signal in subplot (a) and the STFT spectrum of the response corresponding to the strongest aperiodic vibrational resonance in subplot (b). In this figure, the input NLFM signal exhibits an STFT spectrum amplitude of approximately 6×10^{-4} in subplot (a), while the resonance response shows an STFT spectrum amplitude of approximately 10 in subplot (b). Through aperiodic vibrational resonance, the signal energy in the instantaneous frequency curve is significantly enhanced, while preserving the time-frequency characteristics of the original signal.

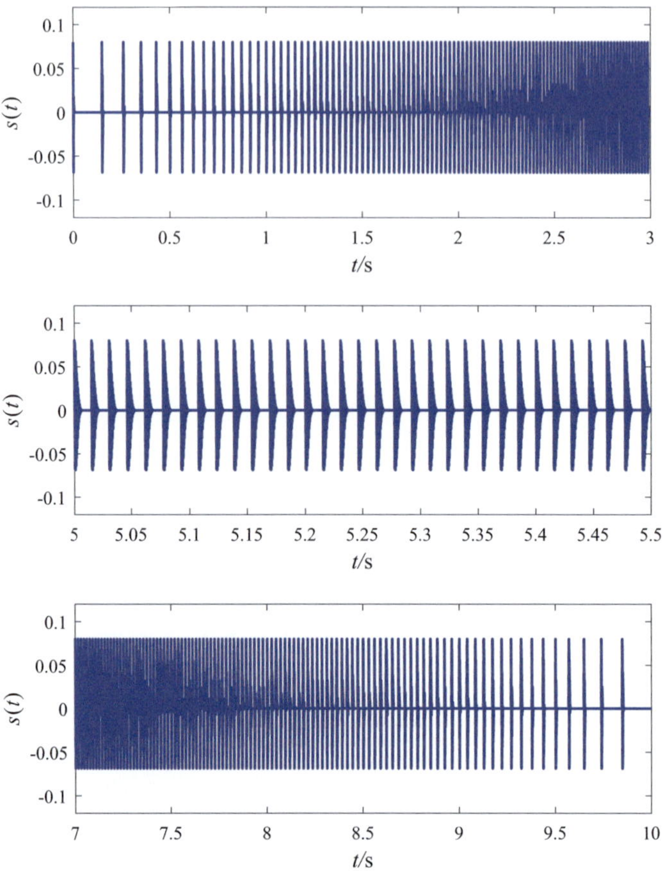

Fig. 7.6 The time series of the input signal from Eq. (7.5) is presented as follows: the top plot illustrates the stage of increasing frequency, the middle plot represents the stage of constant frequency, and the bottom plot depicts the stage of decreasing frequency

7.2.3 Nonlinear Frequency-Modulated Signal with Amplitude and Frequency Modulated Simultaneously

Now, we consider a more complex nonlinear frequency-modulated signal while its amplitude is modulated simultaneously. Its mathematical expression is

$$s(t) = 0.08\left(1 + \frac{1}{3}\cos\frac{2\pi t}{9}\right)\cos\left(70\pi \mathrm{t} + 30\sin\frac{2\pi t}{3} + 20\cos\frac{5\pi t}{3}\right). \tag{7.6}$$

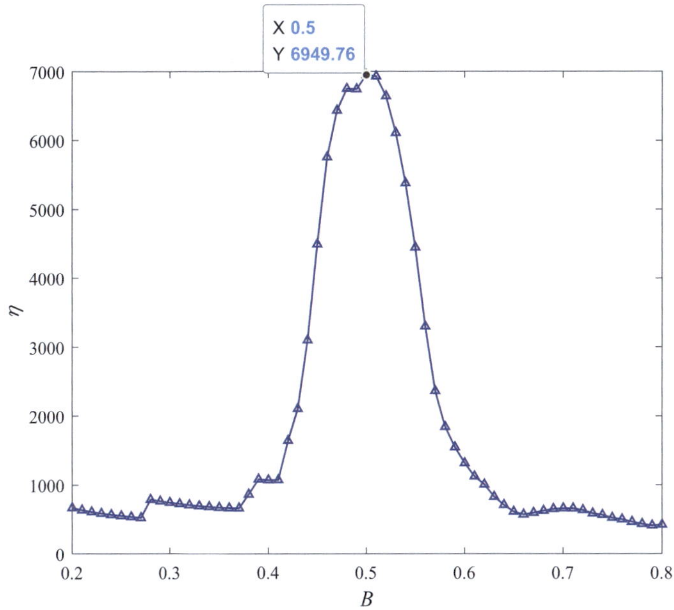

Fig. 7.7 The nonlinear variation of the spectral amplification factor η versus the signal amplitude B. The simulation parameters are $a_1 = 0.1$, $b_1 = 0.001$, $\beta_0 = 2000$, $\Omega_0 = 10$

Its time series is illustrated in Fig. 7.9, demonstrating that its frequency and amplitude are modulated in real time. The modulation forms appear to be relatively complex, with high-energy regions of the signal primarily concentrated in discrete areas. Promoting aperiodic vibrational resonance in such a complex modulation signal presents a technically challenging issue. The real-time rescaling or piecewise rescaling theoretical frameworks will continue to be relevant in this context. We will still use Eq. (7.4) to achieve aperiodic vibrational resonance.

In Fig. 7.10, we compare the aperiodic vibrational resonance achieved through the real-time re-scaled theoretical framework with that realized in the piecewise re-scaled theoretical framework. The spectral amplification factors obtained by both methods are presented. As the amplitude B of the auxiliary signal gradually increases, the spectral amplification factor η initially rises to a maximum value and then decreases. The curves in Fig. 7.10 clearly illustrate the aperiodic vibrational resonance. In the figure, p denotes the number of segments corresponding to the piecewise re-scaled theoretical framework. The number of segments p influences the maximum value of the spectral amplification factor η. Furthermore, the maximum value of η obtained by the real-time re-scaled method also yields satisfactory results. For both the real-time re-scaled method and the piecewise re-scaled method (with $p = 10$ and $p = 20$, the spectral amplification factor η reaches its maximum at $B = 0.7$. Specifically, for the real-time re-scaled method, the maximum value of η is 28.18 at $B = 0.7$. For the

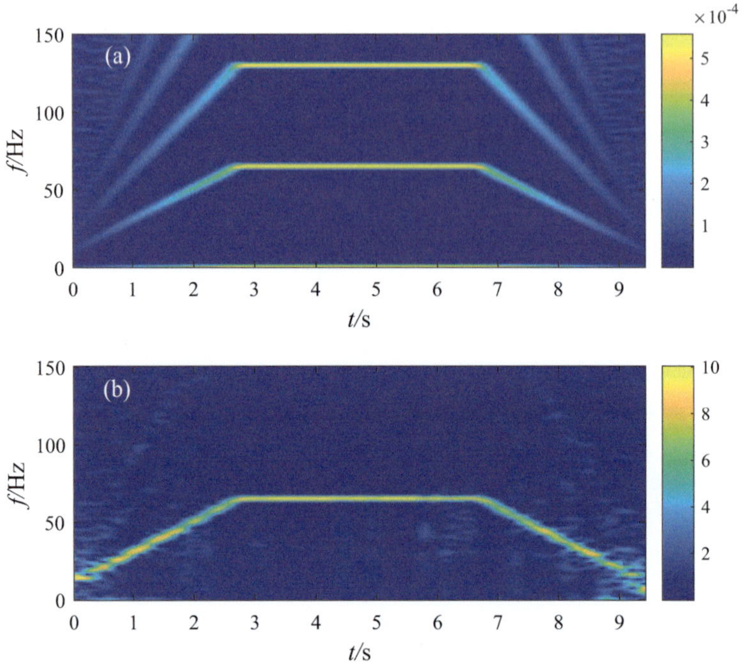

Fig. 7.8 Comparison of the time-frequency spectrum of the input and the output processed by aperiodic vibrational resonance. **a** The time-frequency spectrum of the input signal, **b** the time-frequency spectrum of the output. The simulation parameters are $a_1 = 0.1, b_1 = 0.001, \beta_0 = 2000, \Omega_0 = 10, B = 0.5$

piecewise re-scaled method, the maximum values of η for $p = 5$ and $p = 15$ occur at $B = 0.65$.

In Fig. 7.11, we present the STFT spectrum of the input signal from Eq. (7.6) alongside the STFT spectrum of the output corresponding to aperiodic vibrational resonance under different rescaling methods. In Fig. 7.11a, the black solid line represents the instantaneous frequency curve extracted from the phase, which aligns with the curve in the STFT spectrum. The instantaneous frequency varies in real-time and exhibits complex frequency characteristics. Figure 7.11b displays the output response using the real-time re-scaled method, where the energy values in the STFT spectrum increase from 0.1 to 2.5. Figure 7.11c–f illustrate the "strongest" aperiodic resonance responses for different numbers of segments p in the piecewise re-scaled method. The aperiodic vibrational resonance achieved through the piecewise re-scaled method fails to optimally match the input signal, auxiliary signal, and system parameters, resulting in discontinuous time-frequency features in the STFT spectrum. These results indicate that the real-time re-scaled method outperforms the piecewise re-scaled method.

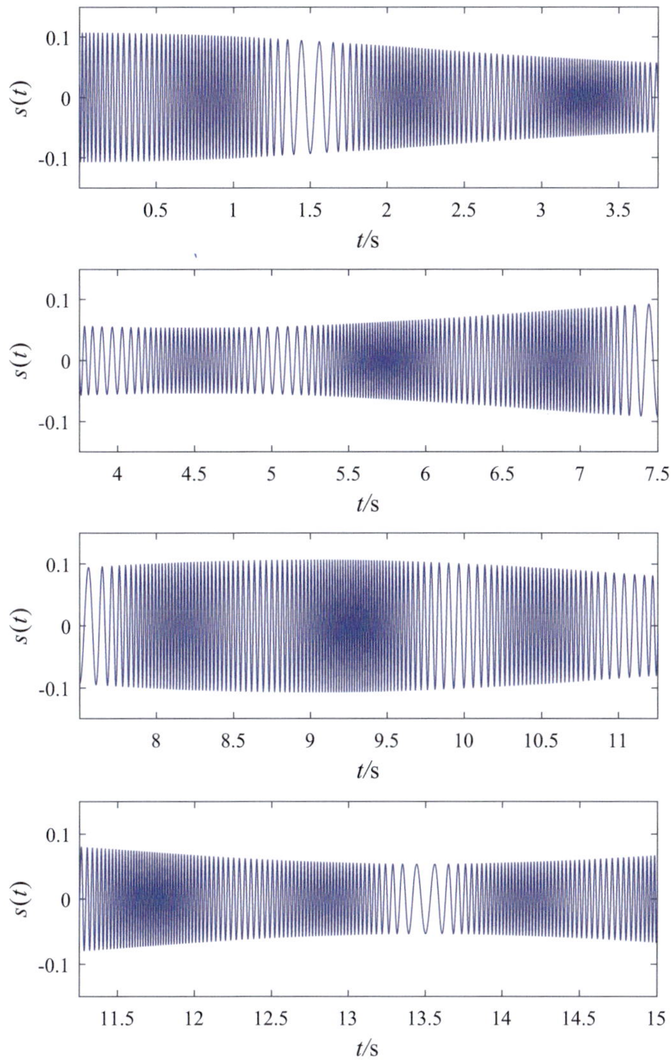

Fig. 7.9 The time series of the input signal from Eq. (7.5) is displayed sequentially across the four subplots

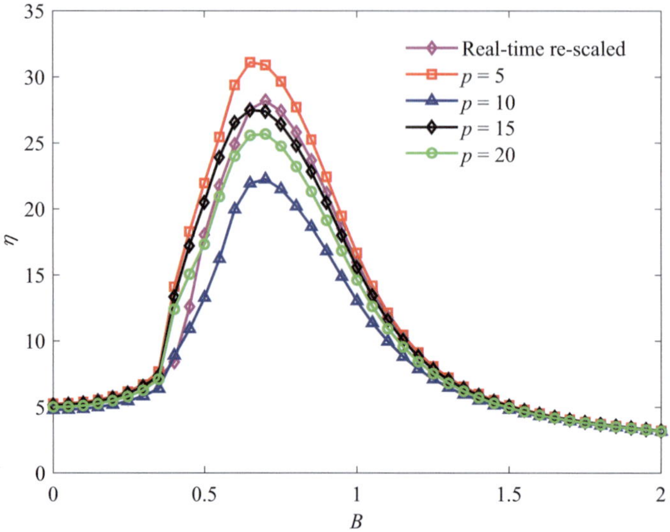

Fig. 7.10 The nonlinear variation of the spectral amplification factor η as a function of the auxiliary signal amplitude B under the real-time re-scaled theoretical framework and piecewise re-scaled theoretical framework, respectively. p is the number of segments corresponding to the piecewise re-scaled theoretical framework. The simulation parameters are $a_1 = 0.1$, $b_1 = 0.01$, $\beta_0 = 200$, $\Omega_0 = 10$

7.3 Concluding Remarks

This chapter discusses the aperiodic vibrational resonance induced by both linear and nonlinear frequency-modulated signals. We present typical phenomena associated with aperiodic vibrational resonance but do not delve into a detailed discussion of the parameters involved. Additionally, we do not explore aperiodic vibrational resonance in other nonlinear models beyond the overdamped bistable system, as many of these have been addressed in our previous papers [9–12]. In terms of engineering applications, we will discuss bearing fault diagnosis under varying speed conditions in Chap. 12.

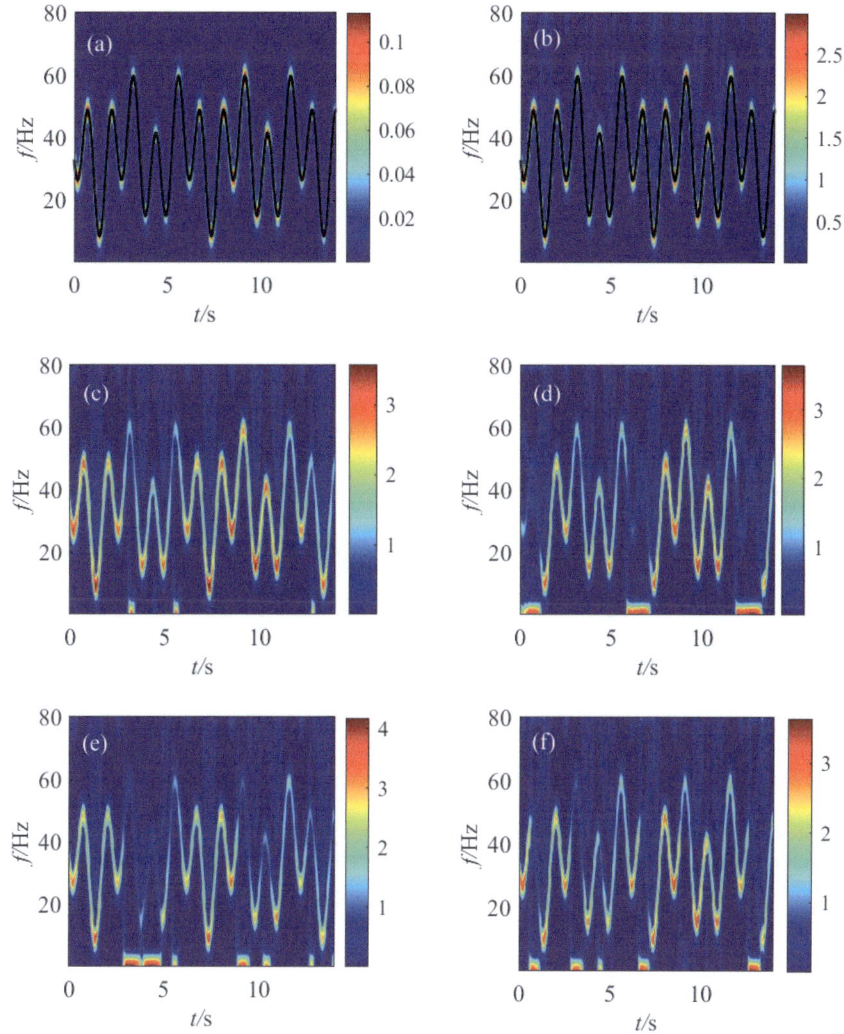

Fig. 7.11 The STFT spectrum of the input signal of Eq. (7.6) and the STFT spectrum of the "strongest" aperiodic vibrational resonance response. **a** The STFT spectrum of the input signal. **b** The STFT spectrum of the output corresponding to the aperiodic vibrational resonance realized by the real-time re-scaled theoretical framework, $B = 0.7$. **c–f** The STFT spectrum of the output corresponding to the aperiodic vibrational resonance realized by the piecewise re-scaled theoretical framework. **c** $p = 5$, $B = 0.65$, **d** $p = 10$, $B = 0.7$, **e** $p = 15$, $B = 0.65$, **f** $p = 20$, $B = 0.7$. The simulation parameters are $a_1 = 0.1$, $b_1 = 0.01$, $\beta_0 = 200$, $\Omega_0 = 10$

References

1. T. Gong, J. Yang, M.A.F. Sanjuán, H. Liu, Z. Shan, Vibrational resonance by using a real-time scale transformation method, *Physica Scripta*, 2022, **97**(4): 045207.
2. J. Yang, C. Yang, X. Zhuang, H. Liu, Z. Wang, Unknown bearing fault diagnosis under time-varying speed conditions and strong noise background. *Nonlinear Dynamics*, 2022, **107**: 2177–C2193.
3. Z. Wang, J. Yang, Y. Guo, Unknown fault feature extraction of rolling bearings under variable speed conditions based on statistical complexity measures, *Mechanical Systems and Signal Processing*, 2022, **172**: 108964.
4. T. Gong, J. Yang, S. Liu, H. Liu, Non-stationary feature extraction by the stochastic response of coupled oscillators and its application in bearing fault diagnosis under variable speed condition, *Nonlinear Dynamics*, 2022, **108**(4): 3839-3857.
5. Z. Wang, J. Yang, Y. Guo, T. Gong, Z. Shan, Nonstationary feature extraction based on stochastic resonance and its application in rolling bearing fault diagnosis under strong noise background, *Review of Scientific Instruments*, 2023, **94**(1): 015110.
6. Z. Shan, Z. Wang, J. Yang, D. Zhou, H. Liu, Extracting non-stationary signal under strong noise background: Time-varying system analysis, *Journal of Vibration and Control*, 2023, **29**(17-18): 4036-4045.
7. Z. Feng, X. Chen, T. Wang, Time-varying demodulation analysis for rolling bearing fault diagnosis under variable speed conditions, *Journal of Sound and Vibration*, 2017, **400**: 71-85.
8. J. Yang, C. Wu, Z. Shan, H. Liu, C. Yang, Extraction and enhancement of unknown bearing fault feature in the strong noise under variable speed condition, *Measurement Science and Technology*, 2021, **32**(10): 105021.
9. P. Jia, J. Yang, Y. Leng, The echo chirp signal amplification by the vibrational information fusion method, *International Journal of Modern Physics B*, 2020, **34**(6): 2050041.
10. P. Jia, J. Yang, X. Zhang, M.A.F. Sanjuán, On the LFM signal improvement by piecewise vibrational resonance using a new spectral amplification factor, *IET Signal Processing*, 2019, **13**(1): 65-69.
11. P. Jia, J. Yang, C. Wu, M.A.F. Sanjuán, Amplification of the LFM signal by using piecewise vibrational methods, *Journal of Vibration and Control*, 2019, **25**(1): 141-150.
12. P. Jia, J. Yang, H. Liu, E. Hu, Improving amplitude-modulated signals by re-scaled and twice sampling vibrational resonance methods, *Pramana*, 2018, **91**(38): 1-11.

Chapter 8
Logical Resonance

Logic resonance is a method for enhancing the accuracy of logical operations through the response of a nonlinear system. The study of logical resonance began with logical stochastic resonance, followed by the introduction of logical vibrational resonance, logical chaotic resonance, and other related concepts. This chapter introduces the concept of logical system resonance. Given the central role of logical operations in technological advancement, improving their accuracy is of paramount importance. The goal of this chapter is to provide insights that can inspire the development of more efficient logical operations. In the order of their introduction, this chapter examines logical stochastic resonance, logical vibrational resonance, and logical system resonance.

8.1 Logical Stochastic Resonance

With the growing research on resonance in nonlinear systems, new types of resonance and their practical applications are constantly being discovered. In 2009, Murali et al. applied the concept of stochastic resonance to the design of new logic gates [1]. They demonstrated that, under the influence of Gaussian white noise, a bistable nonlinear system governed by a set of piecewise functions can perform logical **OR** and **NOR** operations for weak logic input signals. Furthermore, by simply adjusting the parameters of the nonlinear system, the output logic can be switched to **AND** or **NAND**. In the same year, Murali et al. also demonstrated logical stochastic resonance in a circuit composed of linear resistors, capacitors, four CMOS transistors, and a battery. This phenomenon, which closely resembles traditional stochastic resonance [2], shows that the accuracy of the logical output first increases and then decreases with the intensity of the noise. They named this new phenomenon logical stochastic resonance [3].

J. Yang and M. A. F. Sanjuan, *Aperiodic Resonances—Theory and Applications*,
https://doi.org/10.1007/978-981-96-6819-9_8

Logical stochastic resonance offers a promising approach for the design of circuits and new large-scale integrated circuits that operate in environments with strong background noise. In particular, when the influence of noise cannot be ignored, logical stochastic resonance provides an efficient solution. As a result, this concept has attracted significant attention from researchers. The studies in this field primarily focus on two key elements: an effective nonlinear model and the role of noise. For example, Kohar et al. enhanced logical stochastic resonance by varying the amplitude and frequency of the input signals [4]. Lu et al. replaced the bistable system with a tristable system and investigated how the parameters of the tristable system influence the accuracy of logical output in the presence of strong noise [5]. Zhang et al. explored the effects of different types of noise on logical stochastic resonance, specifically inducing logical stochastic resonance using colored noise and $1/f$ noise [6].

The implementation of logical resonance requires the tilt of the potential well of the nonlinear system to determine the corresponding **OR**, **AND**, **NOR**, **NAND**, **XOR**, **NXOR**, logic output. The logical operation relationship between two logical signals I_1 and I_2 is shown in Table 1.1.

In this chapter, we use the logical success probability defined in Eq. (1.21) to evaluate the quality of logical resonance. We set up 100 rounds, each round randomly inputting 4 sets of logic signals at equal intervals, namely (0.5, 0.5), (−0.5, 0.5), (−0.5, −0.5), (0.5, −0.5) and compare the system output of this round with the logical truth table of Table 1.1. Due to the fact that the system logic output has a certain lag, in order to discard the transient process, we discard the top 10% of each logical output time series, and only take 90% of the output time series for comparison. If the system output matches the logical truth table, the number of correct rounds will be added one. If the logical success probability $P = 1$, it can be considered that the system logical output is steady at this time.

We study the logical stochastic resonance in an asymmetric bistable system

$$\frac{dx}{dt} = ax - bx^3 + r + I_1(t) + I_2(t) + N(t), \tag{8.1}$$

where the potential function $V(x)$ has asymmetric bistable potential wells, and

$$V(x) = \frac{b}{4}x^4 - \frac{a}{2}x^2 - rx. \tag{8.2}$$

Unless otherwise specified, $V(x)$ is always used for the same expression in this chapter. The system parameters a and b determine the height of the bistable potential barrier, the bias constant r determines the degree of inclination of the potential well, and also determines the logical output of the system. Usually, when the bias constant $r < 0$, the bistable potential well tilts to the left, and the system achieves reliable **AND** logic operations. When the bias constant $r > 0$, the bistable potential well tilts to the right, and the system achieves reliable **OR** logical operation. In Eq. (8.1) $I_1(t)$ and $I_2(t)$ are square wave signals, which we refer to as logical input signals. Without

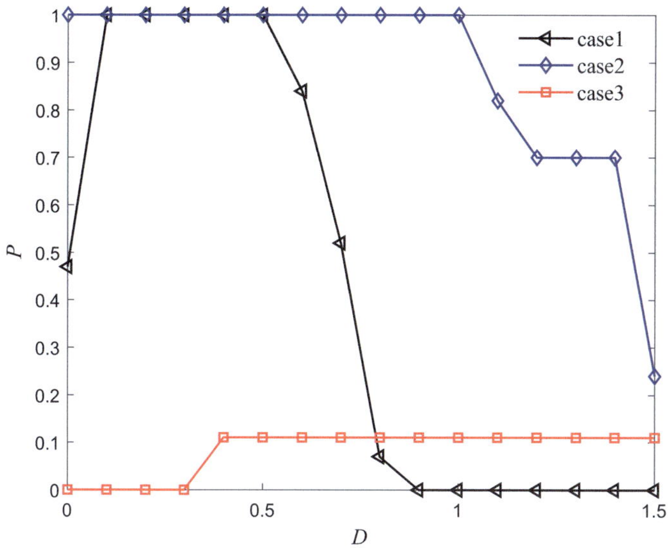

Fig. 8.1 The logical success probability P versus the noise intensity D under the **OR** logical operation for three cases of system parameter settings. The bias constant $r = 0.4$, $a = 1$ and $b = 0.8$ for case 1, $a = 0.15$ and $b = 0.05$ case2, $a = 1$ and $b = 0.05$ for case3

losing generality, the maximum and minimum values of $I_1(t)$ and $I_2(t)$ are set to 0.5 and -0.5, respectively. $N(t)$ is the same with it defined in Eq. (4.2).

Figure 8.1 shows the dependence of the logical success probability P on the noise intensity D under three different cases of a and b. From the figure, for case 1, it can be seen that the logical success probability first increases to 1 and then decreases to 0 with increasing the noise intensity D, which is the typical logical stochastic resonance phenomenon. The positive effect of noise on logical operations is reflected in this situation. For case 2, there exists a wide logical resonance region. Strong noise will reduce the logical success probability. For case 3, the logical success probability has always been low, and noise has not played a positive role. In addition, Fig. 8.1 also indicates that we obtain typical logical stochastic resonance phenomenon only when the system parameters are designed appropriately.

A reliable logical stochastic resonance requires an appropriate noise intensity. However, in practical situations, noise levels are typically either too high or too low. In cases of low noise, adding additional noise can induce logical stochastic resonance, but this approach may reduce the quality of the output signal. On the other hand, in situations with high noise, it becomes difficult, if not impossible, to achieve optimal logical output by merely adjusting the noise. Fortunately, adjusting the system parameters provides an effective alternative solution to these issues. In Fig. 8.2, we demonstrate the achievement of logical stochastic resonance by adjusting the system parameters a and b to find the optimal resonance region. The figure reveals

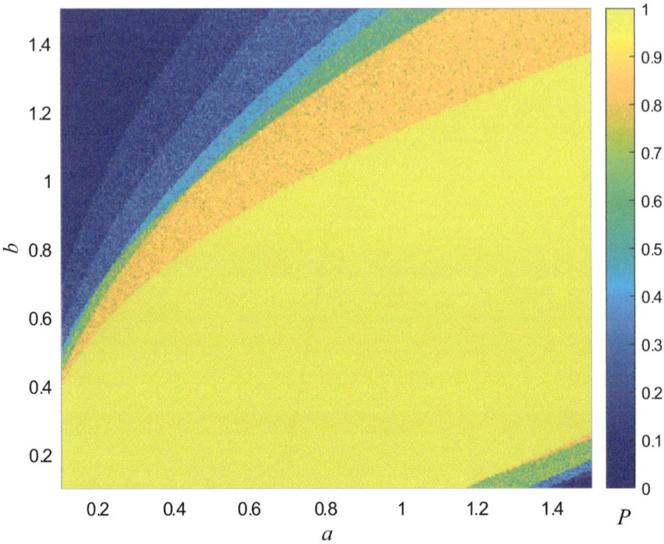

Fig. 8.2 The logical success probability P of the **OR** logical operation versus the system parameters a and b in the two-dimensional plane, $r = 0.4$, $D = 0.3$

a clear and wide logical resonance region. When the parameters a and b are within this region, the logical success probability reaches its optimal value.

To further illustrate the logical stochastic resonance phenomenon, the values of a and b in different regions of Fig. 8.2 are taken to observe the corresponding logical output time series, as shown in Fig. 8.3. When the parameters $a = 1.48$ and $b = 0.76$ locating in the logical resonance region, the logical output time series in Fig. 8.3a salsifies **OR** logic operation completely. When $a = 1.48$ and $b = 0.11$, the system parameters do not lie in the logical resonance region. As a result, the logical output time series in Fig. 8.3b can not satisfy the **OR** logical operation either. In view of system parameter selection, Figs. 8.2 and 8.3 illustrate how to achieve logical stochastic resonance through reasonable parameter selection under fixed external excitations.

In previous works, the logical stochastic resonance occurs always in a nonlinear system with asymmetric potential function. The bias constant r has a key role on the logical operation. Some scholars studied different logical resonances of a nonlinear system by changing the bias constant. We replaced the bias constant by an anharmonic periodic signal to induce the logical stochastic resonance in a symmetric bistable system successfully [7]. We found that when an anharmonic signal with a constant in its Fourier series replaces the bias constant of the system, a symmetric bistable system can also induce logical stochastic resonance, and the amplitude of the anharmonic signal influences the logical operation of the system output. We take the following

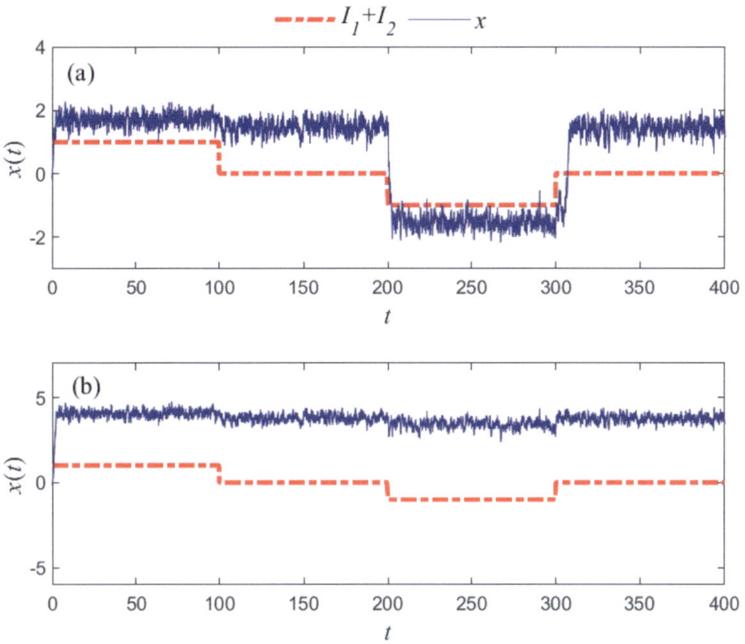

Fig. 8.3 The time series of the output corresponding to different system parameters under **OR** logical operation, $r = 0.4$, $D = 0.3$. (a) $a = 1.48$, $b = 0.76$, (b) $a = 1.48$, $b = 0.11$

symmetric bistable system that excited by logical inputs, noise and a full cosine waveform simultaneously is given as

$$\frac{dx}{dt} = ax - bx^3 + r + I_1(t) + I_2(t) + N(t) + A|\cos(\omega t)|. \tag{8.3}$$

Consider the bias constant r, as the noise increases, the logical success probability first increases and then decreases. Within an appropriate noise intensity, the logical success probability reaches its maximum, as shown in case 1 in Fig. 8.4. When the bias constant r is replaced with a full cosine excitation, there is also logical stochastic resonance presentation, as shown in case 2 in Fig. 8.4. Compared to the previous occurrence of logical stochastic resonance in a symmetric bistable system, Fig. 8.4 indicates that adding an anharmonic periodic excitation (whose Fourier series contains a constant term) can also induce logical stochastic resonance in a symmetric bistable system. For a more detailed discussion on this topic, interested readers can refer to our previous work [7].

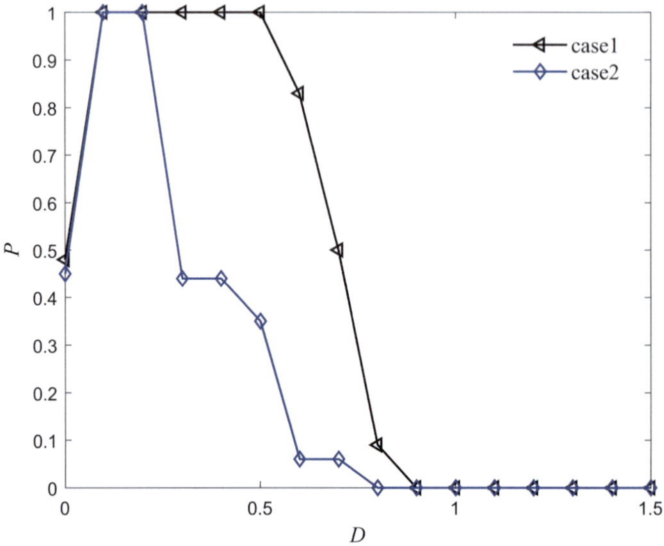

Fig. 8.4 The logical success probability P versus the noise intensity D under the **OR** logical operation for two cases of system parameter settings. The simulation parameter are $a = 1, b = 0.8$, and $r = 0.4$, $A = 0$ for case 1, $r = 0.4$, $A = 0.7$, $\omega = 0.05$ for case2

Logical stochastic resonance can occur not only in ordinary systems, but also in fractional-order systems. We have discussed this issue and revealed the characteristics of logical stochastic resonance in fractional-order systems. Interested readers can refer to our previous work [8]. We will not go into further detail here. Additionally, logical stochastic resonance in coupled bistable systems [9], synthetic genetic networks [10, 11], Hodgkin-Huxley neuron [12], FitzHugh-Nagumo neuron [13], coulomb-coupled quantum-dot rectifier [14], etc., also interested researchers from different research fields. Besides the effect of Gaussian white noise, other kinds of noise such as colored noise [15, 15], Sine-Wiener bounded noise [17] on logical stochastic resonance also have been discussed from different aspects.

Logical stochastic resonance primarily focuses on the positive role of noise. As is well known, noise is a type of random signal. In fact, another type of random-like signal, a chaotic signal, can also induce logical resonance, a phenomenon known as chaotic logical resonance [19, 20]. Therefore, there are many similarities between logical stochastic resonance and chaotic logical resonance, which will not be elaborated upon here. Another new topic, discussed in Chaps. 9 and 10, is the transient nature of chaotic signals. Specifically, a chaotic signal may evolve from chaotic to periodic behavior. What impact does this evolution have on the logical resonance response at different stages of the chaotic signal's evolution? This remains an open question for future research.

8.2 Logical Vibrational Resonance

As research on logical stochastic resonance deepens, some scholars have begun to explore logical resonance in the absence of noise [21, 22]. In this context, logical resonance has been observed even when excited by a single periodic signal. Furthermore, the concept of vibrational resonance has also been introduced into the study of logical resonance [23, 24]. Specifically, two periodic signals with different time scales are used in the excitations, and by adjusting the amplitude and frequency of the slow- and fast-varying signals, the logical success probability can be improved.

We examine the logical response of the system described by the following equation

$$\frac{dx}{dt} = ax - bx^3 + r + I_1(t) + I_2(t) + A\cos(\omega t) + B\cos(\Omega t), \tag{8.4}$$

where $\omega \gg \Omega$, within the framework of vibrational resonance.

Figure 8.5 illustrates an apparent logical resonance region as the amplitude of two harmonic signals is varied. As the amplitude B of the fast-varying signal increases, the logical success probability first rises and then falls. This phenomenon is referred to as *logical vibrational resonance*. The region in which logical vibrational resonance occurs depends on both A and B. Obviously, if the value of A is large, no logical vibrational resonance occurs, and the logical success probability remains zero. In other words, when a weak low-frequency signal is present, adjusting the amplitude

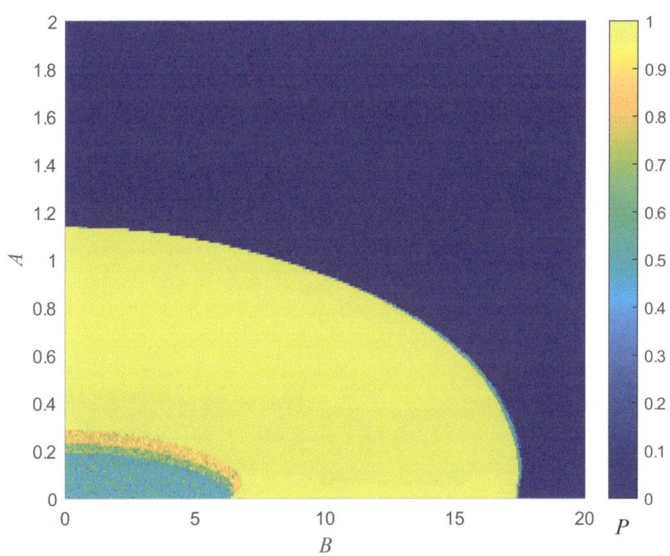

Fig. 8.5 The logical success probability P versus the amplitude of the periodic excitations under the **OR** logical operation. The simulation parameter are $a = 1$, $b = 0.8$, and $r = 0.4$, $\omega = 0.5$, $\Omega = 20$

of a high-frequency auxiliary signal can promote the occurrence of logical vibrational resonance. As shown in Fig. 8.5, logical vibrational resonance appears for weak low-frequency signals and strong high-frequency signals, i.e., $B \ll A$. Similarly, the same phenomenon is observed for weak high-frequency signals and strong low-frequency signals, i.e., $A \gg B$. Thus, careful selection of external harmonic excitations can enhance the logical response of a nonlinear system.

We provide the logic output time series in Fig. 8.6 corresponding to different points in the two-dimensional plane in Fig. 8.5. In Fig. 8.6a, b, d, the time series

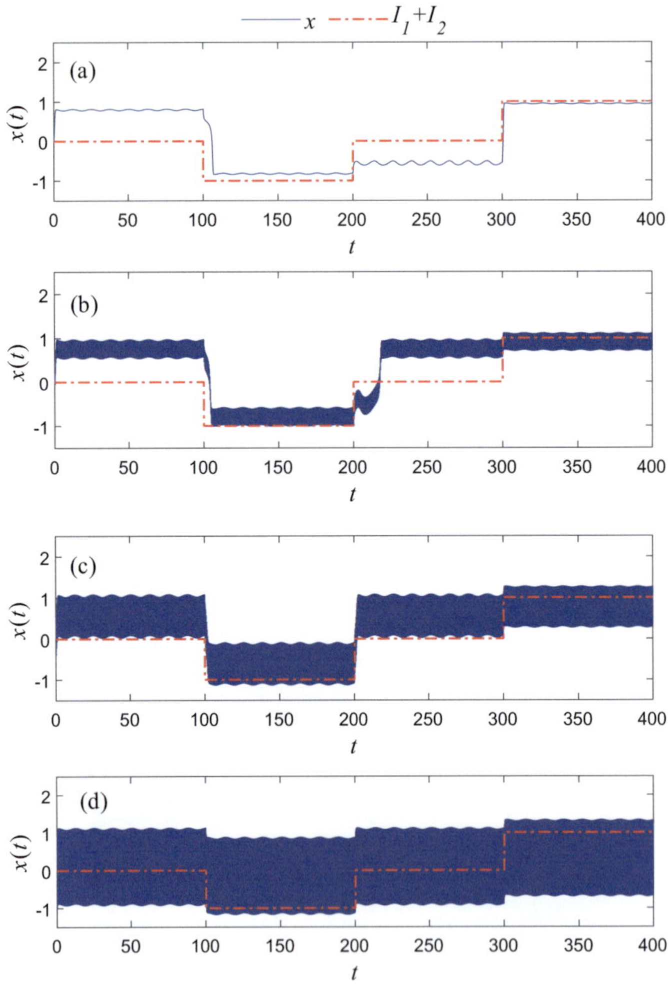

Fig. 8.6 The time series of the output corresponding to different amplitudes of the fast-varying harmonic signal under the **OR** logical operation. The simulation parameter are $a = 1$, $b = 0.8$, $r = 0.4$, $A = 0.1$, $\omega = 0.5$, and $\Omega = 20$. **a** $B = 0$, **b** $B = 4$, **c** $B = 10$, **d** $B = 20$

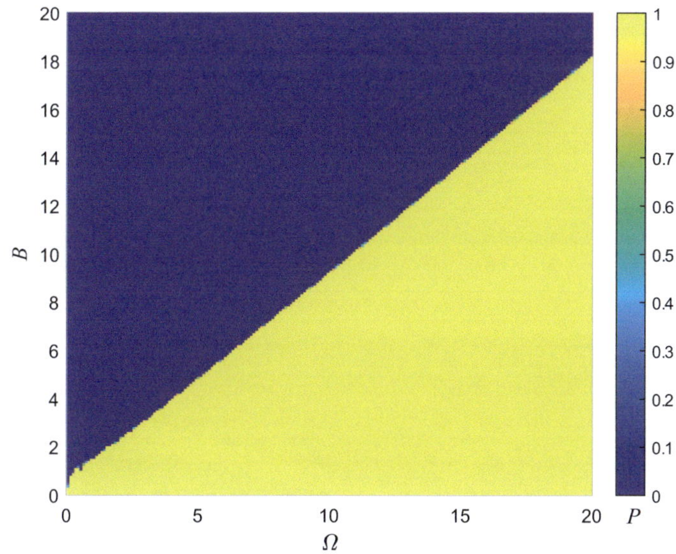

Fig. 8.7 The logical success probability P versus the amplitude and frequency of the fast-varying harmonic signal under the **OR** logical operation. The simulation parameter are $a = 1$, $b = 0.8$, $r = 0.4$, $A = 0.2$ and $\omega = 0.2$

cannot present the specified logical operations. The parameters used in these subplots are located outside the resonance region of Fig. 8.5. In Fig. 8.6c, the output time series presents completely correct logical operations. The value of this subplot is located within the resonance region of Fig. 8.5. As a result, Fig. 8.6 verifies the logical vibrational resonance from the output time series.

When changing the amplitude and frequency of high-frequency harmonic signals simultaneously, the logical resonance region of logical vibrational resonance is shown in Fig. 8.7. If you want to achieve the optimal logical vibrational resonance effect, it is necessary to comprehensively consider selecting the appropriate amplitude and frequency of the fast-varying harmonic signal.

Similar to logical stochastic resonance in a symmetric bistable system, where the bias constant is replaced by an anharmonic periodic signal, logical vibrational resonance can also be induced in a symmetric bistable system by an anharmonic periodic excitation. The nonlinear model discussed here is governed by the following equation

$$\frac{dx}{dt} = ax - bx^3 + I_1(t) + I_2(t) + A|\sin(\omega t)| + B\sin(\Omega t). \tag{8.5}$$

In Eq.(8.5), the role of the bias constant r is replaced by the full sine excitation $A|\sin(\omega t)|$.

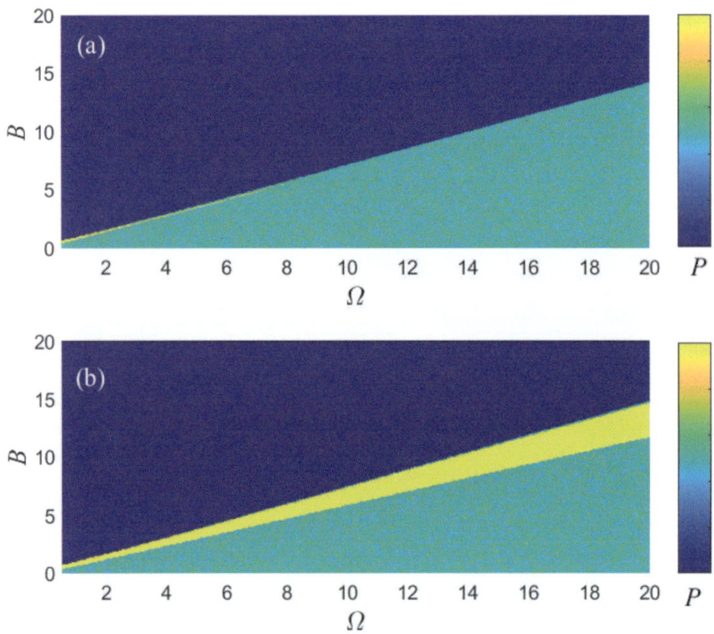

Fig. 8.8 The logical success probability P versus the amplitude and frequency of the fast-varying harmonic signal under the **OR** logical operation. The simulation parameter are $a = 1$, $b = 0.8$, **a** $A = 0.1$, $\omega = 0.2$, **b** $A = 0.2$, $\omega = 0.5$

In the symmetric bistable system of Eq. (8.5), when the amplitude and frequency of high-frequency harmonic signal are changed simultaneously, the optimal resonance region for logical vibrational resonance is shown in Fig. 8.8. The amplitude and frequency of the anharmonic periodic signal affects the region of logical vibrational resonance in the symmetric system. In Fig. 8.8a, the resonance region is very small, and the logical vibrational resonance disappears with the increasing the high-frequency Ω. Whereas, the resonance region is relatively large in Fig. 8.8b, and the region continues to be larger with the increase of Ω.

If we add a noise into Eq. (8.5), then we get

$$\frac{dx}{dt} = ax - bx^3 + I_1(t) + I_2(t) + A|\sin(\omega t)| + B\sin(\Omega t) + N(t), \tag{8.6}$$

where $N(t)$ has been defined in Eq. (4.2), the effect of noise on logical vibrational resonance will be discussed according to the logical response of this equation. In Fig. 8.9, according to the value of the logical success probability, there are four regions in both Fig. 8.9a, b. The noise intensity in Fig. 8.9b is larger than it in Fig. 8.9a, and it leads to the logical vibrational resonance region is larger in Fig. 8.9a than that in Fig. 8.9b. It highlights the positive role of the noise. That is the noise improving the logical success probability. It indicates the occurrence of logical stochastic resonance.

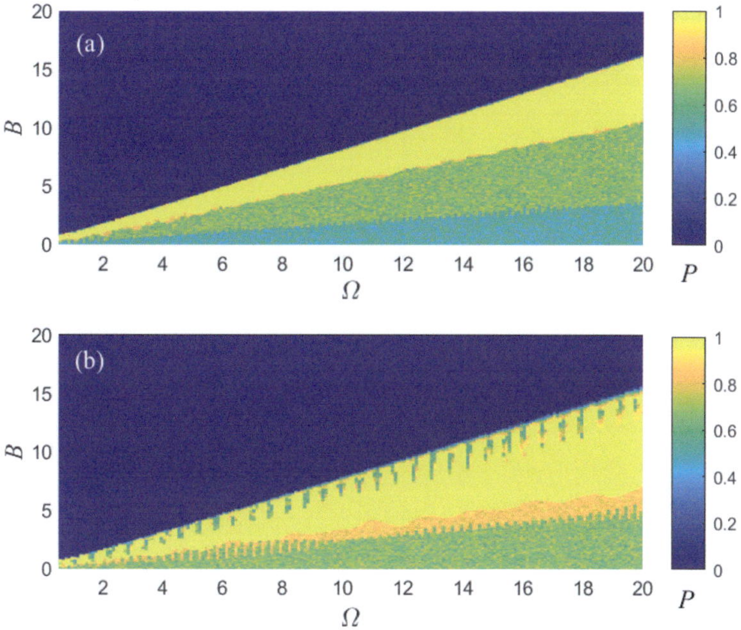

Fig. 8.9 The logical success probability P versus the amplitude and frequency of the fast-varying harmonic signal under the **OR** logical operation in the presence of noise. The simulation parameter are $a = 1$, $b = 0.8$, $A = 0.2$ and $\omega = 0.5$. **a** $D = 0.05$, **b** $D = 0.1$

When there is noise in the excitation, slight burrs appear at the edge of the resonance region. Here, burrs refer to random fluctuations at the edges of the resonance region. From Fig. 8.9a–b, the increase in noise improves the logical success probability, especially in non-resonant regions, where the increase in noise also exacerbates burrs and fluctuations.

8.3 Logical System Resonance

We try to induce logical resonance in the system when it is only stimulated by logical signals and not by any other external stimuli. Here, we treat logical signals as the excitation sources, similar to the system resonance in Chap. 2, by adjusting system parameters to achieve logical resonance of the system.

The system that we will study is given by

$$\frac{dx}{dt} = ax - bx^3 + r + I_1(t) + I_2(t). \tag{8.7}$$

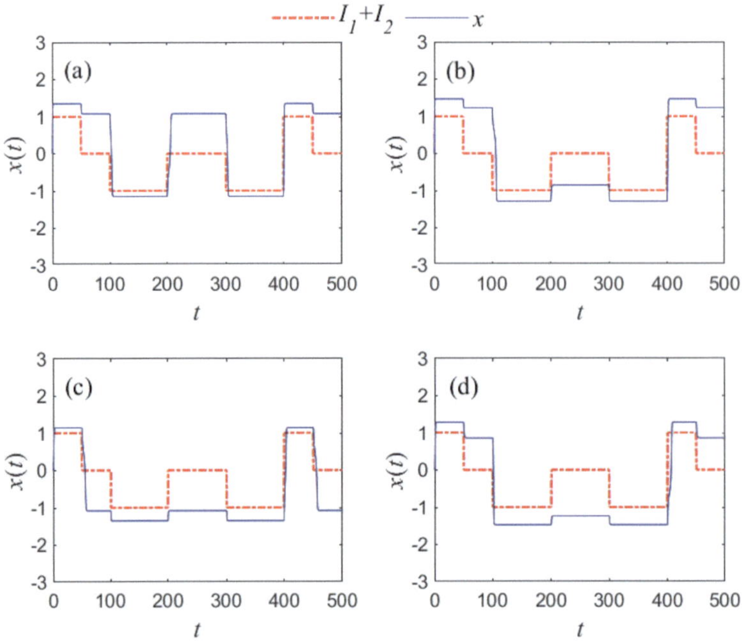

Fig. 8.10 The time series of the output corresponding to different system parameters and different logical operations. **a** $a = 0.8, r = 0.4$, **OR** logical operation, **b** $a = 1.2, r = 0.4$, **OR** logical operation, **c** $a = 0.8, r = -0.4$, **AND** logical operation, **d** $a = 1.2, r = -0.4$, **AND** logical operation. The system parameter $b = 1$

In Fig. 8.10, we observe whether the system has stable **OR** or **AND** logical output only by changing the system parameter a. In Fig. 8.10a, b, the bias constant $r > 0$, the system should be an **OR** logical output. In Fig. 8.10c, d, the bias constant $r < 0$, the system should be the **AND** logical output. It can be seen that the system with $a = 0.8$ in Fig. 8.10a, c has steady logical output, while the system with $a = 1.2$ in Fig. 8.10b, d does not have steady logical output. Therefore, changes in system parameters affect whether the system can have steady logical output.

Taking **OR** logical operation as an example, Fig. 8.11 presents the "strongest" logical resonance region, that is, the region with the logical success probability $P = 1$. This kind of logical resonance is completely caused by changes in system parameter in the absence of external excitation, so we name it as *logical system resonance* here.

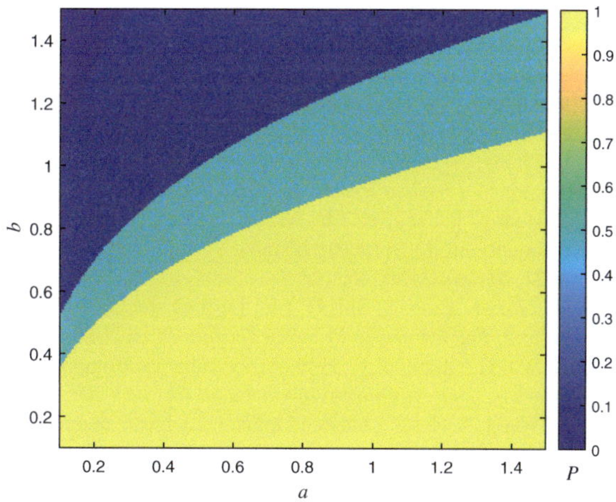

Fig. 8.11 The logical success probability P versus the system parameters a and b under the **OR** logical operation. The bias constant $r = 0.4$

8.4 Concluding Remarks

This chapter focuses on logical stochastic resonance, logical vibrational resonance, and for the first time introduces the concept of logical system resonance. The occurrence of logical resonance directly affects the accuracy of logical operations. Due to the complexity of the logical operation system itself and the complexity of its excitations, in order to achieve high-precision logical operations, it is necessary to comprehensively consider the logical system resonance caused by system parameters, as well as the logical stochastic resonance and the logical vibrational resonance caused by external excitations.

The examples in this chapter are all based on numerical simulations; however, logical resonance can also be studied in hardware circuits [24, 25]. In the future, we look forward to uncovering new phenomena of logical resonance through further research in hardware circuits. Additionally, current studies on logical resonance are mostly focused on theoretical performance, with limited applied research. Although some scholars have explored using logical stochastic resonance for image processing [26], we anticipate further research into the practical applications of logical resonance.

References

1. K. Murali, I. Rajamohamed, S. Sinha, W.L. Ditto, A.R. Bulsara, Realization of reliable and flexible logic gates using noisy nonlinear circuits, *Applied Physics Letters*, 2009, **95**(19):194102.
2. L. Gammaitoni, P. Hänggi, P. Jung, F. Marchesoni, Stochastic resonance, *Reviews of Modern Physics*, 1998, **70**(1): 223-287.
3. K. Murali, S. Sinha, W.L. Ditto, A.R. Bulsara, Reliable logic circuit elements that exploit nonlinearity in the presence of a noise floor, *Physical Rreview Letters*, 2009, **102**(10): 04101.

4. V. Kohar, S. Sinha, Noise-assisted morphing of memory and logic function, *Physics Letters A*, 2012, **376**(8-9): 957-962.
5. S. Lu, Z. Dai, Y. Liu, G. Liu, H. Yang, F. Wang, Decreasing bit error rate by using enhanced tristable logical stochastic resonance in heavy noise condition, *Chinese Journal of Physics*, 2019, **58**: 179-189.
6. L. Zhang, A. Song, J. He, Effect of colored noise on logical stochastic resonance in bistable dynamics, *Physical Review E*, 2010, **82**(5): 051106.
7. S. Huang, J. Yang, H. Liu, M.A.F. Sanjuán, Effect of static bifurcation on logical stochastic resonance in a symmetric bistable system, *International Journal of Bifurcation and Chaos*, 2021, **31**(16): 2150246.
8. M. Hou, J. Yang, S. Shi, H. Liu, Logical stochastic resonance in a nonlinear fractional-order system, *The European Physical Journal Plus*, 2020, **135**(9): 747.
9. H. Wu, H.J. Jiang, Z.H. Hou, Array-enhanced logical stochastic resonance in coupled bistable systems, *Chinese Journal of Chemical Physics*, 2012, 25(1): 70-76.
10. N. Wang, A. Song, Enhanced logical stochastic resonance in synthetic genetic networks, *IEEE Transactions on Neural Networks and Learning Systems*, 2015, **27**(12): 2736-2739.
11. L. Zhang, W. Zheng, A. Song, Adaptive logical stochastic resonance in time-delayed synthetic genetic networks, *Chaos*, 2018, **28**(4): 043117.
12. D. Yu, L. Yang, X. Zhan, Z. Fu, Y. Jia, Logical stochastic resonance and energy consumption in stochastic Hodgkin-Huxley neuron system, *Nonlinear Dynamics*, 2023, **111**(7): 6757-6772.
13. Y. Yao, J. Ma, Logical stochastic and vibrational resonances induced by periodic force in the FitzHugh-Nagumo neuron, *The European Physical Journal Plus*, 2022, **137**(11): 1214.
14. P. Pfeffer, F. Hartmann, S. Höfling, M. Kamp, L. Worschech, Logical stochastic resonance with a coulomb-coupled quantum-dot rectifier, *Physical Review Applied*, 2015, **4**(1): 014011.
15. L. Zhang, A. Song, J. He, Effect of colored noise on logical stochastic resonance in bistable dynamics, *Physical Review E*, 2010, **82**(5): 051106.
16. H. Zhang, Y. Xu, W. Xu, X. Li, Logical stochastic resonance in triple-well potential systems driven by colored noise, *Chaos*, 2012, **22**(4): 043130.
17. G. Cheng, W. Liu, R. Gui, Y. Yao, Sine-Wiener bounded noise-induced logical stochastic resonance in a two-well potential system, *Chaos, Solitons & Fractals*, 2020, **131**: 109514.
18. Y. Yao, J. Ma, Logical chaotic resonance in a bistable system, *International Journal of Bifurcation and Chaos*, 2020, **30**(13): 2050196.
19. Y. Yao, J. Ma, R. Gui, G. Cheng, Enhanced logical chaotic resonance, *Chaos*, 2021, **31**(2): 023103.
20. Y. Yao, Logical chaotic resonance in the FitzHugh-Nagumo neuron, *Nonlinear Dynamics*, 2022, **107**(4): 3887-3901.
21. A. Gupta, A. Sohane, V. Kohar, K. Murali, S. Sinha, Noise-free logical stochastic resonance, *Physical Review E*, 2011, **84**(5): 055201.
22. B. Yang, X. Zhang, M.K. Luo, When noise-free logical stochastic resonance occurs in a bistable system, *Nonlinear Dynamics*, 2017, **87**: 1957-1965.
23. R. Gui, Y. Wang, Y. Yao, G. Cheng, Enhanced logical vibrational resonance in a two-well potential system, *Chaos, Solitons & Fractals*, 2020, **138**: 109952.
24. S. Huang, J. Zhang, J. Yang, H. Liu, M.A.F. Sanjuán, Logical vibrational resonance in a symmetric bistable system: Numerical and experimental studies, *Communications in Nonlinear Science and Numerical Simulation*, 2023, **119**: 107123.
25. A.R. Bulsara, A. Dari, W.L. Ditto, K. Murali, S. Sinha, Logical stochastic resonance, *Chemical Physics*, 2010, **375**(2-3): 424-434
26. N. Wang, B. Zheng, H. Zheng, B. Yang, When underwater degraded images meet logical stochastic resonance, *Nonlinear Dynamics*, 2018, **94**: 295-305.

Chapter 9
Ultrasensitive System Resonance

In this chapter, we study the ultrasensitive resonance of a nonlinear system under a single periodic excitation. The effects of different signal waveforms on the ultrasensitive resonance will be discussed. Over time, the transition from ultrasensitive resonance to sparse ultrasensitive resonance, and then to periodic motion will also be discussed in detail.

9.1 Ultrasensitive System Resonance Caused by the System Parameters

The nonlinear response can be ultrasensitive to very small disturbances in certain situations. These disturbances may arise from slight variations in excitation [1] or from perturbations in the initial conditions of numerical simulations [2]. This phenomenon occurs because the system response is chaotic, making it highly sensitive to minor perturbations. However, the chaos observed here is often transient [3–6], meaning that over time, the system response may gradually transition into a steady periodic motion. In regions of ultrasensitive resonance, the system response is typically chaotic and therefore aperiodic, which justifies the inclusion of this topic in this monograph. The concepts of ultrasensitive resonance, sparse ultrasensitive resonance, ultrasensitive vibrational resonance, and sparse ultrasensitive vibrational resonance represent novel phenomena introduced in recent years.

Ultrasensitive system resonance refers to the resonance response caused by a very small change in system parameters under only a single excitation. We consider a traditional underdamped Duffing system excited by a single harmonic signal

$$\ddot{x} + \delta\dot{x} + ax + bx^3 = A\cos(\omega t), \tag{9.1}$$

where δ is the damping coefficient, and a and b are system parameters. The response amplitude at a given frequency is defined as

J. Yang and M. A. F. Sanjuan, *Aperiodic Resonances—Theory and Applications*,
https://doi.org/10.1007/978-981-96-6819-9_9

$$Q(j\omega) = \sqrt{Q_s^2(j\omega) + Q_c^2(j\omega)} \Big/ A, \tag{9.2}$$

where the Fourier coefficients are calculated by

$$\begin{cases} Q_s(j\omega) = \dfrac{2\Delta t}{mT} \displaystyle\sum_{i=1}^{mT/\Delta t} x(t_i)\sin(j\omega t_i), \\ Q_c(j\omega) = \dfrac{2\Delta t}{mT} \displaystyle\sum_{i=1}^{mT/\Delta t} x(t_i)\cos(j\omega t_i). \end{cases} \tag{9.3}$$

Here, Δt is the time step, and m is an integer. For numerical calculations, the total length of time is labeled as T_L and $T_L = mT$, where T is the period of the excitation and $T = 2\pi/\omega$. The initial conditions are defined as $x(0) = x_0$ and $\dot{x}(0) = \dot{x}_0$. To analyze nonlinear resonance at different frequencies, j can be either an integer or a fractional number, representing various frequency components such as the primary frequency, subharmonic frequency, superharmonic frequency, and combined frequencies.

In Fig. 9.1, the response amplitude $Q(j\omega)$ as a function of the system parameter a exhibits ultrasensitive variations in certain regions. On one hand, even a very small perturbation in a can cause intense fluctuations in the response amplitude. Since this type of resonance is induced by changes in the system parameter, we refer to it as *ultrasensitive system resonance*. Although this resonance occurs at specific values of a, the relationship between its occurrence and variations in a remains unpredictable. Moreover, when ultrasensitive resonance takes place, the system response often exhibits chaotic behavior [1]. On the other hand, resonance occurs not only at the excitation frequency ω but also at the subharmonic frequency $\omega/3$, $\omega/2$, the superharmonic frequency 2ω, 3ω and the combined frequency $5\omega/3$. Based on principles of nonlinear dynamics, we infer that similar resonances are likely to occur at additional frequencies.

We selected a small interval of a within the ultrasensitive resonance region of Fig. 9.1 to examine the resonance behavior at specific frequencies. As shown in Fig. 9.2, ultrasensitive resonance continues to appear frequently. To explore the structure of the $Q(j\omega)$ curve in greater detail, we present Fig. 9.3, which illustrates the irregular and frequent fluctuations of $Q(j\omega)$ as a function of the system parameter a over a much smaller interval. From Fig. 9.3, an infinitely self-similar pattern is clearly visible, indicating a typical fractal structure. Evidently, if we further reduce the interval of a and refine the step size Δa, this structure would persist, which is an essential characteristic of fractals.

From Figs. 9.1, 9.2 and 9.3, we gradually narrow down the range of values for a, and find the highly fractalized characteristic of the $Q(j\omega)$ curve in the ultrasensitive system resonance regions. We observe that ultrasensitive resonance differs significantly from traditional resonance. To the best of our knowledge, while extensive

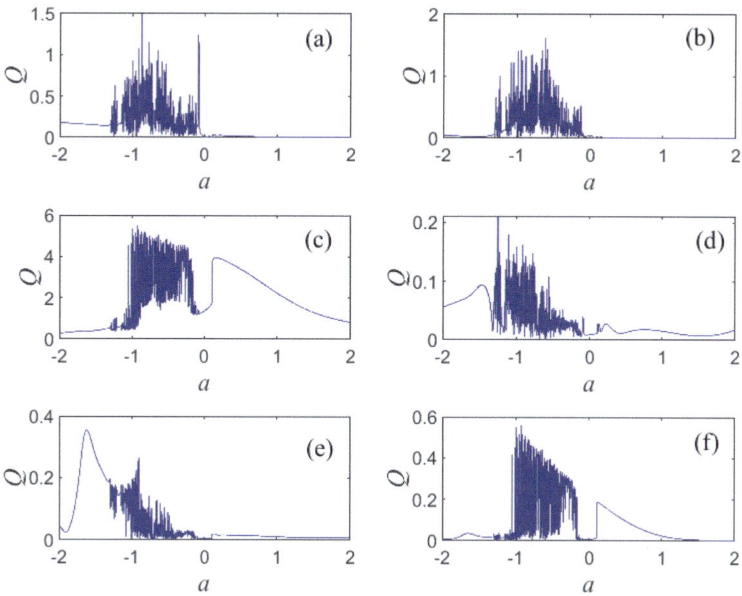

Fig. 9.1 The response amplitude $Q(j\omega)$ as a function of the system parameter a exhibits ultrasensitive system resonance in some intervals. **a** $j = 1/3$, **b** $j = 1/2$, **c** $j = 1$, **d** $j = 5/3$, **e** $j = 2$, **f** $j = 3$. The simulation parameters are $\delta = 0.15$, $b = 1$, $A = 0.245$, $\omega = 0.9$, $m = 50$, the step $\Delta a = 0.001$. The initial conditions are $x_0 = 0$ and $\dot{x}_0 = 0$. The total calculation time starts from the very beginning

research exists on traditional resonance, studies on ultrasensitive resonance remain relatively scarce.

Not only the parameter a but also the parameter b can induce ultrasensitive system resonance, as shown in Fig. 9.4. Because the parameter b determines the nonlinear term of the system, an appropriate value of b can induce strong resonance at a nonlinear frequency as given in Fig. 9.4a, b. In these two subplots, the ultrasensitive system resonance occurs at subharmonic frequencies $\omega/3$ and $\omega/2$. An appropriate value of parameter b can make the response amplitude at a nonlinear frequency much larger than that at the primary frequency.

To investigate the parameter-induced ultrasensitive system resonance further, we provide Fig. 9.5, which shows different resonance pattern in the two-dimensional plane. Several distinct regions can be identified: ultrasensitive resonant regions, traditional resonant regions, and non-resonant regions. Ultrasensitive system resonance occurs only within specific ranges of the system parameters. Due to its unpredictability, the ultrasensitive resonance region is of particular interest. To reduce computational time, we have slightly increased the calculation step size compared to Figs. 9.1 and 9.4. This adjustment does not affect the fundamental laws governing the resonance response.

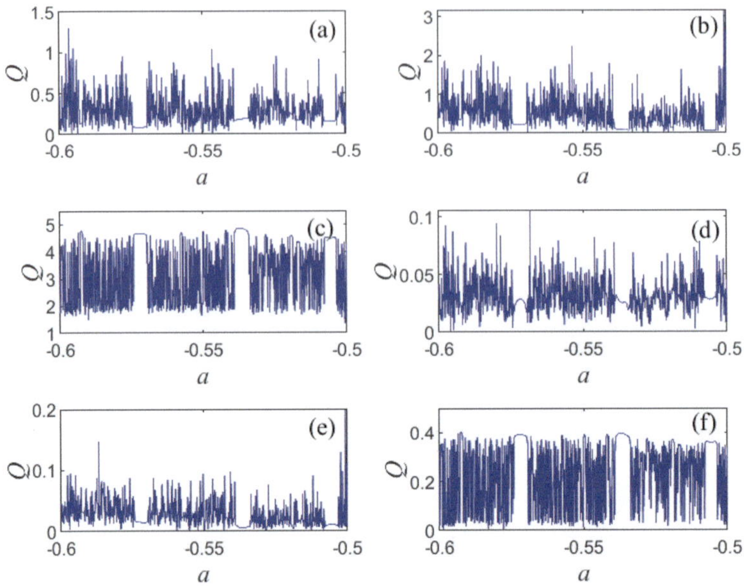

Fig. 9.2 The response amplitude $Q(j\omega)$ as a function of the system parameter a exhibits ultrasensitive system resonance in some small intervals. **a** $j = 1/3$, **b** $j = 1/2$, **c** $j = 1$, **d** $j = 5/3$, **e** $j = 2$, **f** $j = 3$. The simulation parameters are $\delta = 0.15$, $b = 1$, $A = 0.245$, $\omega = 0.9$, $m = 50$, the step $\Delta a = 0.0001$. The initial conditions are $x_0 = 0$ and $\dot{x}_0 = 0$. The total calculation time starts from the very beginning

9.2 Ultrasensitive System Resonance Caused by the Excitation

In addition to system parameters and initial conditions, the amplitude and frequency of excitation-being the source of vibration-can also induce ultrasensitive resonance in certain cases. Examples of this phenomenon are presented in Figs. 9.6 and 9.7, respectively. As the amplitude f or the frequency ω, the response amplitude $Q(j\omega)$) exhibits ultrasensitive resonance in specific regions. This behavior differs significantly from traditional resonance in the frequency response curve. In Fig. 9.6, for different response frequencies $j\omega$, the regions corresponding to the "strongest" resonance typically occur at different intervals of the amplitude f. Nevertheless, as seen in both Figs. 9.6 and 9.7, the regions of ultrasensitive variation in $Q(j\omega)$ may coincide for certain values of f and ω. This is expected, as the system response consists of multiple frequency components. If the response at one frequency component exhibits ultrasensitive variation with respect to a control parameter, similar variations are likely to occur at other frequency components as well.

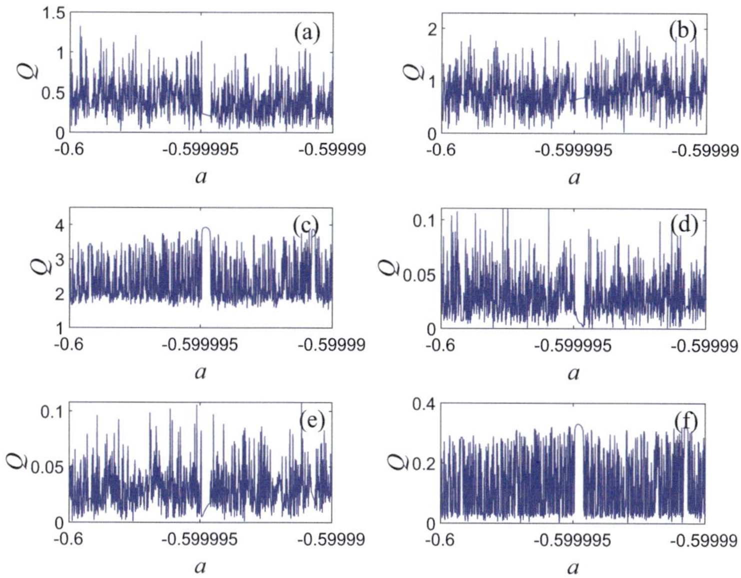

Fig. 9.3 The response amplitude $Q(j\omega)$ versus the system parameter a presents ultrasensitive resonance in a very small interval. **a** $j = 1/3$, **b** $j = 1/2$, **c** $j = 1$, **d** $j = 5/3$, **e** $j = 2$, **f** $j = 3$. The simulation parameters are $\delta = 0.15$, $b = 1$, $A = 0.245$, $\omega = 0.9$, $m = 50$, the step $\Delta a = 0.00000001$. The initial conditions are $x_0 = 0$ and $\dot{x}_0 = 0$. The total calculation time starts from the very beginning

9.3 Ultrasensitive System Resonance Caused by the Initial Conditions

9.3.1 *A Cosine Harmonic Excitation*

Although we choose the same initial conditions of x_0 and $\dot{x}_0$ in Figs. 9.1, 9.2, 9.3, 9.4 and 9.5, in fact, like the effect of the system parameters in the ultrasensitive resonant region, the initial condition of x_0 and $\dot{x}(0)$ can also induce this interesting ultrasensitive system resonance. As shown in Fig. 9.8, even a very small disturbance in the initial condition x_0 causes the response amplitude $Q(j\omega)$ to fluctuate significantly. As is well known, the sensitivity of a system's response to initial conditions is one of the most characteristic features of chaotic behavior.

To further reveal the effect of the initial conditions of x_0 and $\dot{x}_0$ on the resonance behavior, we provide Fig. 9.9 showing $Q(\omega)$, $Q(\omega/3)$ and $Q(3\omega)$ versus x_0 and $\dot{x}_0$ respectively in three subplots. There are obvious traditional (stable) resonant region and ultrasensitive resonant region. In Fig. 9.9a, there is no obvious pure non-resonant region. It is different from Fig. 9.5. Specifically, there are resonant region, non-resonant region, and ultrasensitive resonant region in Fig. 9.5, but there are only a resonant region and an ultrasensitive resonant region in Fig. 9.9a. Hence, different

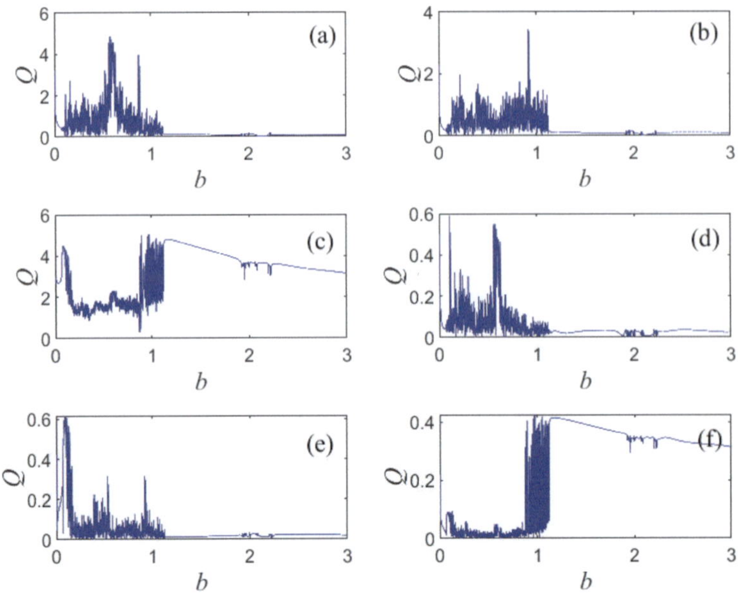

Fig. 9.4 The response amplitude $Q(j\omega)$ versus the system parameter b presents ultrasensitive system resonance in some intervals. **a** $j = 1/3$, **b** $j = 1/2$, **c** $j = 1$, **d** $j = 5/3$, **e** $j = 2$, **f** $j = 3$. The simulation parameters are $\delta = 0.15$, $a = -0.6$, $A = 0.245$, $\omega = 0.9$, $m = 50$, the step $\Delta b = 0.001$. The initial conditions are $x_0 = 0$ and $\dot{x}_0 = 0$. The total calculation time starts from the very beginning

resonance patterns caused by the initial conditions are clearly presented in Fig. 9.9a. It is interesting that in the ultrasensitive resonant region, the value of $Q(\omega)$ has a highly fractal characteristic. In Figs. 9.9b, c, the plots depict the relationship between $Q(\omega/3)$ and $Q(3\omega)$, i.e., the response amplitude at the nonlinear frequency components versus the initial conditions in the two-dimensional plane. Interestingly, the resonance patterns in Fig. 9.9a are almost identical with those in Fig. 9.9c, except for their different values. However, the resonance basins in Fig. 9.9b are exactly opposite to those in Fig. 9.9a, c. This indicates that the frequency component $\omega/3$ is in an antagonistic relationship with the other two frequency components ω and 3ω. Specifically, when the amplitude corresponding to the frequency component $\omega/3$ increases, the amplitudes corresponding to the two frequency components ω and 3ω are suppressed. On the contrary, when the frequency component $\omega/3$ is suppressed, other two frequency components ω and 3ω are enhanced.

Besides the curve of the response amplitude, we also can use the phase trajectory and the spectrum to reflect the characteristic of the response, as shown in Figs. 9.10 and 9.11. Specifically, there is a periodic orbit in the phase portrait and there is a single line in Figs. 9.10a, 9.11a, respectively, which demonstrates the periodic motion of the response. The phase portraits in Fig. 9.10b–d and the corresponding response amplitudes in Fig. 9.11b–d display the aperiodicity of the response.

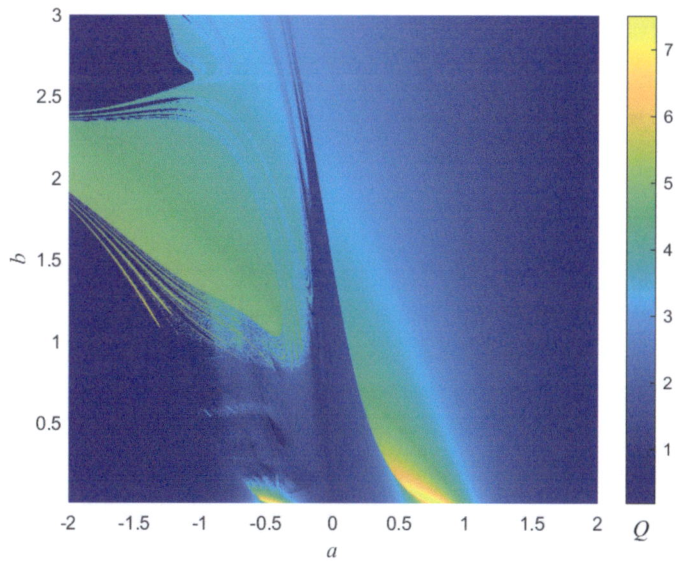

Fig. 9.5 The response amplitude $Q(\omega)$ versus the system parameters a and b presents different resonance patterns in the two-dimensional plane. The color encodes the value of $Q(\omega)$. The simulation parameters are $\delta = 0.15$, $A = 0.245$, $\omega = 0.9$, $m = 50$, the step $\Delta a = 0.002$, the step $\Delta b = 0.005$. The initial conditions are $x_0 = 0$ and $\dot{x}_0 = 0$. The total calculation time starts from the very beginning

9.3.2 A Square Waveform Excitation

When the underdamped Duffing system is excited by a single square waveform signal, i.e.,

$$\ddot{x} + \delta\dot{x} + ax + bx^3 = s_{qua}(t), \tag{9.4}$$

where $s_{qua}(t)$ is a square waveform excitation defined in Eq. (B1) of Appendix B, the response amplitudes of $Q(\omega)$, $Q(\omega/3)$, $Q(3\omega)$ in the $x_0 - \dot{x}_0$ planes are given in Fig. 9.12. In this figure, the resonance patterns of Fig. 9.12a, c are still the same, but opposite to the resonance patterns of Fig. 9.12b. This phenomenon is the same as that in Fig. 9.9.

9.3.3 A Full Sine Waveform Excitation

If the excitation signal is in a full sine waveform, the nonlinear equation is given by

$$\ddot{x} + \delta\dot{x} + ax + bx^3 = s_{fsi}(t), \tag{9.5}$$

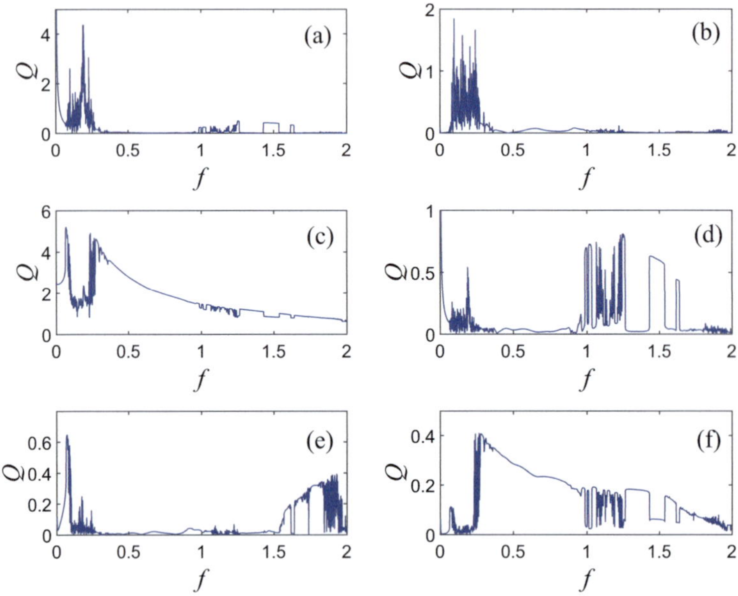

Fig. 9.6 The response amplitude $Q(j\omega)$ versus the signal amplitude f presents ultrasensitive system resonance. **a** $j = 1/3$, **b** $j = 1/2$, **c** $j = 1$, **d** $j = 5/3$, **e** $j = 2$, **f** $j = 3$. The simulation parameters are $\delta = 0.15, a = -0.6, b = 1, A = 0.245, \omega = 0.9, m = 50$, the step $\Delta f = 0.001$. The initial conditions are $x_0 = 0$ and $\dot{x}_0 = 0$. The total calculation time starts from the very beginning

where $s_{fsi}(t)$ is a full sine waveform excitation defined in Eq. (B3) of Appendix B. The response amplitudes at the frequencies $Q(\omega)$, $Q(\omega/3)$, and $Q(2\omega)$ versus the initial conditions x_0 and $\dot{x}_0$ are given in Fig. 9.13.

9.3.4 A Rectified Sine Waveform Excitation

If the excitation is a rectified sine waveform signal, then the underdamped Duffing system is given by

$$\ddot{x} + \delta\dot{x} + ax + bx^3 = s_{rsi}(t), \tag{9.6}$$

where $s_{rsi}(t)$ is a rectified sine waveform excitation defined in Eq. (B5) of Appendix B. The structures of the initial conditions inducing resonance basins of the response amplitudes $Q(\omega)$, $Q(2\omega/3)$ and $Q(2\omega)$ are shown in Fig. 9.14.

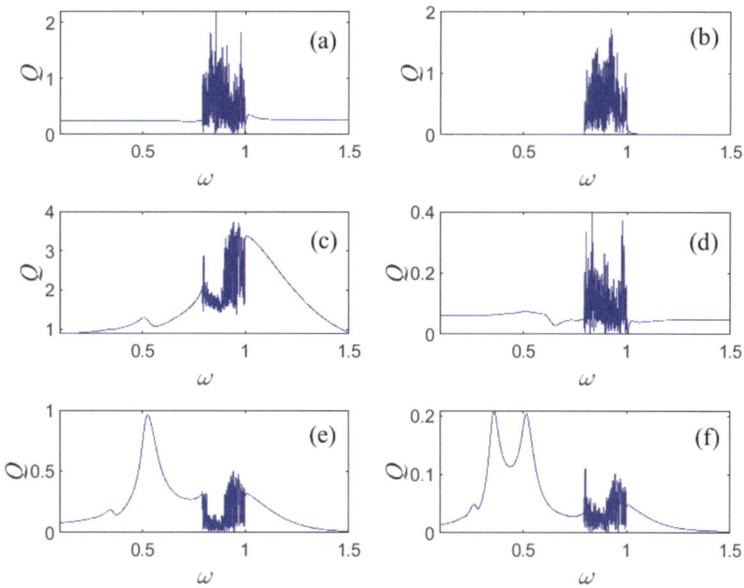

Fig. 9.7 The response amplitude $Q(j\omega)$ versus the signal frequency ω presents ultrasensitive system resonance. **a** $j = 1/3$, **b** $j = 1/2$, **c** $j = 1$, **d** $j = 5/3$, **e** $j = 2$, **f** $j = 3$. The simulation parameters are $\delta = 0.15, a = -0.6, b = 1, A = 0.245, f = 0.1, m = 50$, the step $\Delta\omega = 0.004$. The initial conditions are $x_0 = 0$ and $\dot{x}_0 = 0$. The total calculation time starts from the very beginning

9.3.5 A Symmetric Triangular Waveform Excitation

We consider the Duffing system excited by a symmetric triangular waveform signal given by

$$\ddot{x} + \delta\dot{x} + ax + bx^3 = s_{tri}(t), \tag{9.7}$$

where $s_{tri}(t)$ is a symmetric triangular waveform excitation defined in Eq. (B7) of Appendix B.

The structures of the resonance basins for the response amplitudes at frequencies $Q(\omega)$, $Q(\omega/3)$ and $Q(2\omega)$ are shown in Fig. 9.15. The resonance structures in Fig. 9.15a, c are completely opposite to each other. In contrast, the structural distribution in Fig. 9.15b is neither identical nor opposite to those in Fig. 9.15a, c; instead, it exhibits a distinct pattern.

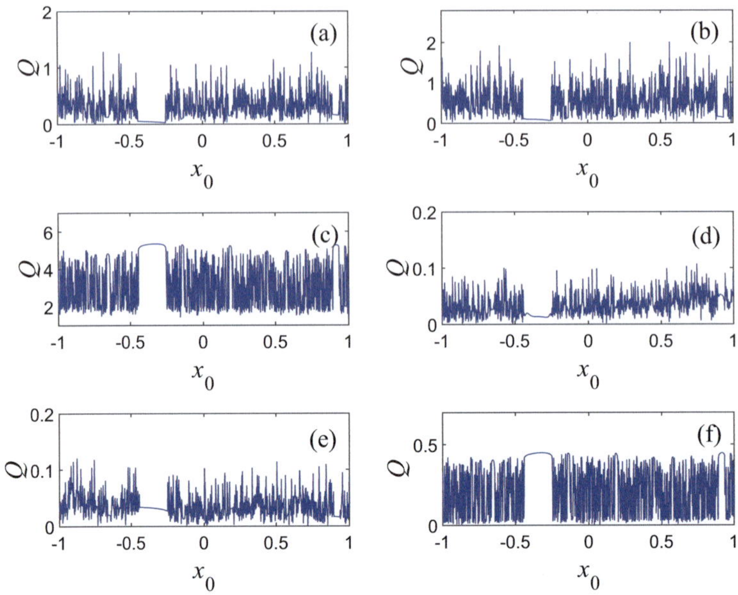

Fig. 9.8 The response amplitude $Q(j\omega)$ versus the initial condition x_0 presents ultrasensitive system resonance. **a** $j = 1/3$, **b** $j = 1/2$, **c** $j = 1$, **d** $j = 5/3$, **e** $j = 2$, **f** $j = 3$. The simulation parameters are $\delta = 0.15$, $a = -0.6$, $b = 1$, $A = 0.245$, $\omega = 0.9$, $m = 50$, the step $\Delta x_0 = 0.002$. The initial condition $\dot{x}_0 = 0$. The total calculation time starts from the very beginning

9.4 Ultrasensitive System Resonance for A Long Time

9.4.1 A Cosine Harmonic Excitation

For the numerical examples mentioned earlier in this chapter, we always let $m = 50$. In other words, the total calculation time we used is 50 periods of the excitation signal from the very beginning. In most cases, we are more concerned about the dynamic behavior of the system for a long time. To achieve this goal, we use 300 periods of the excitation signal for the computations. The specific approach is to first divide the entire time process into an average of 6 parts, each of which still has 50 signal cycles. Then calculate the value of $Q(\omega)$ separately for each time period, and we obtain Figs. 9.16 and 9.17.

In Fig. 9.16, when the computation was carried out using the first 50 cycles, as illustrated in Fig. 9.16a, the ultrasensitive resonant region is the largest among all these subplots, and the amplitude oscillates the most densely within the ultrasensitive resonant region. When we use the second 50 cycles for the computation, as given in Fig. 9.16b, the ultrasensitive resonant region decreases, and the amplitude oscillates slightly sparsely within the ultrasensitive resonant region. From Fig. 9.16a–d, a clear evolutionary trend is observed over time. Specifically, the ultrasensitive resonant

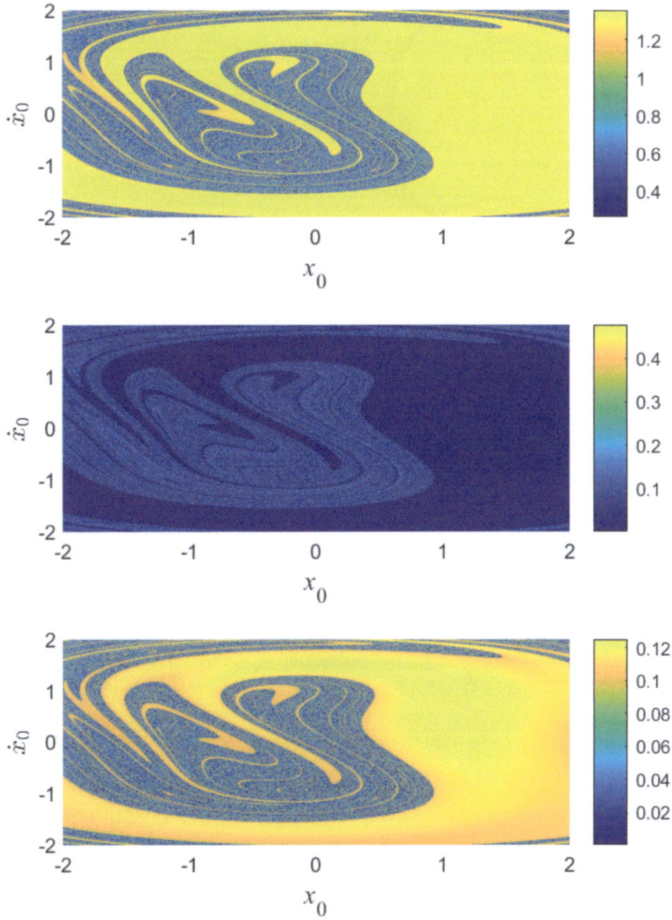

Fig. 9.9 The response amplitude $Q(j\omega)$ caused by a harmonic excitation versus the initial conditions x_0 and $\dot{x}_0$ presents different resonance patterns in the two-dimensional planes. The color bars encode the value of $Q(\omega)$, $Q(\omega/3)$ and $Q(3\omega)$ from the top subplot to the bottom subplot respectively. The simulation parameters are $\delta = 0.15$, $a = -0.6$, $b = 1$, $A = 0.245$, $\omega = 0.9$, $m = 50$, the step $\Delta x_0 = \Delta \dot{x}_0 = 0.004$. The total calculation time starts from the very beginning

region gradually shrinks, and the amplitude fluctuations become increasingly sparse. However, this pattern does not persist indefinitely. As seen in Fig. 9.16e, f, the curves remain almost identical. On one hand, this indicates that the system's response has reached a steady state and no longer evolves with time. On the other hand, it confirms that ultrasensitive resonance persists and does not vanish over time.

In Fig. 9.17, we show the change of the ultrasensitive variation of the response amplitude $Q(\omega)$ in a much small interval of the system parameter a. With the time growing, the emergence of the amplitude fluctuation reduces sharply from Fig. 9.17a–e. In Fig. 9.17b–e, with a slight change of a, the value of $Q(\omega)$ may be severely

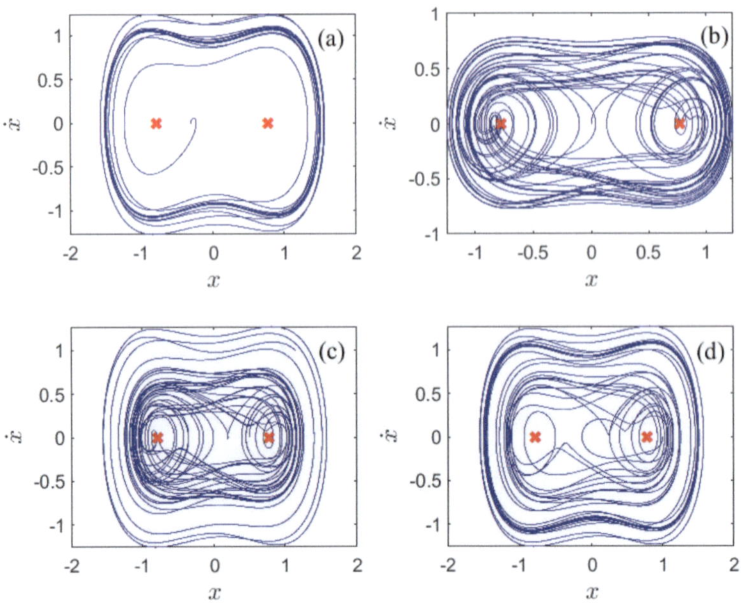

Fig. 9.10 The phase portrait of the output at different initial conditions x_0. **a** $x_0 = -0.3$, **b** $x_0 = 0$, **c** $x_0 = 0.5$, **d** $x_0 = 1$. The symbols $\times$ are the two stable equilibria. The simulation parameters are $\delta = 0.15$, $a = -0.6$, $b = 1$, $A = 0.245$, $\omega = 0.9$, $m = 50$, $\dot{x}_0 = 0$

decreased, and the degree of resonance is suppressed. In addition, in Fig. 9.17f, the response amplitude $Q(\omega)$ changes continuously with the system parameter a. Certainly, it does not mean that the ultrasensitive resonance phenomenon will disappear throughout the entire parameter range, as shown in Fig. 9.16. However, it is possible that the ultrasensitive phenomenon may no longer exist over time within this interval of a.

Like Fig. 9.16, the ultrasensitive system resonance caused by the initial condition x_0 and $\dot{x}_0$ also becomes sparse over time, as shown in Figs. 9.18 and 9.19, respectively. Sparsity does not mean that this ultrasensitive system resonance will completely disappear, but rather that the probability of its occurrence will decrease, but this occurrence is still unpredictable.

To further understand the effect of initial conditions on the system's response, we present Fig. 9.20, which illustrates the evolution of the frequency spectrum over time. In Fig. 9.20a, only two distinct straight lines are visible throughout, indicating the presence of just two frequency components. In contrast, Fig. 9.20b–d reveal a more complex spectral structure in the initial stages, with numerous spectral components. Over time, however, the spectrum simplifies, and the frequency components become more distinct and well-defined.

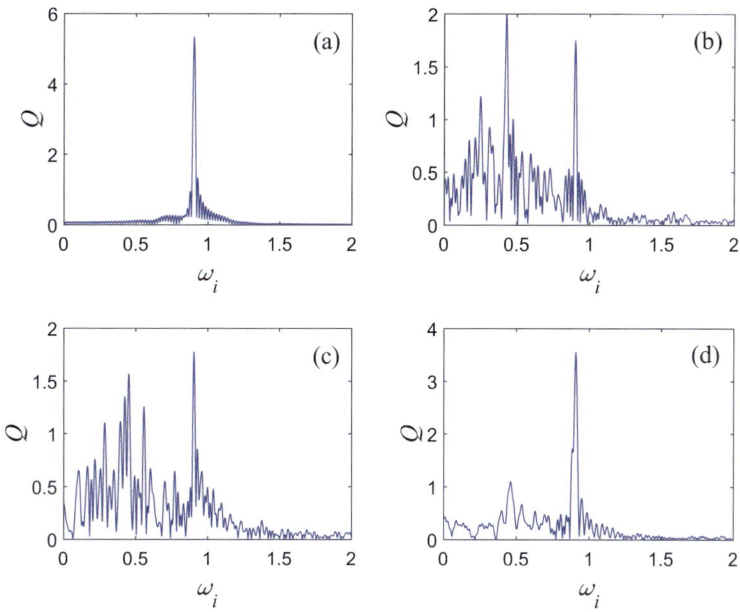

Fig. 9.11 The response amplitude of the output at a series of discrete frequencies under different initial conditions x_0. **a** $x_0 = -0.3$, **b** $x_0 = 0$, **c** $x_0 = 0.5$, **d** $x_0 = 1$. The simulation parameters are $\delta = 0.15$, $a = -0.6$, $b = 1$, $A = 0.245$, $\omega = 0.9$, $m = 50$, $\dot{x}_0 = 0$

9.4.2 A Square Waveform Excitation

We consider the Duffing oscillator excited by a square waveform signal, as shown in Eq. (9.4). The long term behavior of the response is presented in Fig. 9.21. In order to better describe the change in system response over time, we select 40 signal cycles as one time interval in the following discussions, and study a total of 8 consecutive time intervals. Therefore, these plots are more refined than the curves in the previous subsection. If x_0 is the control parameter, as shown in the figure, the response transits from the ultrasensitive state to the steady state. Specifically, only in Fig. 9.21a, the fluctuation appears with the change of the initial condition x_0. For other curves in this figure, the response amplitude will be independent of the initial condition x_0 when the time is exceeds $40T$. There is no transition stage "ultrasensitive response—sparse sensitive response—steady response" in the system response versus the initial condition x_0, as clearly shown in Fig. 9.21b–f. Therefore, we believe that under the excitation of square waveform signal, the transient chaotic state is very short-lived, and the system response quickly enters a steady state.

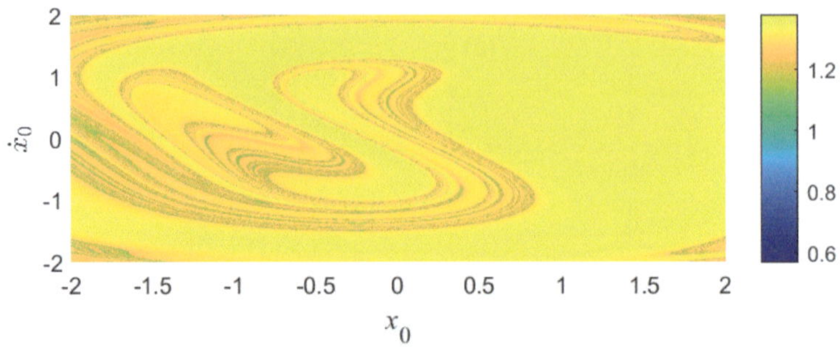

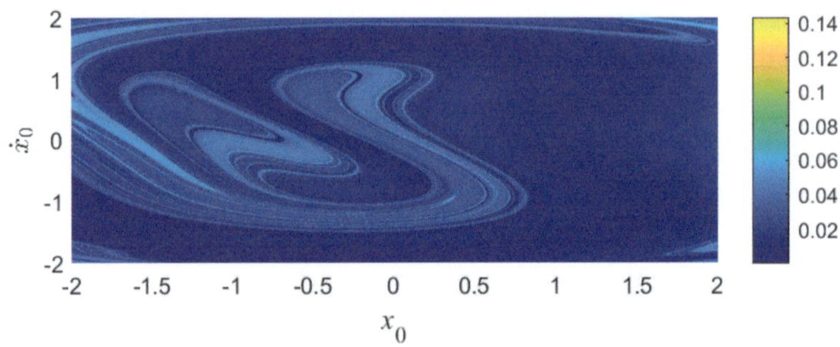

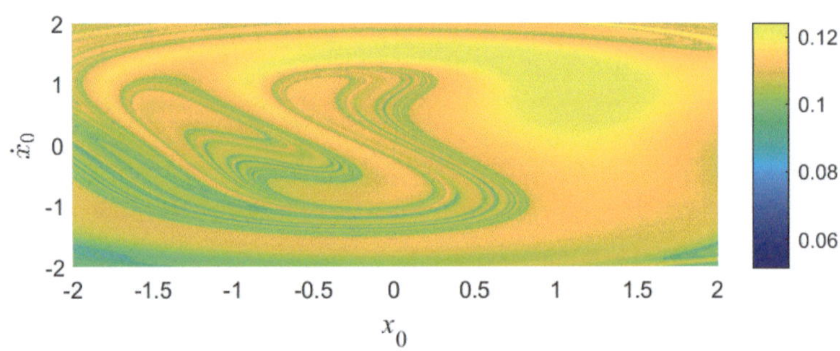

Fig. 9.12 The response amplitude $Q(j\omega)$ caused by a square waveform excitation versus the initial conditions x_0 and $\dot{x}_0$ presents different resonance patterns in the two-dimensional planes. The color bars encode the value of $Q(\omega)$, $Q(\omega/3)$ and $Q(3\omega)$ from the top subplot to the bottom subplot respectively. The simulation parameters are $\delta = 0.15, a = -0.6, b = 1, A = 0.245, \omega = 0.9, m = 50$, the step $\Delta x_0 = \Delta \dot{x}_0 = 0.004$. The total calculation time starts from the very beginning

9.4.3 A Full Sine Waveform Excitation

In Fig. 9.22, the long-term behavior of the response amplitude as a function of the initial condition is presented. Unlike Fig. 9.21, the ultrasensitive resonance caused by the initial condition x_0 persists even after an extended period. As shown in Fig. 9.22a, during the first forty periods T, the system exhibits a transient chaotic response. This subplot differs slightly from the subsequent subplots in the figure, reflecting the system's evolving dynamics over time.

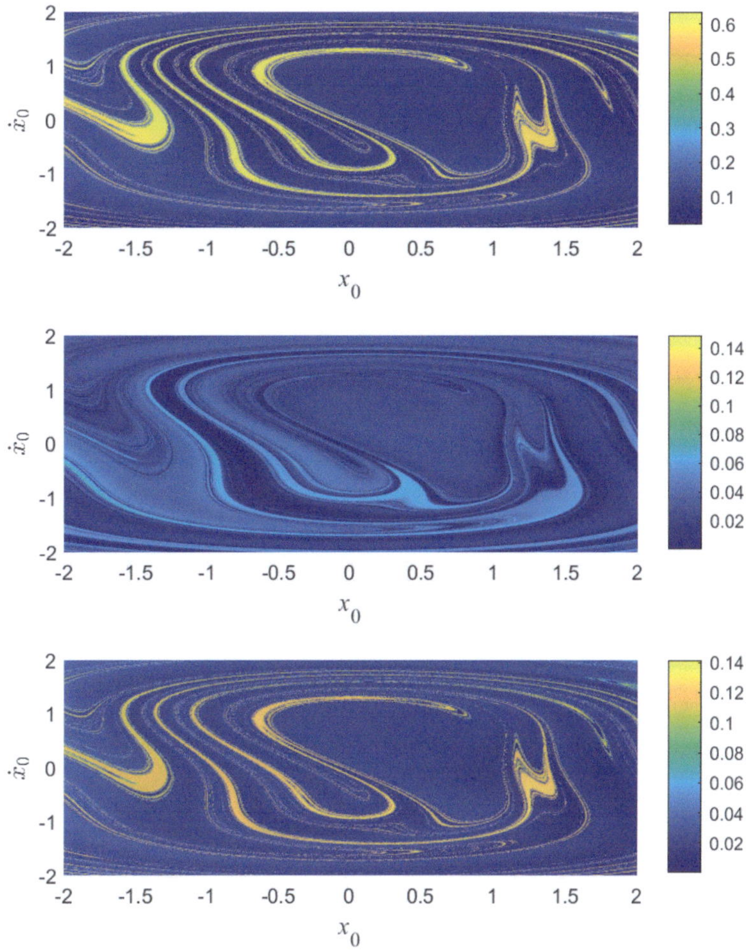

Fig. 9.13 The response amplitude $Q(j\omega)$ caused by a full sine waveform excitation versus the initial conditions x_0 and $\dot{x}_0$ presents different resonance patterns in the two-dimensional planes. The color bars encode the value of $Q(\omega)$, $Q(\omega/3)$ and $Q(2\omega)$ from the top subplot to the bottom subplot respectively. The simulation parameters are $\delta = 0.15$, $a = -0.6$, $b = 1$, $A = 0.245$, $\omega = 0.9$, $m = 50$, the step $\Delta x_0 = \Delta \dot{x}_0 = 0.004$. The total calculation time starts from the very beginning

9.4.4 A Rectified Sine Waveform Excitation

If the excitation is a rectified sine waveform, the long behavior of the response is shown in Fig. 9.23. It also presents the transition state of "ultrasensitive response—sparse sensitive response—steady response" as time increases.

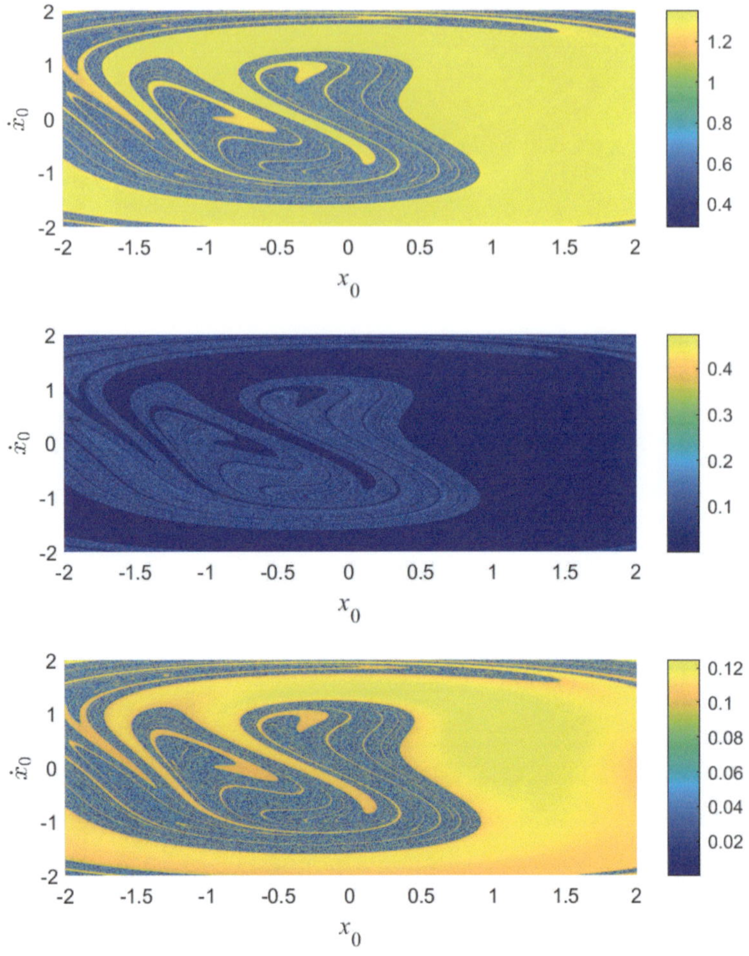

Fig. 9.14 The response amplitude $Q(j\omega)$ caused by a rectified sine waveform excitation versus the initial conditions x_0 and $\dot{x}_0$ presents different resonance patterns in the two-dimensional planes. The color bars encode the value of $Q(\omega)$, $Q(2\omega/3)$ and $Q(2\omega)$ from the top subplot to the bottom subplot, respectively. The simulation parameters are $\delta = 0.15$, $a = -0.6$, $b = 1$, $A = 0.245$, $\omega = 0.9$, $m = 50$, the step $\Delta x_0 = \Delta \dot{x}_0 = 0.004$. The total calculation time starts from the very beginning

9.4.5 A Symmetric Triangular Waveform Excitation

In Fig. 9.24, the long term behavior of the response under the excitation of a symmetric triangular waveform signal is shown. In Fig. 9.24a, it mainly presents the ultrasenstive response. In Fig. 9.24b, c, the response displays the sparse sensitive characteristic. In Fig. 9.24d–h, the response shows steady response and it is independent of the initial condition x_0.

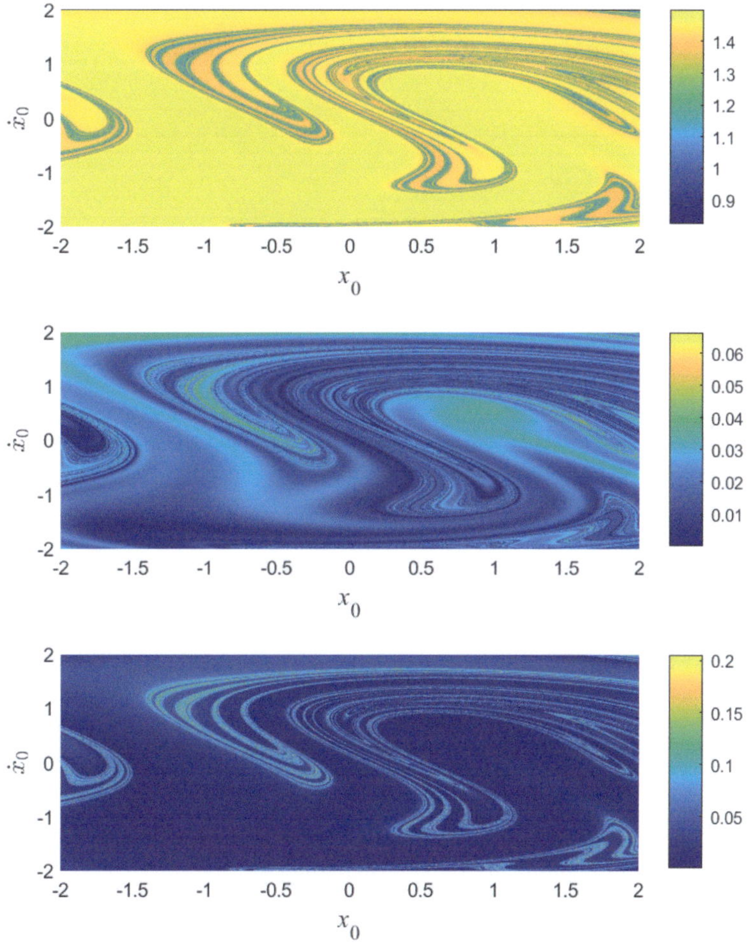

Fig. 9.15 The response amplitude $Q(j\omega)$ caused by a symmetric triangular waveform excitation versus the initial conditions x_0 and $\dot{x}_0$ presents different resonance patterns in the two-dimensional planes. The color bars encode the value of $Q(\omega)$, $Q(\omega/3)$ and $Q(2\omega)$ from the top subplot to the bottom subplot, respectively. The simulation parameters are $\delta = 0.15, a = -0.6, b = 1, A = 0.245$, $\omega = 0.9$, $m = 50$, the step $\Delta x_0 = \Delta \dot{x}_0 = 0.004$. The total calculation time starts from the very beginning

9.5 Concluding Remarks

This chapter explores the phenomenon of ultrasensitive system resonance. Various disturbances, such as external excitation, system parameters, and initial conditions, can induce ultrasensitive resonance due to the presence of chaotic responses within the system. Additionally, the chapter examines the characteristics of ultrasensitive

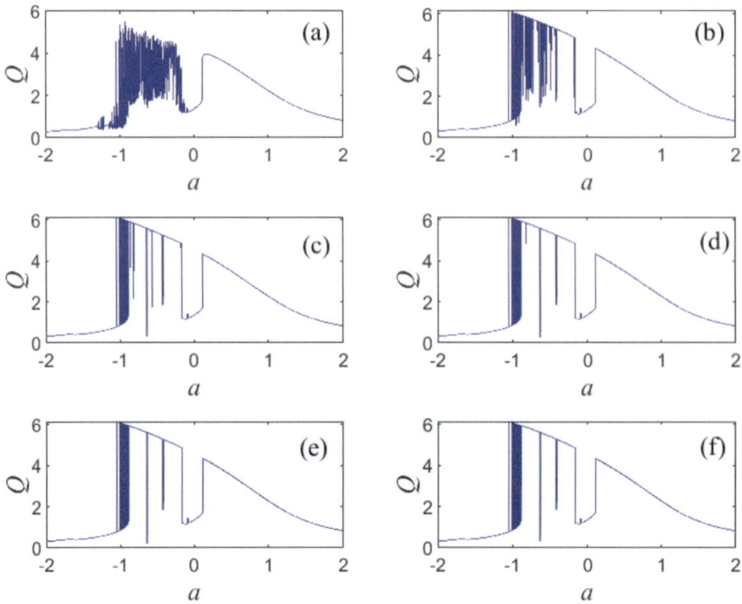

Fig. 9.16 The response amplitude $Q(\omega)$ versus the system parameter a presents ultrasensitive system resonance in some intervals. The total calculation time starts from the very beginning of $(50m+1)T$ to the end of $50jT$, $m = j-1$ and $j = 1 \sim 6$ in turn from **a** to **f**, and T is the period of the excitation. The simulation parameters are $\delta = 0.15$, $b = 1$, $A = 0.245$, $\omega = 0.9$, $m = 50$, the step $\Delta a = 0.001$. The initial conditions are $x_0 = 0$ and $\dot{x}_0 = 0$

resonance under different periodic excitations. Over time, the resonance behavior typically transitions through distinct stages, including ultrasensitive resonance, sparse ultrasensitive resonance, and ultimately, a periodic response. This progression occurs because the system evolves from a transient chaotic state to a steady-state periodic response.

In general, there is a direct relationship between ultrasensitive resonance and chaos, an enduring topic of study. This chapter presents only a preliminary discussion, and much remains to be explored in greater depth. For instance, we have introduced the concept of sparse resonance, but important questions remain: How can sparse resonance be quantified? What are the necessary conditions for its occurrence? These are intriguing topics for further investigation. Additionally, given the vast number of chaotic systems, each with unique characteristics, the study of ultrasensitive resonance within these systems presents a promising avenue for future research.

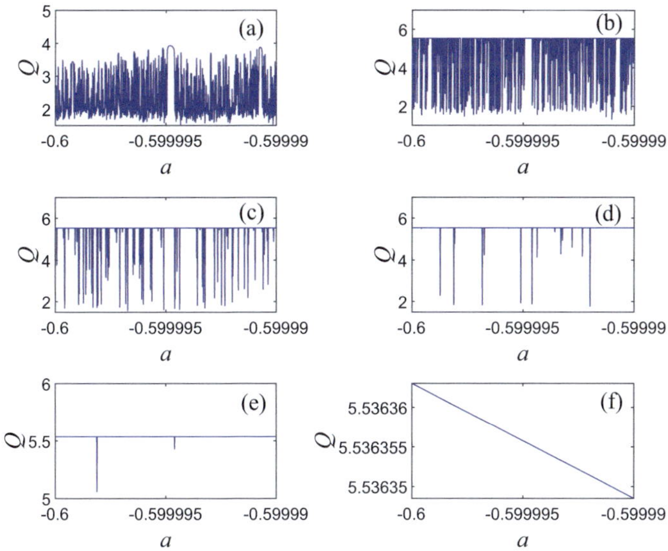

Fig. 9.17 The response amplitude $Q(\omega)$ versus the system parameter a presents ultrasensitive system resonance in some intervals. The total calculation time starts from the very beginning of $(50m + 1)T$ to the end of $50jT$, $m = j - 1$ and $j = 1 \sim 6$ in turn from **a** to **f**, and T is the period of the excitation. The simulation parameters are $\delta = 0.15, b = 1, A = 0.245, \omega = 0.9, m = 50$, the step $\Delta a = 0.00000001$. The initial conditions are $x_0 = 0$ and $\dot{x}_0 = 0$

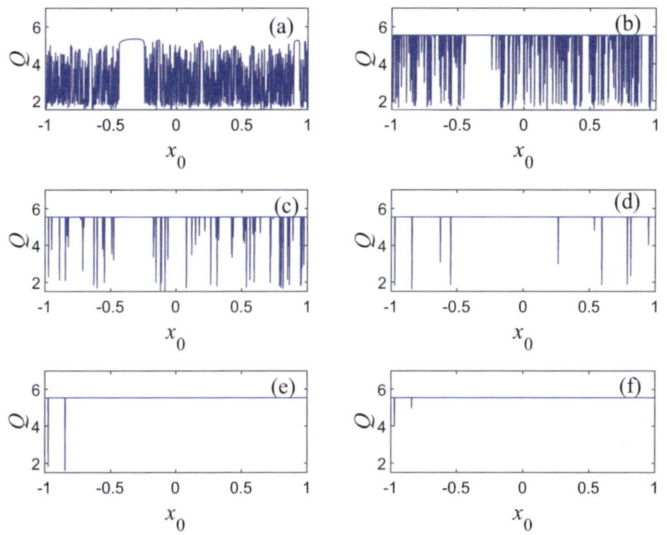

Fig. 9.18 The response amplitude $Q(\omega)$ versus the initial condition x_0 presents ultrasensitive system resonance in some intervals. The total calculation time starts from the very beginning of $(50m + 1)T$ to the end of $50jT$, $m = j - 1$ and $j = 1 \sim 6$ in turn from **a** to **f**, and T is the period of the excitation. The simulation parameters are $\delta = 0.15$, $a = -0.6$, $b = 1$, $A = 0.245$, $\omega = 0.9$, the step $\Delta x_0 = 0.002$, $\dot{x}_0 = 0$

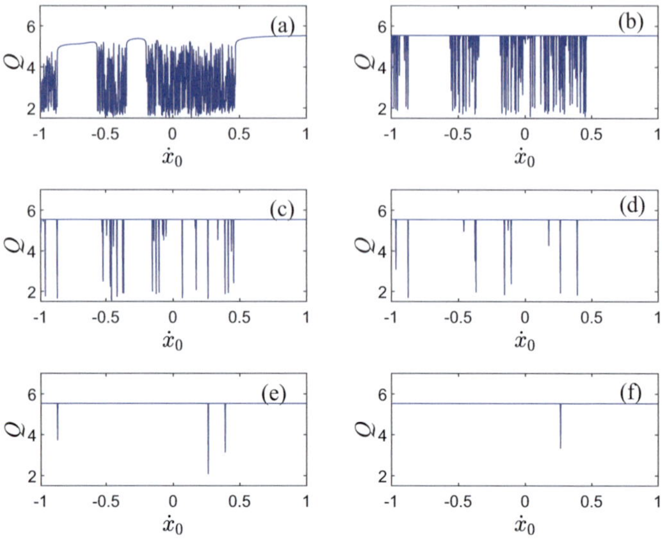

Fig. 9.19 The response amplitude $Q(\omega)$ versus the initial condition $\dot{x}_0$ presents ultrasensitive system resonance in some intervals. The total calculation time starts from the very beginning of $(50m + 1)T$ to the end of $50jT$, $m = j - 1$ and $j = 1 \sim 6$ in turn from **a** to **f**, and T is the period of the excitation. The simulation parameters are $\delta = 0.15$, $a = -0.6$, $b = 1$, $A = 0.245$, $\omega = 0.9$, the step $\Delta x_0 = 0.002$, $x_0 = 0$

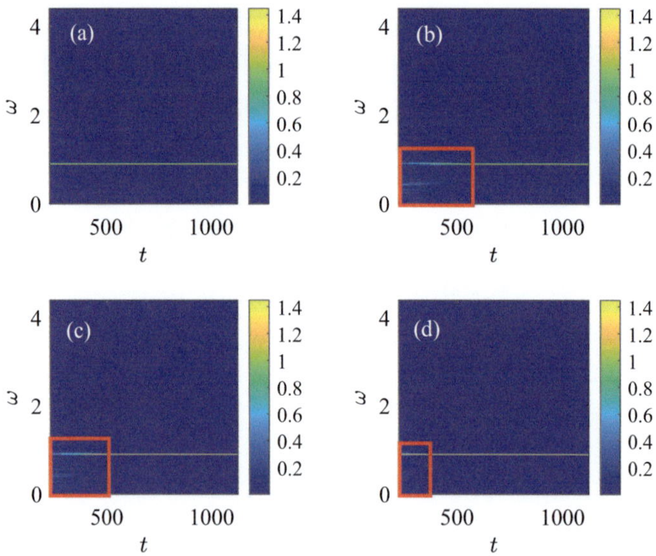

Fig. 9.20 The frequency components of the response change with time under different initial conditions. **a** $x_0 = -0.3$, **b** $x_0 = 0$, **c** $x_0 = 0.5$, **d** $x_0 = 1$. The simulation parameters are $\delta = 0.15$, $a = -0.6$, $b = 1$, $A = 0.245$, $\omega = 0.9$, $m = 200$, $\dot{x}_0 = 0$. The total calculation time starts from the very beginning

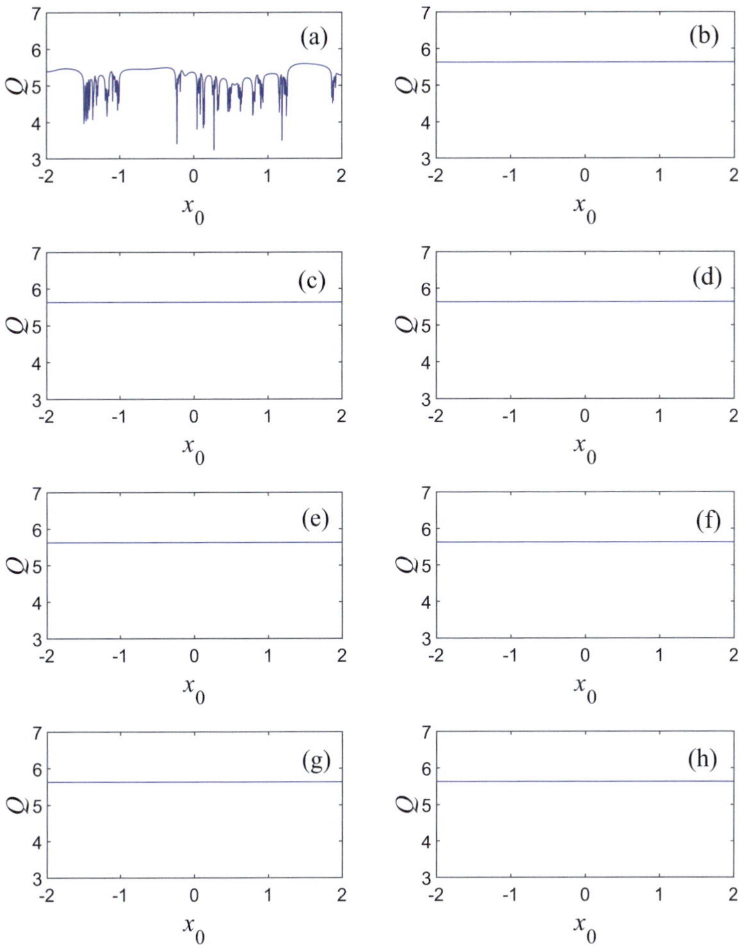

Fig. 9.21 The response amplitude $Q(\omega)$ caused by a square waveform excitation versus the initial condition x_0 presents ultrasensitive system resonance in some intervals. The total calculation time starts from the very beginning of $(40m + 1)T$ to the end of $40jT$, $m = j - 1$ and $j = 1 \sim 8$ in turn from **a** to **f**, and T is the period of the excitation. The simulation parameters are $\delta = 0.15$, $a = -0.6$, $b = 1$, $A = 0.245$, $\omega = 0.9$, the step $\Delta x_0 = 0.004$, $\dot{x}_0 = 0$

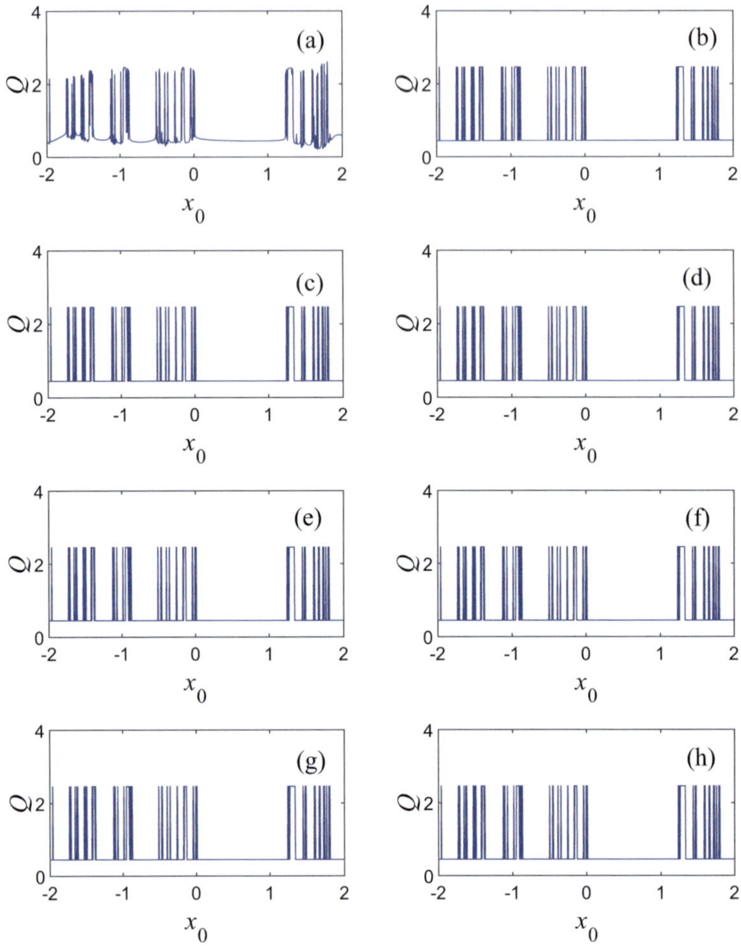

Fig. 9.22 The response amplitude $Q(\omega)$ caused by a full sine waveform excitation versus the initial condition x_0 presents ultrasensitive system resonance in some intervals. The total calculation time starts from the very beginning of $(40m + 1)T$ to the end of $40jT$, $m = j - 1$ and $j = 1 \sim 8$ in turn from **a** to **f**, and T is the period of the excitation. The simulation parameters are $\delta = 0.15$, $a = -0.6$, $b = 1$, $A = 0.245$, $\omega = 0.9$, the step $\Delta x_0 = 0.004$, $\dot{x}_0 = 0$

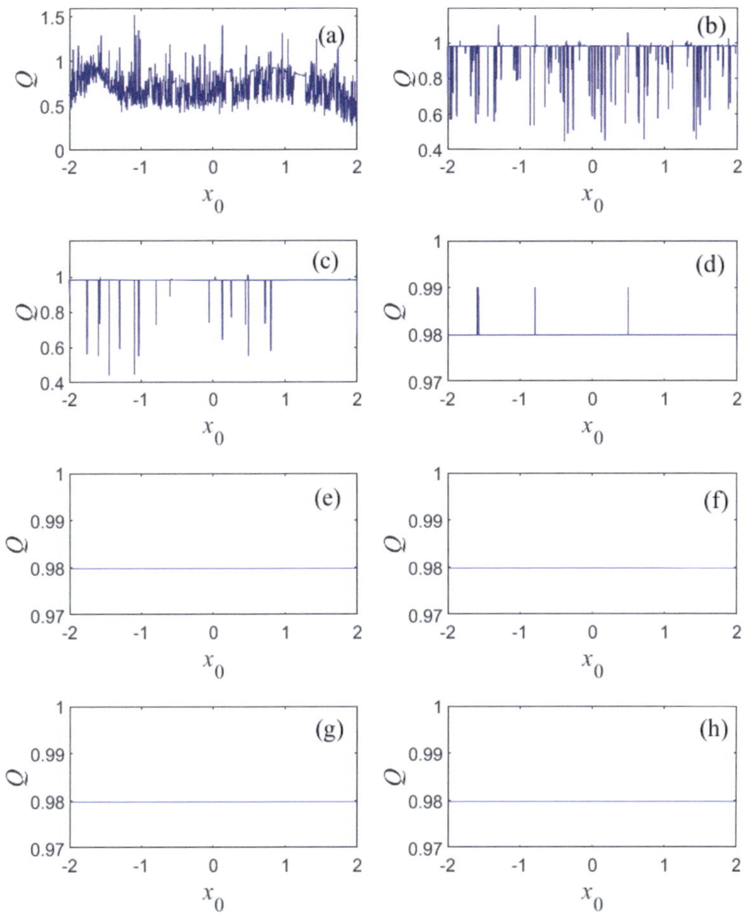

Fig. 9.23 The response amplitude $Q(\omega)$ caused by a rectified sine waveform excitation versus the initial condition x_0 presents ultrasensitive system resonance in some intervals. The total calculation time starts from the very beginning of $(40m+1)T$ to the end of $40jT$, $m = j - 1$ and $j = 1 \sim 8$ in turn from **a** to **f**, and T is the period of the excitation. The simulation parameters are $\delta = 0.15$, $a = -0.6$, $b = 1$, $A = 0.245$, $\omega = 0.9$, the step $\Delta x_0 = 0.004$, $\dot{x}_0 = 0$

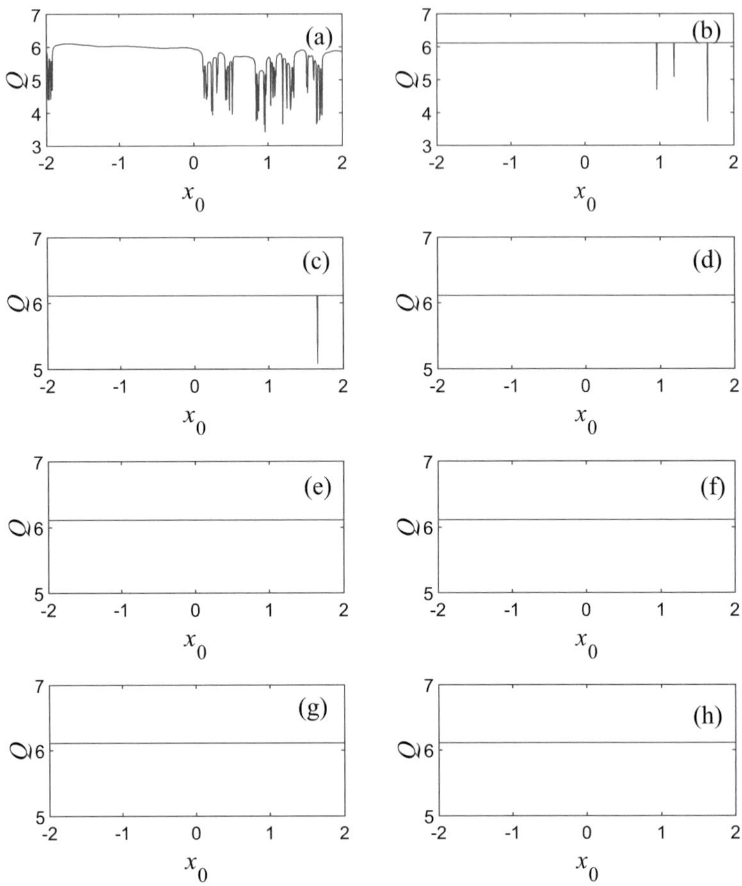

Fig. 9.24 The response amplitude $Q(\omega)$ caused by a symmetric triangular waveform excitation versus the initial condition x_0 presents ultrasensitive system resonance in some intervals. The total calculation time starts from the very beginning of $(40m + 1)T$ to the end of $40jT$, $m = j - 1$ and $j = 1 \sim 8$ in turn from **a** to **f**, and T is the period of the excitation. The simulation parameters are $\delta = 0.15$, $a = -0.6$, $b = 1$, $A = 0.245$, $\omega = 0.9$, the step $\Delta x_0 = 0.004$, $\dot{x}_0 = 0$

References

1. A. Daza, A. Wagemakers, M.A.F. Sanjuán, Strong sensitivity of the vibrational resonance induced by fractal structures, *International Journal of Bifurcation and Chaos*, 2013, **23**(7): 1350129.
2. S. Li, Z. Wang, J. Yang, M.A.F. Sanjuán, S. Huang, L. Lou, Ultrasensitive vibrational resonance induced by small disturbances, *Chaos*, 2023, **33**(12):123111.
3. C. Grebogi, E. Ott, J.A. Yorke, Crises, sudden changes in chaotic attractors, and transient chaos, *Physica D*, 1983, **7**(1-3):181-200.
4. A.S. de Paula, M.A. Savi, F.H.I. Pereira-Pinto, Chaos and transient chaos in an experimental nonlinear pendulum, *Journal of Sound and Vibration*, 2006, **294**(3):585-595.
5. T. Tél, Y.C. Lai, Chaotic transients in spatially extended systems, *Physics Reports*, 2008, **460**(6): 245-275.
6. Y.C. Lai, T. Tél, *Transient Chaos: Complex Dynamics on Finite Time Scales* (Springer Science & Business Media 2011)

Chapter 10
Ultrasensitive Vibrational Resonance

In this chapter, we discuss ultrasensitive vibrational resonance in detail. Unlike ultrasensitive system resonance, this phenomenon involves two simultaneous excitations: one slow-varying and the other fast-varying. Specifically, we examine how different periodic excitations and initial conditions contribute to ultrasensitive vibrational resonance. Additionally, we explore the evolution of vibrational resonance patterns over time.

10.1 Ultra-Sensitive Vibrational Resonance Caused by Different Periodic Excitations

10.1.1 Two Harmonic Excitations

Now, we will discuss the ultrasensitive vibrational resonance in an underdamped Duffing system excited by different periodic excitations, containing harmonic excitations and anharmonic periodic excitations. Under different periodic excitations, a nonlinear system may exhibit very complicated dynamics phenomena. Up to now, there are related research works on stochastic resonance [1–3], vibrational resonance [2, 4], logical resonance [5, 6], bifurcation [7], and so on.

Next, we consider the underdamped Duffing system subjected to both low-frequency and high-frequency excitations

$$\ddot{x} + \delta\dot{x} + ax + bx^3 = A\cos(\omega t) + B\cos(\Omega t), \tag{10.1}$$

where $\omega \ll \Omega$.

In Fig. 10.1, a small disturbance in the amplitude of the high-frequency signal induces ultrasensitive vibrational resonance. On one hand, even a very small excitation, i.e., a minimal value of parameter B, can trigger strong vibrational resonance at the primary frequency or nonlinear vibrational resonance at certain nonlinear

J. Yang and M. A. F. Sanjuan, *Aperiodic Resonances—Theory and Applications*,
https://doi.org/10.1007/978-981-96-6819-9_10

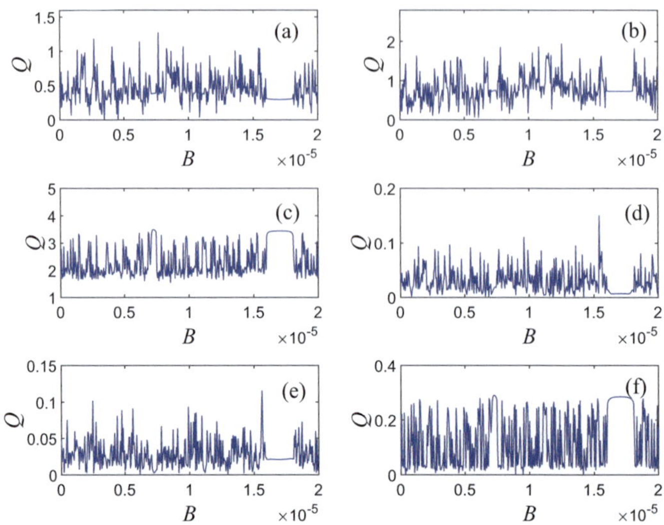

Fig. 10.1 The response amplitude $Q(j\omega)$ versus the signal amplitude B presents ultrasensitive vibrational resonance in a very small interval. **a** $j = 1/3$, **b** $j = 1/2$, **c** $j = 1$, **d** $j = 5/3$, **e** $j = 2$, **f** $j = 3$. The simulation parameters are $\delta = 0.15$, $a = -0.6$, $b = 1$, $A = 0.245$, $\omega = 0.9$, $\Omega = 9.8$, $m = 50$, the step $\Delta B = 0.00000005$. The initial conditions are $x_0 = 0$ and $\dot{x}_0 = 0$. The total calculation time starts from the very beginning

frequency components. On the other hand, a slight change in B causes $Q(j\omega)$ to undergo rapid variations. This significant change in ultrasensitive vibrational resonance occurs not only at the excitation frequency but also at certain nonlinear frequency components.

Figure 10.2 illustrates ultrasensitive vibrational resonance over a relatively large range of B, resembling parameter-caused ultrasensitive system resonance. A slight change in the auxiliary signal amplitude B leads to intense and dense variations in the response amplitude Q at certain existing frequencies.

Not only the amplitude but also the frequency of the high-frequency signal can induce ultrasensitive vibrational resonance, as illustrated in Fig. 10.3. As Ω varies, the response amplitude $Q(j\omega)$ undergoes significant changes at certain frequencies, leading to intense vibrational resonance effects.

10.1.2 A Harmonic Excitation and a Square Waveform Excitation

The first kind of periodic excitation we considered is a square waveform signal. Then, Eq. (10.1) changes to

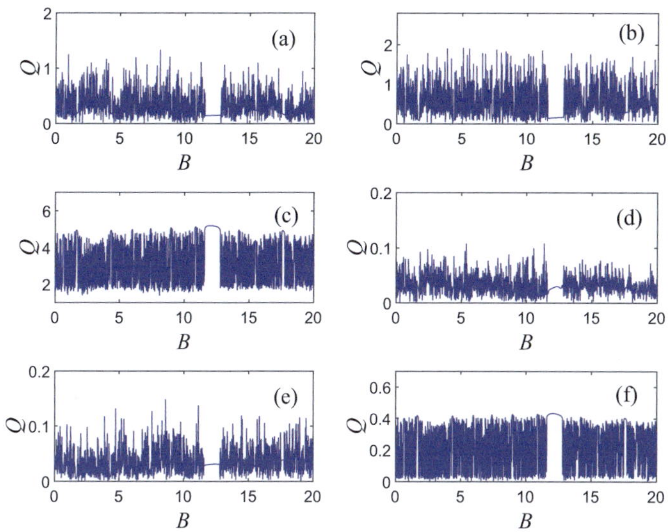

Fig. 10.2 The response amplitude $Q(j\omega)$ versus the signal amplitude B presents ultrasensitive vibrational resonance in a relatively large interval. **a** $j = 1/3$, **b** $j = 1/2$, **c** $j = 1$, **d** $j = 5/3$, **e** $j = 2$, **f** $j = 3$. The simulation parameters are $\delta = 0.15$, $a = -0.6$, $b = 1$, $A = 0.245$, $\omega = 0.9$, $\Omega = 9.8$, $m = 50$, the step $\Delta B = 0.01$. The initial conditions are $x_0 = 0$ and $\dot{x}_0 = 0$. The total calculation time starts from the very beginning

$$\ddot{x} + \delta\dot{x} + ax + bx^3 = s_{qua}(t) + B\cos(\Omega t). \tag{10.2}$$

The expression of the square waveform signal $s_{qua}(t)$ is given in Eq. (B1) of the Appendix B.

In Fig. 10.4, the ultrasensitive vibrational resonance is displayed at different frequency components. Among them, in Fig. 10.4a, b, d, f, these frequency components are not contained in the excitation that can be obtained directly from the Fourier series in Eq. (B2) of Appendix B. Consequently, they are nonlinear vibrational resonance in these subplots. In Fig. 10.4c, e, the vibrational resonance occurs at ω and 3ω, and these two frequency components are contained in the Fourier series of the square wave excitation. Figure 10.4 illustrates the ultrasensitive vibrational resonance caused by the square waveform in the slow-varying signal and the harmonic waveform in the fast-varying signal.

Now, we make the fast-varying signal $S_{qua}(t)$ in a square waveform, i.e.,

$$\ddot{x} + \delta\dot{x} + ax + bx^3 = A\cos(\omega t) + S_{qua}(t), \tag{10.3}$$

where the signal parameters A and ω in Eq. (B1) are replaced by B and Ω which are large parameters.

We still investigate the ultrasensitive vibrational resonance in Fig. 10.5 at the frequency components that the same as Fig. 10.4. The difference is that all frequency

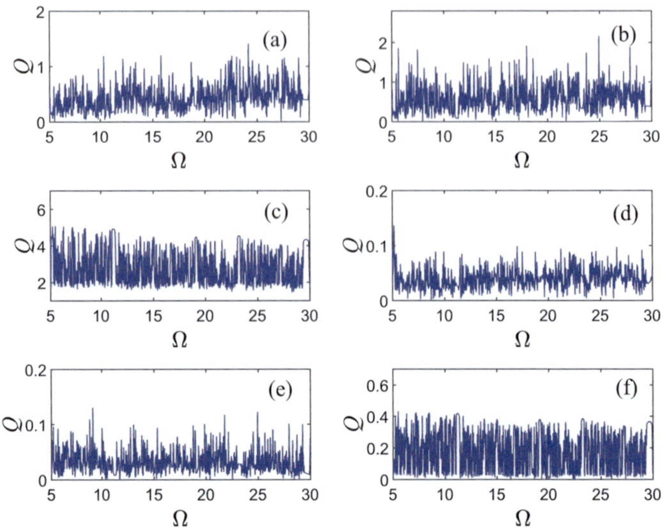

Fig. 10.3 The response amplitude $Q(j\omega)$ versus the frequency Ω presents ultrasensitive vibrational resonance in a relatively large interval. **a** $j = 1/3$, **b** $j = 1/2$, **c** $j = 1$, **d** $j = 5/3$, **e** $j = 2$, **f** $j = 3$. The simulation parameters are $\delta = 0.15$, $a = -0.6$, $b = 1$, $A = 0.245$, $\omega = 0.9$, $B = 5$, $m = 50$, the step $\Delta\Omega = 0.03$. The initial conditions are $x_0 = 0$ and $\dot{x}_0 = 0$. The total calculation time starts from the very beginning

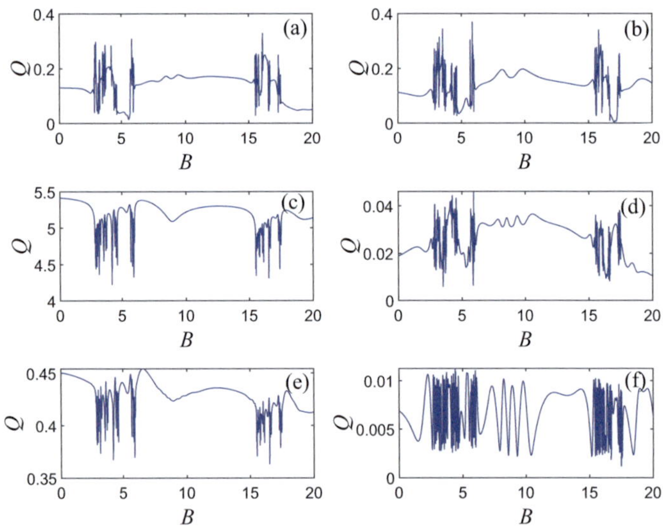

Fig. 10.4 The response amplitude $Q(j\omega)$ versus the signal amplitude B presents ultrasensitive vibrational resonance in a relatively large interval. The slow-varying signal is a square wave excitation. **a** $j = 1/3$, **b** $j = 1/2$, **c** $j = 1$, **d** $j = 2$, **e** $j = 3$, **f** $j = 4$. The simulation parameters are $\delta = 0.15$, $a = -0.6$, $b = 1$, $A = 0.245$, $\omega = 0.9$, $\Omega = 9.8$, $m = 50$, the step $\Delta B = 0.01$. The initial conditions are $x_0 = 0$ and $\dot{x}_0 = 0$. The total calculation time starts from the very beginning

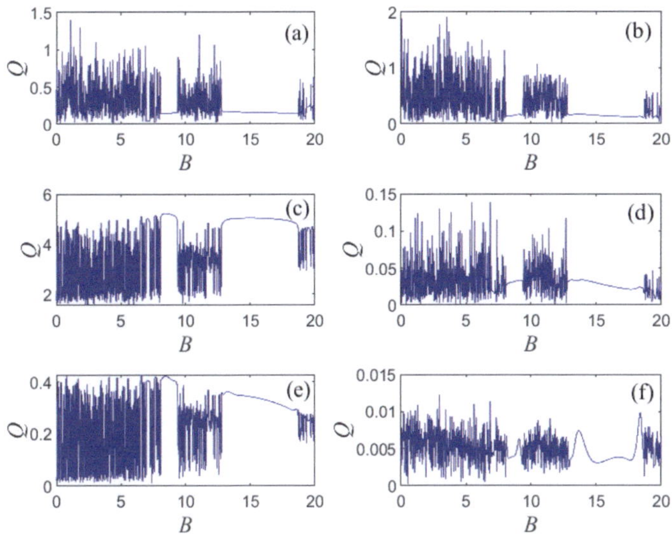

Fig. 10.5 The response amplitude $Q(j\omega)$ versus the signal amplitude B presents ultrasensitive vibrational resonance in a relatively large interval. The fast-varying signal is a square wave excitation. **a** $j = 1/3$, **b** $j = 1/2$, **c** $j = 1$, **d** $j = 2$, **e** $j = 3$, **f** $j = 4$. The simulation parameters are $\delta = 0.15$, $a = -0.6$, $b = 1$, $A = 0.245$, $\omega = 0.9$, $\Omega = 9.8$, $m = 50$, the step $\Delta B = 0.01$. The initial conditions are $x_0 = 0$ and $\dot{x}_0 = 0$. The total calculation time starts from the very beginning

components except ω are nonlinear frequencies. Further, compared Fig. 10.5 with Fig. 10.4, the response amplitude is much larger when the fast varying signal is in the square wave from. Although the amplitude of the slow excitation signal in both Figs. 10.4 and 10.5 are 0.245, from the Fourier series in Eq. (B2), the amplitude of the slow square waveform at ω is smaller than the harmonic signal with the same amplitude.

10.1.3 A Harmonic Excitation and a Full Sine Waveform Excitation

When the slow-varying signal in a full sine waveform, the dynamical equation is given by

$$\ddot{x} + \delta\dot{x} + ax + bx^3 = s_{fsi}(t) + B\cos(\Omega t). \tag{10.4}$$

The expression of the full sine waveform $s_{fsi}(t)$ is given in Eq. (B3) of the Appendix B.

In Fig. 10.6, the ultrasensitive vibrational resonance of Eq. (10.4) is displayed at different frequency components. The ultrasensitive vibrational resonance occurs in a small range and small value of B. In other words, a high-frequency harmonic signal

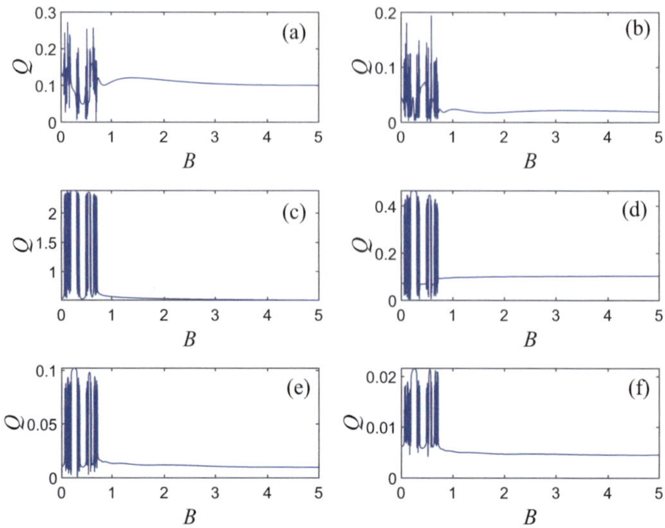

Fig. 10.6 The response amplitude $Q(j\omega)$ versus the signal amplitude B presents ultrasensitive vibrational resonance in a relatively large interval. The slow-varying signal is a full sine wave excitation. **a** $j = 1/3$, **b** $j = 1/2$, **c** $j = 1$, **d** $j = 2$, **e** $j = 3$, **f** $j = 4$. The simulation parameters are $\delta = 0.15$, $a = -0.6$, $b = 1$, $A = 0.245$, $\omega = 0.9$, $\Omega = 9.8$, $m = 50$, the step $\Delta B = 0.01$. The initial conditions are $x_0 = 0$ and $\dot{x}_0 = 0$. The total calculation time starts from the very beginning

with relative large amplitude can reduce the vibration caused by a slow varying full sine signal. Evenly, in Fig. 10.6c, the response amplitude $Q(\omega)$ is almost zero. That is to say, the amplitude at frequency ω is very weak and basically disappears. In fact, we can consider it as the anti-resonance phenomenon. There are also volumes of references on anti-resonance [8–11]. Anti-resonance refers to the phenomenon which the amplitude at the considered frequency almost disappears. In addition, the subharmonic frequency components $\omega/3$ and $\omega/2$ in Fig. 10.6a, b are nonlinear frequency components in the nonlinear dynamical framework. The frequency components ω, 2ω, 3ω and 4ω are inherent frequency components in the signal.

When the fast varying signal is in the full sine waveform, the dynamical equation is

$$\ddot{x} + \delta\dot{x} + ax + bx^3 = A\cos(\omega t) + S_{fsi}(t), \tag{10.5}$$

where S_{fsi} is the same as Eq. (B3) but the signal parameters A and ω are replaced by B and Ω.

Corresponding to Eq. (10.5), Fig. 10.7 gives the response amplitude curves at different frequency components. In each subplot, there is ultrasensitive variation with change of the signal amplitude B. In Fig. 10.7a–c, e, we can consider the response as ultrasensitive vibrational resonance due to the response amplitude can achieve some large values. Nevertheless, the ultrasensitive variation of the response amplitude in Fig. 10.7d, f are small and the resonance occurs at a traditional resonance region.

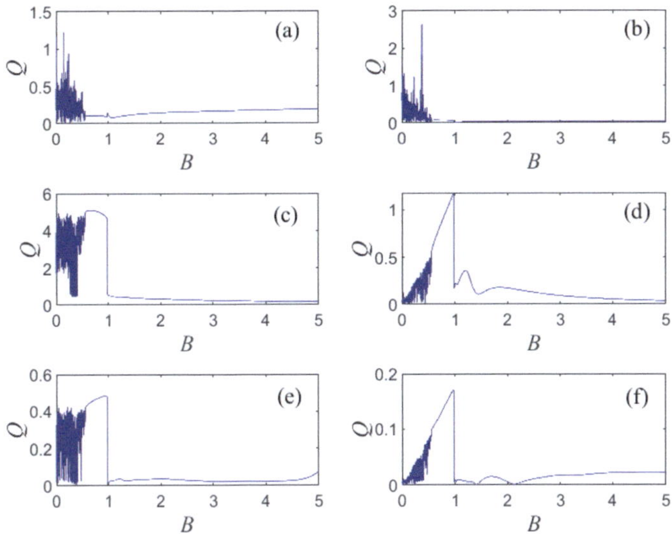

Fig. 10.7 The response amplitude $Q(j\omega)$ versus the signal amplitude B presents ultrasensitive vibrational resonance in a relatively large interval. The fast-varying signal is a full sine wave excitation. **a** $j = 1/3$, **b** $j = 1/2$, **c** $j = 1$, **d** $j = 2$, **e** $j = 3$, **f** $j = 4$. The simulation parameters are $\delta = 0.15$, $a = -0.6$, $b = 1$, $A = 0.245$, $\omega = 0.9$, $\Omega = 9.8$, $m = 50$, the step $\Delta B = 0.01$. The initial conditions are $x_0 = 0$ and $\dot{x}_0 = 0$. The total calculation time starts from the very beginning

It extends the ultrasensitive vibrational resonance in the former works to a general ultrasensitive response.

10.1.4 A Harmonic Excitation and a Rectified Sine Waveform Excitation

When the slow varying signal is a rectified sine waveform excitation, the governing equation is given by

$$\ddot{x} + \delta\dot{x} + ax + bx^3 = s_{rsi}(t) + B\cos(\Omega t). \tag{10.6}$$

The expression of the rectified sine waveform $s_{rsi}(t)$ is given in Eq. (B5) of the Appendix B.

In the response of Eq. (10.6), the ultrasensitive vibrational resonance presents intermittently. The linear resonance and the nonlinear resonance occur almost in the same interval, because the chaotic response is for all frequencies, while periodic vibration is only for a few frequencies. Although the amplitude oscillations in each subplot are very severe, the resonance is stronger at the primary resonance and 1/3 subharmonic resonance, as shown in Fig. 10.8a, c.

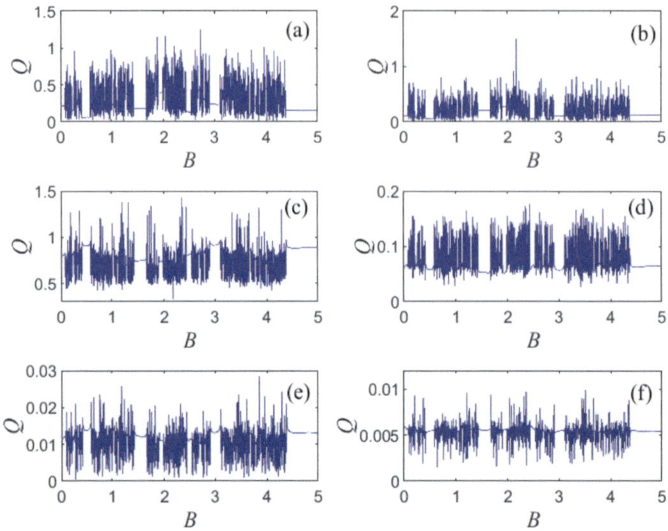

Fig. 10.8 The response amplitude $Q(j\omega)$ versus the signal amplitude B presents ultrasensitive vibrational resonance in a relatively large interval. The slow-varying signal is a rectified sine wave excitation. **a** $j = 1/3$, **b** $j = 1/2$, **c** $j = 1$, **d** $j = 2$, **e** $j = 3$, **f** $j = 4$. The simulation parameters are $\delta = 0.15$, $a = -0.6$, $b = 1$, $A = 0.245$, $\omega = 0.9$, $\Omega = 9.8$, $m = 50$, the step $\Delta B = 0.01$. The initial conditions are $x_0 = 0$ and $\dot{x}_0 = 0$. The total calculation time starts from the very beginning

When the fast-varying signal is in the rectified sine waveform and the slow-varying signal is in the harmonic form, the dynamic equation changes to

$$\ddot{x} + \delta\dot{x} + ax + bx^3 = A\cos(\omega t) + S_{rsi}(t), \tag{10.7}$$

where S_{rsi} is also the rectified sine waveform in Eq. (B5) but with the large parameters B and Ω replacing small parameters A and ω.

In Fig. 10.9, when the fast-varying signal changes to the rectified sine wave excitation, there are some reduction in the scopes of the ultrasensitive vibrational resonance. Interestingly, for the primary resonance and superharmonic resonance, there are obvious and wide resonance bands, as illustrated in Fig. 10.9c–f.

10.1.5 A Harmonic Excitation and a Symmetric Triangular Waveform Excitation

When the slow-varying signal is a symmetric triangular waveform excitation, the nonlinear equation is

$$\ddot{x} + \delta\dot{x} + ax + bx^3 = s_{tri}(t) + B\cos(\Omega t). \tag{10.8}$$

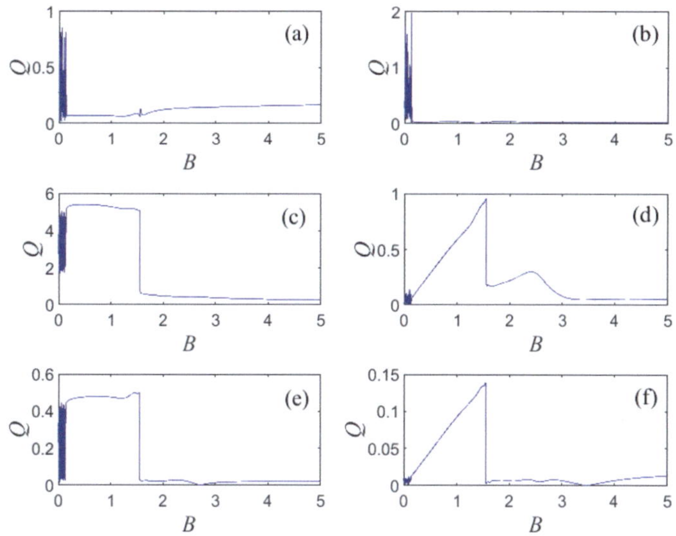

Fig. 10.9 The response amplitude $Q(j\omega)$ versus the signal amplitude B presents ultrasensitive vibrational resonance in a relatively large interval. The fast-varying signal is a rectified sine wave excitation. **a** $j = 1/3$, **b** $j = 1/2$, **c** $j = 1$, **d** $j = 2$, **e** $j = 3$, **f** $j = 4$. The simulation parameters are $\delta = 0.15$, $a = -0.6$, $b = 1$, $A = 0.245$, $\omega = 0.9$, $\Omega = 9.8$, $m = 50$, the step $\Delta B = 0.01$. The initial conditions are $x_0 = 0$ and $\dot{x}_0 = 0$. The total calculation time starts from the very beginning

Here, the symmetric triangular waveform is given in Eq. (B7) of the Appendix B.

When the slow-varying signal is the symmetric triangular waveform, the ultrasensitive vibrational resonance at some frequencies are given in Fig. 10.10. In the figure, $Q(j\omega)$ shows a high sensitivity with the variation of B, and the curve shows ultrasensitive vibrational resonance and ultrasensitive nonlinear vibrational resonance. In addition, for this case, the variation of $Q(j\omega)$ presents a higher sensitivity compared to Figs. 10.4, 10.6, and 10.8.

When the fast-varying signal is a symmetric triangular waveform excitation

$$\ddot{x} + \delta\dot{x} + ax + bx^3 = A\cos(\omega t) + S_{tri}(t), \tag{10.9}$$

where S_{tri} is defined by Eq. (B7) but the signal parameters A and ω are replaced by B and Ω.

The ultrasensitive vibrational resonance caused by the fast-varying symmetric triangular waveform excitation is given in Fig. 10.11. There is still high sensitivity of the response amplitude $Q(j\omega)$ with the variation of the amplitude B of the fast varying symmetric triangular waveform excitation. The sensitivity is higher than that in Figs. 10.5, 10.7, and 10.9.

In addition to the four types anharmonic periodic signals discussed above, there are also a large number of other forms of periodic signals, even aperiodic signals, which can cause ultrasensitive vibrational resonance or ultrasensitive aperiodic vibrational

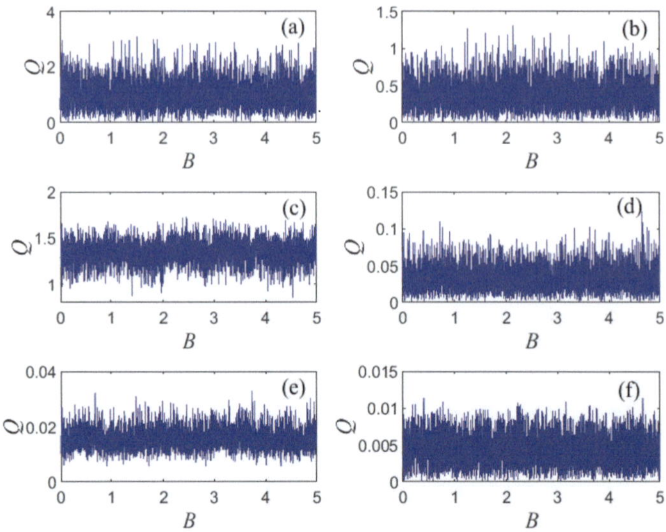

Fig. 10.10 The response amplitude $Q(j\omega)$ versus the signal amplitude B presents ultrasensitive vibrational resonance in a relatively large interval. The slow varying signal is a symmetric triangular wave excitation. **a** $j = 1/3$, **b** $j = 1/2$, **c** $j = 1$, **d** $j = 2$, **e** $j = 3$, **f** $j = 4$. The simulation parameters are $\delta = 0.15$, $a = -0.6$, $b = 1$, $A = 0.245$, $\omega = 0.9$, $\Omega = 9.8$, $m = 50$, the step $\Delta B = 0.01$. The initial conditions are $x_0 = 0$ and $\dot{x}_0 = 0$. The total calculation time starts from the very beginning

resonance. This will not be repeated here. Interested readers can conduct in-depth research on their own.

10.2 Ultra-Sensitive Vibrational Resonance Caused by the Initial Conditions

When the two excitations are in harmonic form, as given in Fig. 10.12, the ultrasensitive vibrational resonance of Eq. (10.1) can also be caused by the initial condition $x(0)$. It is also a prominent feature of ultrasensitive system resonance and ultrasensitive vibrational resonance. The strongest resonance appears in Fig. 10.12c, i.e., the resonance at the primary frequency ω. At the subharmonic frequency components $\omega/3$, $\omega/2$, there are also strong vibrational resonance. Compared with the subharmonic and superharmonic vibrational resonance in the overdamped Duffing oscillator realized by tuning the system parameter or the high-frequency signal, the nonlinear vibrational resonance herein can be realized only by choosing the initial conditions of the system. It is a double-edged sword. On the one hand, the resonance can be achieved much more easily. On the other hand, the ultrasensitivity of the response to the initial conditions makes it difficult to control the response, and a very small disturbance may cause a significant change in the system response.

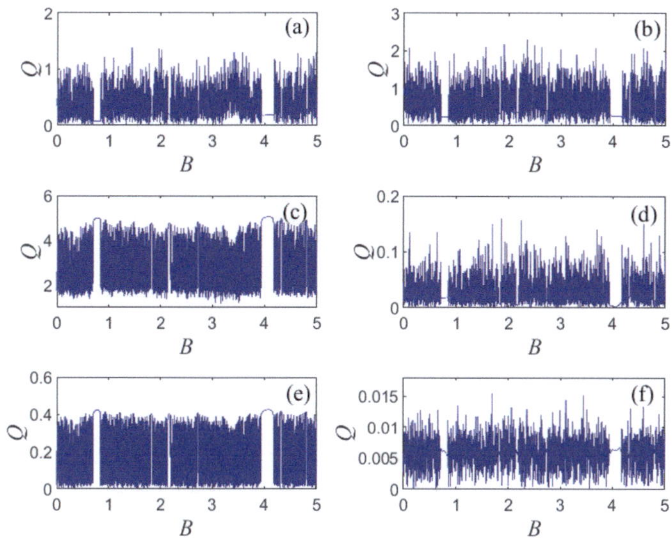

Fig. 10.11 The response amplitude $Q(j\omega)$ versus the signal amplitude B presents ultrasensitive vibrational resonance in a relatively large interval. The fast-varying signal is a symmetric triangular wave excitation. **a** $j = 1/3$, **b** $j = 1/2$, **c** $j = 1$, **d** $j = 2$, **e** $j = 3$, **f** $j = 4$. The simulation parameters are $\delta = 0.15$, $a = -0.6$, $b = 1$, $A = 0.245$, $\omega = 0.9$, $k = 1$, $\Omega = 9.8$, $m = 50$, the step $\Delta B = 0.01$. The initial conditions are $x_0 = 0$ and $\dot{x}_0 = 0$. The total calculation time starts from the very beginning

In Fig. 10.13, the regions are given in the two-dimensional plane which shows the resonance at the frequency $Q(\omega)$, $Q(\omega/3)$ and $Q(3\omega)$ caused by the initial displacement and initial velocity, respectively. Both the initial displacement and the initial velocity can induce resonance. In the subplots of Fig. 10.13a, c, the resonance patterns of the response amplitude versus the initial conditions are almost the same except the values are different. In the subplot of Fig. 10.13(b), the resonance patterns are exactly opposite to the situation in the two subplots of Fig. 10.13a, c. In addition, there are many traditional vibrational resonance regions and ultrasensitive vibrational resonance regions in the figure. In the traditional vibrational resonance region, the response amplitude of the system is stable in a large interval of the initial condition x_0 or $\dot{x}_0$. In the ultrasensitive resonance region, the response amplitude fluctuates sensitively with the tiny change of the initial conditions. Figure 10.13 can serve as a basis for selecting stable resonant regions for the components ω and 3ω or non-resonant region for the subharmonic frequency $\omega/3$. In addition, comparing Fig. 10.13 with Fig. 9.9, we find that the two figures are very similar. Except for the difference in the presence or absence of the high-frequency signal in the excitation, other simulation parameters in these two figures are the same. It indicates that there is a trivial effect of the high-frequency auxiliary signal on the sensitive response of the low-frequency ω under the simulation parameters of these two figures.

To further illustrate the dependence of the response on the initial conditions in the initial stage of the time, Fig. 10.14 shows the phase trajectories of the response in

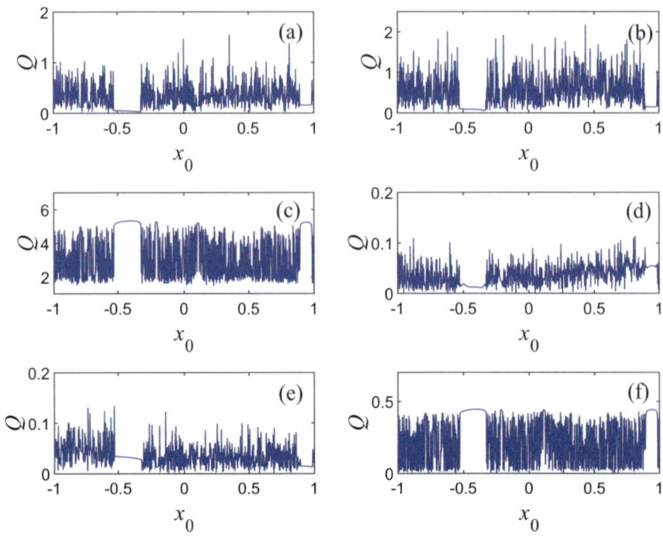

Fig. 10.12 The response amplitude $Q(j\omega)$ versus the initial condition x_0 presents ultrasensitive resonance. **a** $j = 1/3$, **b** $j = 1/2$, **c** $j = 1$, **d** $j = 5/3$, **e** $j = 2$, **f** $j = 3$. The simulation parameters are $\delta = 0.15$, $a = -0.6$, $b = 1$, $A = 0.245$, $\omega = 0.9$, $B = 5$, $\Omega = 9.8$, $m = 50$, the step $\Delta x_0 = 0.002$. The initial condition $\dot{x}_0 = 0$. The total calculation time starts from the very beginning

the first 50 cycles under different initial conditions. The difference in phase diagram caused by initial conditions is significant. They visually presented the differences in system response. Compared to Fig. 9.10, the addition of high-frequency signals in the excitation makes the phase diagram more complex. Corresponding to Fig. 10.14, the results in Fig. 10.15 show the response amplitude spectrum in the low frequency range. These response amplitude spectra exhibit continuous characteristics in some frequency bands. It is the reason for the ultrasensitive response of the system. Besides the frequency of the signal, many other interference frequency components also exist.

10.3 Ultrasensitive Vibrational Resonance for A Long Time

Here, we will discuss the evolution from the ultrasensitive vibrational resonance to the steady periodic motion for a long time.

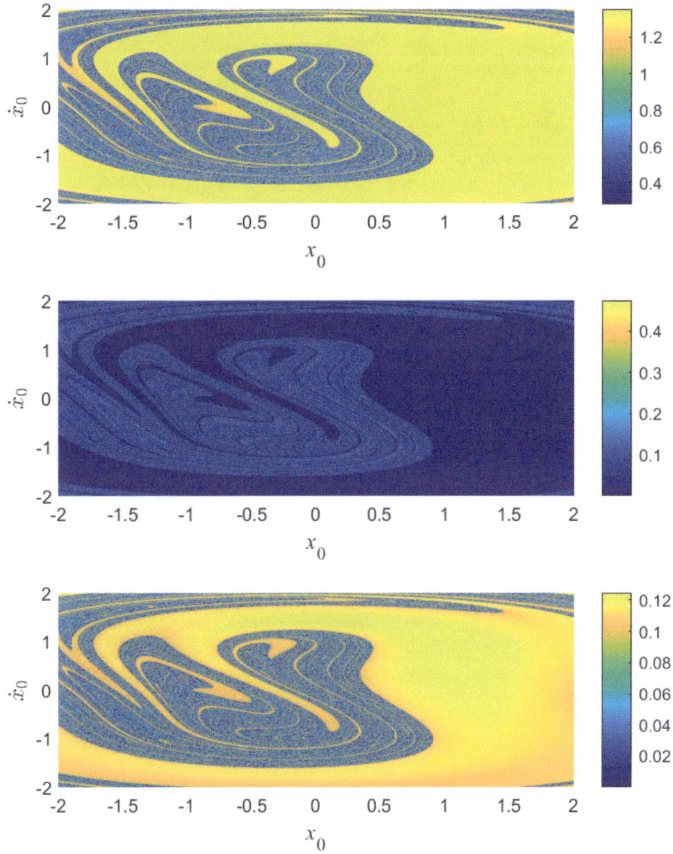

Fig. 10.13 The response amplitude $Q(j\omega)$ that caused by two harmonic excitations versus the initial conditions x_0 and $\dot{x}_0$ presents different resonance patterns in the two-dimensional planes. The color bars encode the value of $Q(\omega)$, $Q(\omega/3)$ and $Q(3\omega)$ from the top subplot to the bottom subplot respectively. The simulation parameters are $\delta = 0.15$, $a = -0.6$, $b = 1$, $A = 0.245$, $\omega = 0.9$, $B = 5$, $\Omega = 9.8$, $m = 50$, the step $\Delta x_0 = \Delta \dot{x}_0 = 0.004$. The total calculation time starts from the very beginning

10.3.1 Two Harmonic Excitations

When the underdamped Duffing system excited by two cosine waveforms, as described in Eq. (10.1), the evolution of the response amplitude $Q(j\omega)$ versus the signal amplitude B over a long time is given in Fig. 10.16. During the response of the first 50 cycles of the low-frequency signal, the amplitude of the response at the frequency ω versus the amplitude of the fast-varying signal exhibits ultrasensitive vibrational resonance. With a very slight change in the amplitude of the fast-varying signal, the value of amplitude of the system response changes dramatically. It is interesting that as time increases, for example, in the second 50 cycles, the change

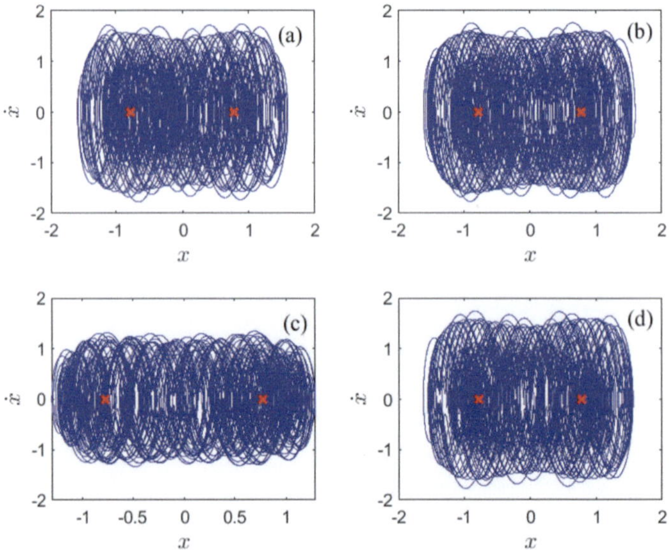

Fig. 10.14 The phase portraits of the output at different initial conditions x_0. **a** $x_0 = -0.3$, **b** $x_0 = 0$, **c** $x_0 = 0.5$, **d** $x_0 = 1$. The symbols $\times$ are the two stable equilibria. The simulation parameters are $\delta = 0.15$, $a = -0.6$, $b = 1$, $A = 0.245$, $\omega = 0.9$, $B = 5$, $\Omega = 9.8$, $m = 50$, $\dot{x}_0 = 0$

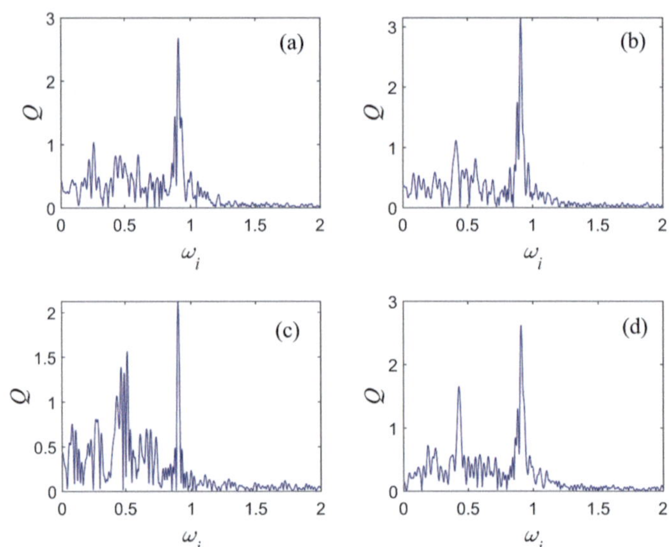

Fig. 10.15 The response amplitude of the output at a series of discrete frequencies under different initial conditions x_0. **a** $x_0 = -0.3$, **b** $x_0 = 0$, **c** $x_0 = 0.5$, **d** $x_0 = 1$. The simulation parameters are $\delta = 0.15$, $a = -0.6$, $b = 1$, $A = 0.245$, $\omega = 0.9$, $B = 5$, $\Omega = 9.8$, $m = 50$, $\dot{x}_0 = 0$

of Q relative to B becomes sparser than the corresponding situation in the first 50 cycles. As time goes on, this phenomenon becomes increasingly apparent, as shown

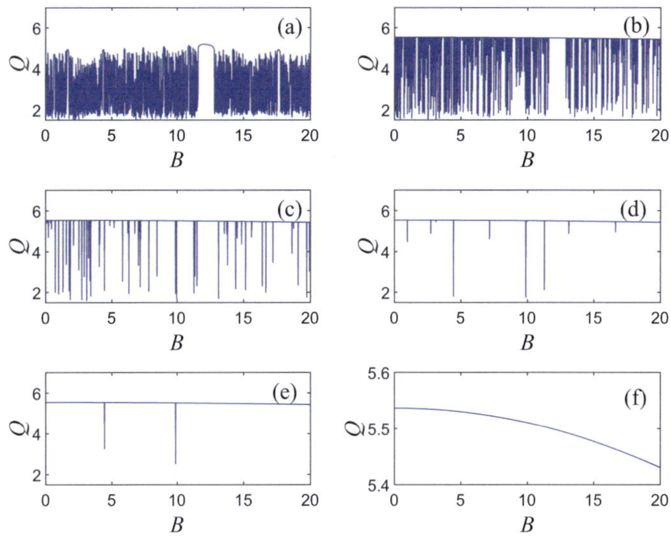

Fig. 10.16 The evolution of the response amplitude $Q(\omega) - B$ curve over a long time. The total calculation time starts from the very beginning of $(50m + 1)T$ to the end of $50jT$, $m = j - 1$ and $j = 1 \sim 6$ in turn from **a** to **f**, and T is the period of the excitation. The simulation parameters are $\delta = 0.15$, $a = -0.6$, $b = 1$, $A = 0.245$, $\omega = 0.9$, $\Omega = 9.8$, $m = 50$, the step $\Delta B = 0.01$. The initial conditions are $x_0 = 0$ and $\dot{x}_0 = 0$

in Fig. 10.16c–e. As time continues to goes on further, as shown in Fig. 10.16f, the ultrasensitive variation in response amplitude caused by the small disturbance will disappear. The reason for this phenomenon is that the system response is transient chaotic [12–15], and as time goes on, finally, the system response gradually transits from chaotic to periodic or quasi-periodic motion. Combining the study of transient chaos with vibrational resonance is a new research direction, and there are many phenomena that have not been further revealed, which are worth conducting in-depth research. When studying ultrasensitive vibrational resonance, the dynamic behavior of the system response over a long time should be considered, and attention should be paid to the change from ultrasensitive vibrational resonance to the steady periodic response. In [16, 17], this issue was not addressed, but in [18], this issue has already been addressed. Again in Fig. 10.16f, the response amplitude Q decreases with the increase of B, there is not a vibrational resonance phenomenon. It is mainly because the system resonance has already occurred only under the excitation of the slow varying signal. Similar to the results in Figs. 9.17, 9.18, 9.19, 9.23, 9.24, especially in Fig. 10.16c–e, the curve presents sparse ultrasensitive vibrational resonance.

Similarly, as time goes on, the system response caused by the change in initial condition x_0 undergoes a process from ultrasensitive vibrational resonance to the steady periodic or quasi-periodic response, as presented in Fig. 10.17. As time goes on, the chaotic characteristics of the system gradually weaken until the chaos disappears, which is a gradual process of change.

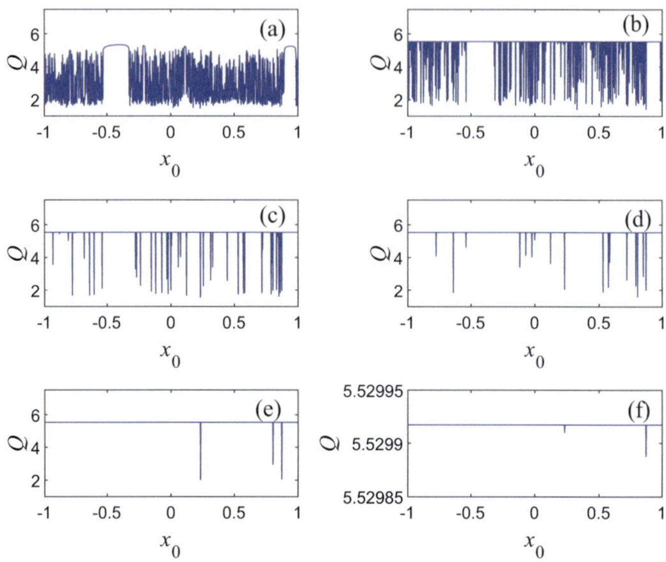

Fig. 10.17 The evolution of the response amplitude $Q(\omega) - x_0$ curve over a long time. The total calculation time starts from the very beginning of $(50m + 1)T$ to the end of $50jT$, $m = j - 1$ and $j = 1 \sim 6$ in turn from **a** to **f**, and T is the period of the excitation. The simulation parameters are $\delta = 0.15, a = -0.6, b = 1, A = 0.245, \omega = 0.9, B = 5, \Omega = 9.8, m = 50$, the step $\Delta x_0 = 0.002$, $\dot{x}_0 = 0$

To further reveal the long-term response behavior of the system, in Fig. 10.18, we have provided phase diagrams of the system response at $251T$-$300T$ under different initial conditions of x_0. Apparently, for different initial conditions, these phase diagrams are almost identical. It indicates that the influence of the transient response has disappeared, mainly remains the motion presented by the steady-state response.

In Fig. 10.19, we once again demonstrate from the perspective of the response spectrum that the system response has transitioned from transient chaos to steady-state periodic motion over a long period of time. There are only a few discrete spectral lines on the spectrum, indicating that the system cannot be in a chaotic state. In addition, it can be observed that the amplitude of the system response at the low-frequency ω has been significantly amplified, while the amplitude at the high-frequency Ω has been suppressed.

In Fig. 10.20, we continue to discuss this issue on the time-frequency plot. This figure can reflect more intuitively how the system response changes over time. Within the boxes in the figure, there are multiple and complex frequency components, and the spectrum is continuously changing. The system response presents a chaotic state. Outside the boxes, the chaotic characteristics have basically disappeared, and the frequency components are relatively seldom, making frequency identification easy.

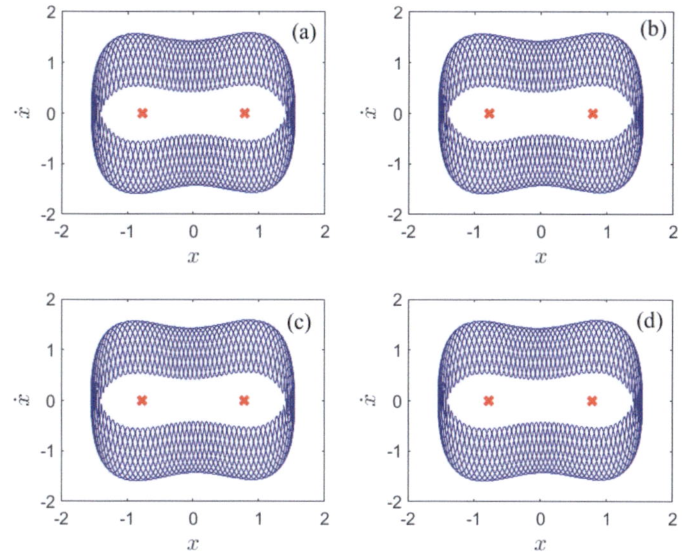

Fig. 10.18 The phase portraits of the output at different initial conditions x_0. **a** $x_0 = -0.3$, **b** $x_0 = 0$, **c** $x_0 = 0.5$, **d** $x_0 = 1$. The symbols $\times$ are the two stable equilibria. The simulation parameters are $\delta = 0.15$, $a = -0.6$, $b = 1$, $A = 0.245$, $\omega = 0.9$, $B = 5$, $\Omega = 9.8$, $m = 50$, $\dot{x}_0 = 0$. The total calculation time starts from the beginning of $251T$ to $300T$

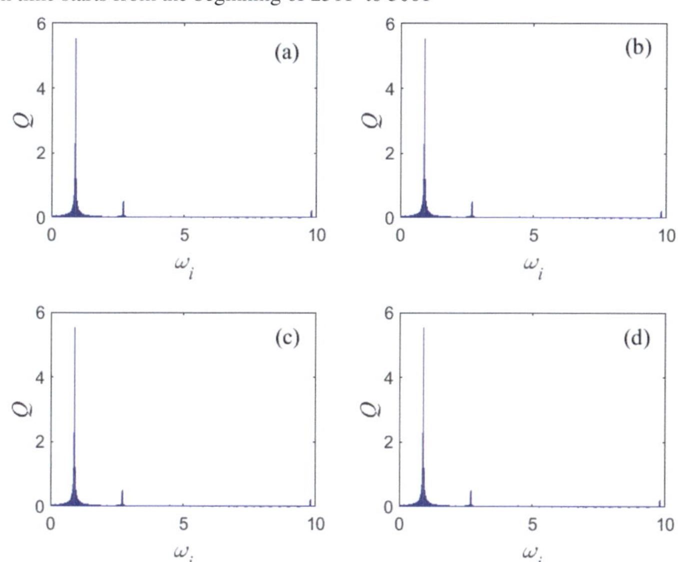

Fig. 10.19 The response amplitude of the output at a series of discrete frequencies under different initial conditions x_0. **a** $x_0 = -0.3$, **b** $x_0 = 0$, **c** $x_0 = 0.5$, **d** $x_0 = 1$. The simulation parameters are $\delta = 0.15$, $a = -0.6$, $b = 1$, $A = 0.245$, $\omega = 0.9$, $B = 5$, $\Omega = 9.8$, $m = 50$, $\dot{x}_0 = 0$. The total calculation time starts from the beginning of $251T$ to $300T$

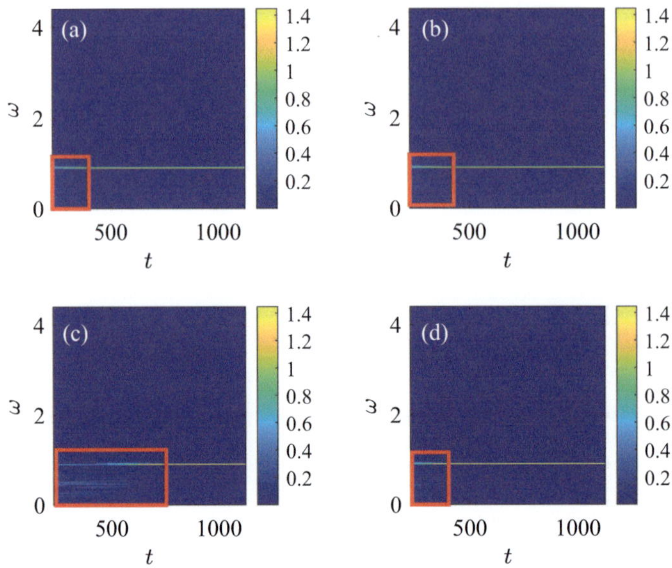

Fig. 10.20 The frequency components of the response change with time under different initial conditions. **a** $x_0 = -0.3$, **b** $x_0 = 0$, **c** $x_0 = 0.5$, **d** $x_0 = 1$. The simulation parameters are $\delta = 0.15$, $a = -0.6$, $b = 1$, $A = 0.245$, $\omega = 0.9$, $B = 5$, $\Omega = 9.8$, $m = 200$, $\dot{x}_0 = 0$

10.3.2 A Harmonic Excitation and a Square Waveform Excitation

When the slow varying excitation is a square waveform, as described in Eq. (B1), the long-term behavior of the system response is presented in Fig. 10.21. Only certain ultrasensitive vibrational resonance characteristics are shown in Fig. 10.21a. In Fig. 10.21b–e, Q varies less with the change of B. Whereas in Fig. 10.21f, the response amplitude Q decreases with the increase of B.

If the fast-varying signal is replaced by a square waveform, as the system in Eq. (10.3), in Fig. 10.22, the system response gradually transitions from ultrasensitive vibrational resonance to a steady response state in a long period, similar to the phenomenon reflected in Fig. 10.16.

10.3.3 A Harmonic Excitation and a Full Sine Waveform Excitation

The response amplitude Q of Eq. (10.4) versus the signal amplitude B for a long time is given in Fig. 10.23. In this figure, although it has been a long time, the ultrasensitive vibrational resonance still exists when the value of B is in an interval of small value.

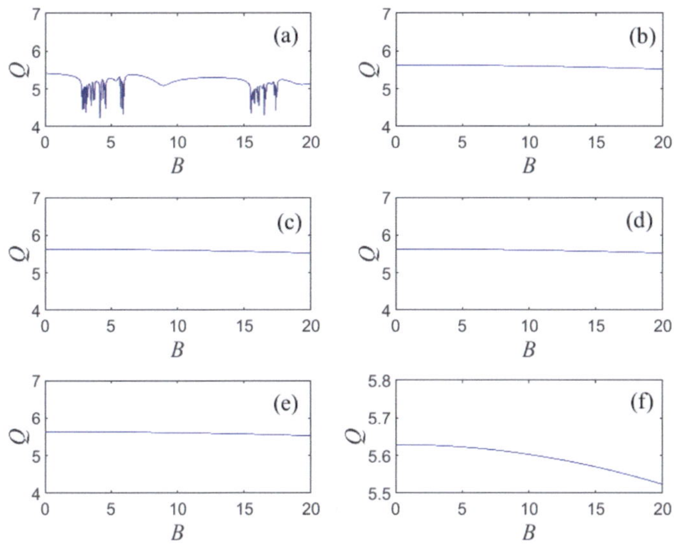

Fig. 10.21 The evolution of the response amplitude $Q(\omega) - B$ curve over a long time. The slow-varying signal is a square waveform excitation. The total calculation time starts from the very beginning of $(50m + 1)T$ to the end of $50jT$, $m = j - 1$ and $j = 1 \sim 6$ in turn from **a** to **f**, and T is the period of the excitation. The simulation parameters are $\delta = 0.15$, $a = -0.6$, $b = 1$, $A = 0.245$, $\omega = 0.9$, $\Omega = 9.8$, $m = 50$, the step $\Delta B = 0.01$. The initial conditions are $x_0 = 0$ and $\dot{x}_0 = 0$

When the value of B is large, the ultrasensitive vibrational resonance disappears, which is different from the results in Figs. 10.16, 10.21, and 10.22.

If the fast-varying signal is a full sine wave excitation, as given in Eq. (10.4), the evolution of the response amplitude as a function of the signal amplitude B is shown in Fig. 10.24. Ultrasensitive vibrational resonance also occurs within a narrow range of small B values. Despite the long time scale considered, the response amplitude in the ultrasensitive resonance region does not become sparse over time. In both Figs. 10.23 and 10.24, compared to the $Q(\omega) - B$ curves observed in other cases, the effect of full sine wave excitation on the $Q(\omega) - B$ curves is clearly evident.

10.3.4 A Harmonic Excitation and a Symmetric Triangular Waveform Excitation

In Fig. 10.25, corresponding to Eq. (10.8) for a long time, the ultrasensitive vibrational resonance is shown when the slow varying signal is a symmetric triangular waveform excitation. As time goes on, the ultrasensitive vibrational resonance persists and does not weaken. It indicates that the system response is in a chaotic state for a long

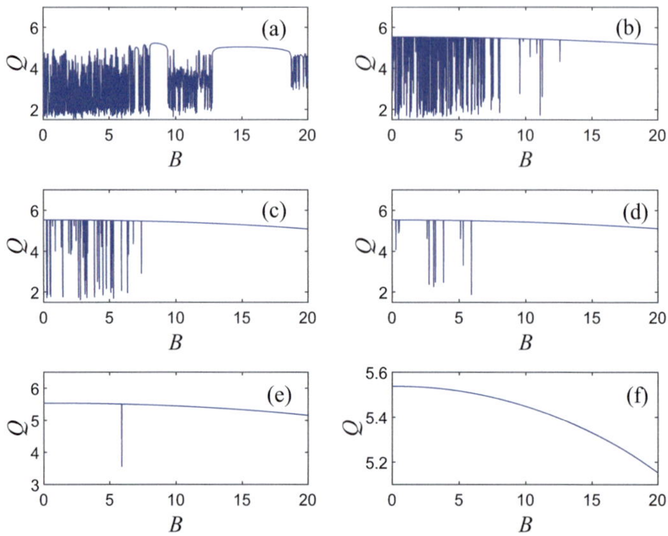

Fig. 10.22 The evolution of the response amplitude $Q(\omega) - B$ curve over a long time. The fast-varying signal is a square waveform excitation. The total calculation time starts from the very beginning of $(50m + 1)T$ to the end of $50jT$, $m = j - 1$ and $j = 1 \sim 6$ in turn from **a** to **f**, and T is the period of the excitation. The simulation parameters are $\delta = 0.15$, $a = -0.6$, $b = 1$, $A = 0.245$, $\omega = 0.9$, $\Omega = 9.8$, $m = 50$, the step $\Delta B = 0.01$. The initial conditions are $x_0 = 0$ and $\dot{x}_0 = 0$

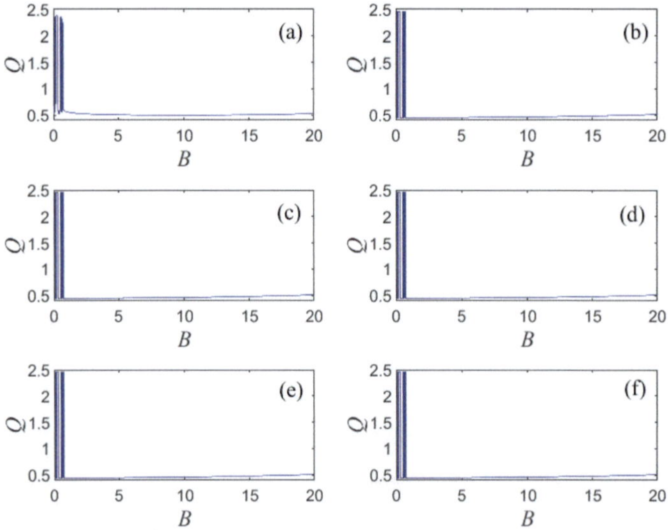

Fig. 10.23 The evolution of the response amplitude $Q(\omega) - B$ curve over a long time. The slow-varying signal is a full sine waveform excitation. The total calculation time starts from the very beginning of $(50m + 1)T$ to the end of $50jT$, $m = j - 1$ and $j = 1 \sim 6$ in turn from **a** to **f**, and T is the period of the excitation. The simulation parameters are $\delta = 0.15$, $a = -0.6$, $b = 1$, $A = 0.245$, $\omega = 0.9$, $\Omega = 9.8$, $m = 50$, the step $\Delta B = 0.01$. The initial conditions are $x_0 = 0$ and $\dot{x}_0 = 0$

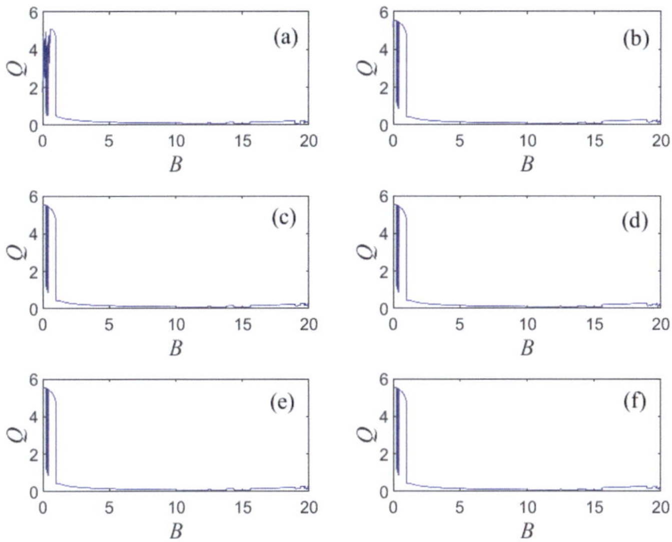

Fig. 10.24 The evolution of the response amplitude $Q(\omega) - B$ curve over a long time. The fast-varying signal is a full sine waveform excitation. The total calculation time starts from the very beginning of $(50m + 1)T$ to the end of $50jT$, $m = j - 1$ and $j = 1 \sim 6$ in turn from **a** to **f**, and T is the period of the excitation. The simulation parameters are $\delta = 0.15$, $a = -0.6$, $b = 1$, $A = 0.245$, $\omega = 0.9$, $\Omega = 9.8$, $m = 50$, the step $\Delta B = 0.01$. The initial conditions are $x_0 = 0$ and $\dot{x}_0 = 0$

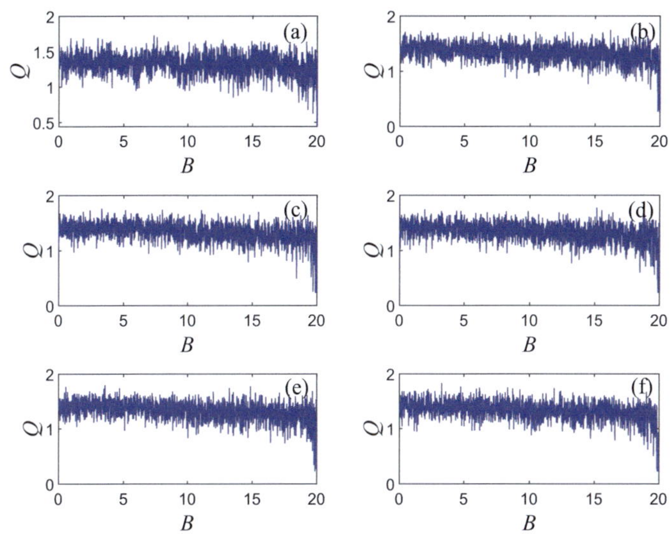

Fig. 10.25 The evolution of the response amplitude $Q(\omega) - B$ curve over a long time. The slow varying signal is a symmetric triangular waveform excitation. The total calculation time starts from the very beginning of $(50m + 1)T$ to the end of $50jT$, $m = j - 1$ and $j = 1 \sim 6$ in turn from **a** to **f**, and T is the period of the excitation. The simulation parameters are $\delta = 0.15$, $a = -0.6$, $b = 1$, $A = 0.245$, $\omega = 0.9$, $\Omega = 9.8$, $m = 50$, the step $\Delta B = 0.01$. The initial conditions are $x_0 = 0$ and $\dot{x}_0 = 0$

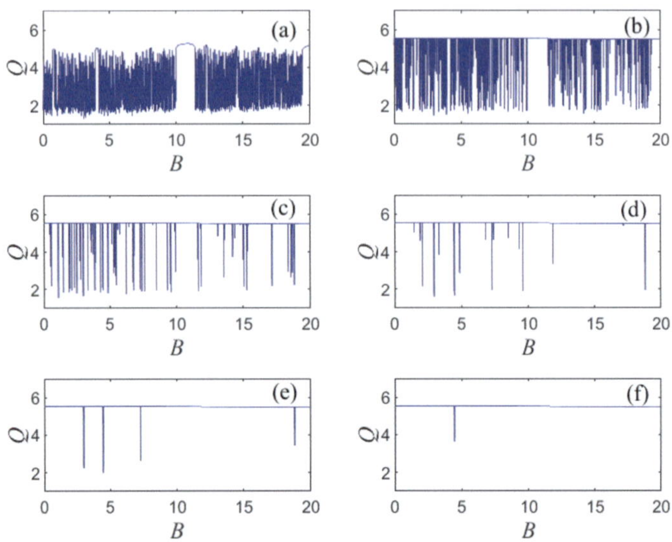

Fig. 10.26 The evolution of the response amplitude $Q(\omega) - B$ curve over a long time. The fast-varying signal is a symmetric triangular waveform excitation. The total calculation time starts from the very beginning of $(50m + 1)T$ to the end of $50jT$, $m = j - 1$ and $j = 1 \sim 6$ in turn from **a** to **f**, and T is the period of the excitation. The simulation parameters are $\delta = 0.15$, $a = -0.6$, $b = 1$, $A = 0.245$, $\omega = 0.9$, $\Omega = 9.8$, $m = 50$, the step $\Delta B = 0.01$. The initial conditions are $x_0 = 0$ and $\dot{x}_0 = 0$

time under the excitation of slow symmetric triangular waveform and fast harmonic waveform.

When the fast-varying signal is a symmetric triangular waveform excitation, as given in Eq. (10.9), the evolution of the $Q - B$ curve from ultrasensitive vibrational resonance to sparse ultrasensitive vibrational resonance clearly shown in Fig. 10.26. The results in this figure is very similar to that in Figs. 10.16 and 10.22.

10.4 Concluding Remarks

This chapter provides a detailed analysis of ultrasensitive vibrational resonance, examining the effects of external excitations in different forms and the influence of initial conditions on the phenomenon. Additionally, various combinations of periodic excitations significantly impact the corresponding response of ultrasensitive vibrational resonance. As time progresses, the response transitions from ultrasensitive vibrational resonance to sparse vibrational resonance and eventually to steady vibrational resonance over long durations, following a pattern similar to ultrasensitive vibrational resonance. In this chapter, we focus exclusively on the undamped Duffing system. However, many other nonlinear systems can also exhibit ultrasensitive vibrational resonance, making their study worthwhile. Furthermore, future research should explore potential engineering applications of ultrasensitive vibrational resonance.

References

1. V.M. Gandhimathi, K. Murali, S. Rajasekar, Stochastic resonance with different periodic forces in overdamped two coupled anharmonic oscillators, *Chaos, Solitons & Fractals*, 2006, **30**(5): 1034-1047.
2. V.M. Gandhimathi, S. Rajasekar, J. Kurths, Vibrational and stochastic resonances in two coupled overdamped anharmonic oscillators, *Physics Letters A*, 2006, **360**(2): 279-286.
3. S. Arathi, S. Rajasekar, Stochastic resonance in a single-well anharmonic oscillator with coexisting attractors, *Communications in Nonlinear Science and Numerical Simulation*, 2014, **19**(12): 4049-4056.
4. J. Zhang, J. Yang, Z. Zhu, G Shen, M.A.F. Sanjuán, Effects of different fast periodic excitations on the pitchfork bifurcation and vibrational resonance, *International Journal of Bifurcation and Chaos*, 2020, **30**(06): 2050092.
5. S. Huang, J. Yang, H. Liu, M.A.F. Sanjuán, Effect of static bifurcation on logical stochastic resonance in a symmetric bistable system, *International Journal of Bifurcation and Chaos*, 2021, **31**(16): 2150246.
6. S. Huang, J. Zhang, J. Yang, H. Liu, M.A.F. Sanjuán, Logical vibrational resonance in a symmetric bistable system: Numerical and experimental studies, *Communications in Nonlinear Science and Numerical Simulation*, 2023, **119**: 107123.
7. K. Srinivasan, K. Thamilmaran, A. Venkatesan, Effect of nonsinusoidal periodic forces in Duffing oscillator: Numerical and analog simulation studies, *Chaos, Solitons & Fractals*, 2009, **40**(1): 319-330.
8. F. Dohnal, Damping by parametric stiffness excitation: resonance and anti-resonance, *Journal of Vibration and Control*, 2008, **14**(5): 669-688.
9. R. Jothimurugan, K. Thamilmaran, S. Rajasekar, M.A.F. Sanjuán, Multiple resonance and anti-resonance in coupled Duffing oscillators, *Nonlinear Dynamics*, 2016, **83**: 1803-1814.
10. F. Dohnal, Damping by parametric stiffness excitation: resonance and anti-resonance, *Journal of Vibration and Control*, 2008, **14**(5): 669-688.
11. P.J. Gonçalves, A.D. Shaw, M.J. Brennan, B. Tang, G. Gatti, An investigation into the interaction between a nonlinear resonance and a fixed anti-resonance in a harmonically excited system, *International Journal of Non-Linear Mechanics*, 2023, **157**: 104531.
12. C. Grebogi, E. Ott, J.A. Yorke, Crises, sudden changes in chaotic attractors, and transient chaos, *Physica D*, 1983, **7**(1-3):181-200.
13. A.S. de Paula, M.A. Savi, F.H.I. Pereira-Pinto, Chaos and transient chaos in an experimental nonlinear pendulum, *Journal of Sound and Vibration*, 2006, **294**(3):585-595.
14. T. Tél, Y.C. Lai, Chaotic transients in spatially extended systems, *Physics Reports*, 2008, **460**(6): 245-275.
15. Y.C. Lai, T. Tél, *Transient Chaos: Complex Dynamics on Finite Time Scales* (Springer Science & Business Media 2011)
16. A. Daza, A. Wagemakers, M.A.F. Sanjuán, Strong sensitivity of the vibrational resonance induced by fractal structures, *International Journal of Bifurcation and Chaos*, 2013, **23**(7): 1350129.
17. S. Li, Z. Wang, J. Yang, M.A.F. Sanjuán, S. Huang, L. Lou, Ultrasensitive vibrational resonance induced by small disturbances, *Chaos*, 2023, **33**(12):123111.
18. J. Yang, S. Rajasekar, M.A.F. Sanjuán, Vibrational resonance: A review, *Physics Reports*, 2024, **1067**: 1-62.

Part II
Applications of Aperiodic Resonances

Chapter 11
Image Processing by Aperiodic Resonances

This chapter focuses on the application of aperiodic resonances in image processing. Images are often decomposed into aperiodic signals and then further processed. The resonance that occurs when such signals are input into a nonlinear system is known as aperiodic resonance. These non-traditional image processing methods include image denoising through aperiodic stochastic resonance, weak image feature extraction via logical stochastic resonance, and image contour perception using aperiodic vibrational resonance. These approaches offer new perspectives in the field of image processing.

11.1 Image Noise Reduction by Aperiodic Stochastic Resonance

Digital images are ubiquitous in both human society and industrial applications, serving as essential carriers of information. However, during imaging, compression, and transmission, these images often encounter various forms of noise [1–7]. When the noise is particularly strong, it can severely hinder the interpretation and analysis of image content. Therefore, effective image noise reduction is crucial for accurately extracting and understanding detailed information [8].

Due to the importance of image noise reduction, a variety of image denoising methods have always garnered significant attention [9, 10]. Numerous methods have been proposed to address this challenge. Among them, the traditional methods contain spatial domain filtering [11, 12], as well as transform domain methods such as Fourier transform [13, 14], discrete cosine transform [15], and wavelet transform [16]. Additionally, the block matching 3D (BM3D) method integrates concepts from both the spatial and transform domains [17, 18]. Compressed sensing leverages prior knowledge of sparsity to distinguish between images and noise [19, 20]. More recently, both data-driven and model-driven deep learning methods have been

J. Yang and M. A. F. Sanjuan, *Aperiodic Resonances—Theory and Applications*,
https://doi.org/10.1007/978-981-96-6819-9_11

widely employed for image noise reduction [21–23]. Despite their effectiveness, these approaches primarily focus on noise elimination rather than harnessing the constructive role of noise. However, excessive noise removal often leads to the loss of important image details [24–26], and their performance tends to deteriorate as noise intensity increases, especially when the characteristic information is weak and the noise is strong.

Stochastic resonance is a phenomenon in which a weak signal can be enhanced by leveraging the beneficial role of noise. With the expanding applications of stochastic resonance, researchers have begun exploring its potential in image processing [27–40]. Given the non-periodicity of signals in image transformations, it is more appropriate to refer to this phenomenon as aperiodic stochastic resonance. While aperiodic stochastic resonance-based methods can partially mitigate image noise, there remains a need for more effective and efficient denoising techniques, particularly for images with substantial noise. Recent studies suggest that aperiodic stochastic resonance in a periodic potential system (APSR) offers significant advantages over aperiodic stochastic resonance in a bistable system (ABSR) for signal denoising [41–43]. Moreover, APSR demonstrates particular benefits in image noise reduction [32].

An overdamped nonlinear system with periodic potential is described by the following Langevin equation

$$\frac{dx}{dt} = -\frac{\partial V}{\partial x} + s(t) + N(t). \tag{11.1}$$

In Eq. (11.1), $s(t)$ and $N(t)$ are the image signal and the noise, respectively. Here, the periodic potential function is given by

$$V(x) = a\cos(bx), \quad a > 0, \quad b > 0. \tag{11.2}$$

The configuration of the periodic potential function is shown in Fig. A.8 of the Appendix A.4.

To apply aperiodic stochastic resonance (ASR) for image denoising, the image must first be transformed into a one-dimensional signal. Since the nonlinear system requires bipolar input signals, the reduced-dimensional signal undergoes bipolarization to generate an aperiodic square waveform signal. This processed signal is then used as the input $s(t)$ for computation. The effectiveness of APSR in image denoising is validated through the system's output.

Here, the aperiodic square waveform signal, denoted as $L(t)$, is generated based on a pseudo-random sequence. Gaussian white noise with a $L(t)$ to obtain $s(t)$), which simulates the noisy image signal. Gaussian white noise is one of the most common types of image noise, making it a suitable choice for the image background noise in this study.

Based on Eq. (11.1), optimal aperiodic stochastic resonance is achieved when the random noise, input signal, and system parameters are optimally matched. However, obtaining this optimal resonance solely by adjusting the noise intensity is challenging,

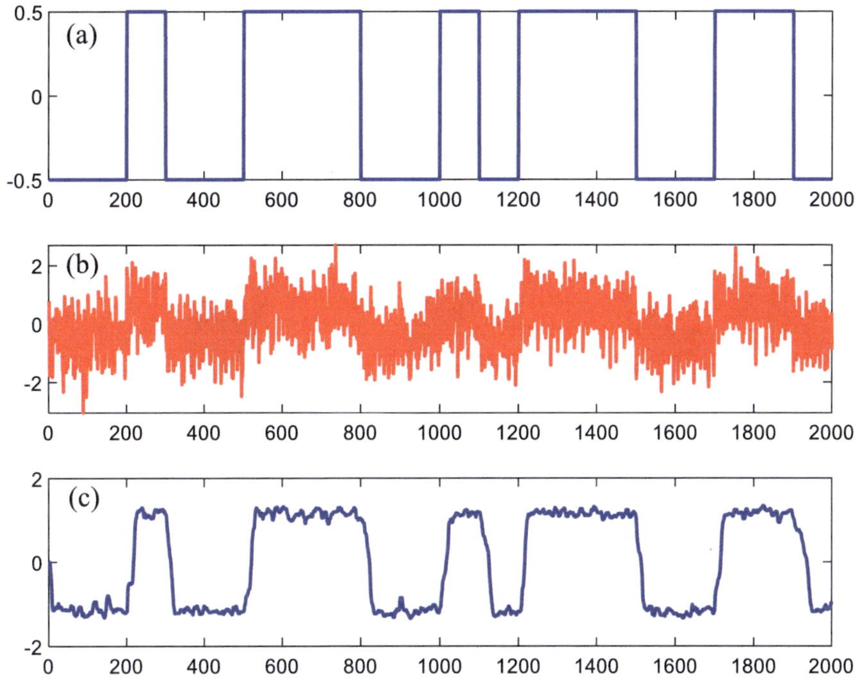

Fig. 11.1 Aperiodic stochastic resonance realizes noise reduction. **a** The pure signal $L(t)$, **b** the input noisy signal $s(t)$, **c** the recovered signal $x(t)$. The system parameters are $a = 2$, $b = 1$

as the optimal noise level cannot always be predicted in real-world applications. Therefore, for a given noisy input signal, we achieve optimal aperiodic stochastic resonance by fine-tuning the system parameters. Unless otherwise specified, we consider the noisy signal $s(t)$ as the input image signal and do not introduce additional random noise.

Input the signal $s(t)$ into the periodic potential system described in Eq. (11.1) to evaluate its denoising performance. As shown in Fig. 11.1, the pure signal $L(t)$ is successfully recovered from the noisy signal $s(t)$, with its amplitude also amplified. This demonstrates that aperiodic stochastic resonance in a nonlinear system with a periodic potential can effectively recover an aperiodic square waveform signal from a noisy background. Essentially, this phenomenon results from aperiodic stochastic resonance induced by an aperiodic binary signal, a topic previously discussed in Chap. 4. Therefore, we do not revisit the theoretical details here.

Fig. 11.2 presents the results of denoising Cameraman noise images using the ABSR and APSR methods, respectively. The denoised images are affected by Gaussian white noise, with variations in parameters such as the statistical mean μ and variance σ. From this figure, it is evident that the APSR method outperforms the ABSR method, offering superior visual quality and achieving a higher peak signal-to-noise ratio (PSNR) [32].

Fig. 11.2 Noise reduction effects by using the ABSR method or the APSR method. The first column on the right shows the noisy images, and the noise parameters (μ, σ) are (0, 0.1), (0.5, 0.1), (1, 0.1), (0, 1), (0.5, 1) and (1, 1) from top to bottom. The first column on the left and the middle column display the denoised images obtained using the APSR and ABSR methods, respectively. From the top row to the bottom row, the PSNR values of the noisy images in the first column on the right are 9.8, −11.1, −16.7, −6.97, −11.5, and −12.6. The PSNR values of the denoised images in the middle column (ABSR method) are −3.9, 3.7, −1.7, −6.5, 1.3, and −5.9, while the PSNR values of the denoised images in the first column on the left (APSR method) are −1.1, 7.5, 1.3, −3.1, 3.2, and −0.8

PSNR is usually applied to quantify image equality due to its good performance in comparing each pixel. Assuming a two-dimensional monochromatic image with size m (row)$\times$ n (column), $S(i, j)$ is the original reference image and $X(i, j)$ is the image to be evaluated. Then, the mean square error (MSE) between the two images is

$$MSE = \frac{1}{m \cdot n} \sum_{i=1}^{m} \sum_{j=1}^{n} [S(i, j) - X(i, j)]^2. \tag{11.3}$$

Then, PSNR is defined by the formula

$$PSNR = 10\log_{10}\left(\frac{MAX_s^2}{MSE}\right), \tag{11.4}$$

where MAX_s is the maximum gray level of the images. If the images are pixel normalized, the gray scale range of images lies in $[0 - 1]$, and $MAX_s = 1$. Therefore, the PSNR of the pixel normalization image is calculated by

$$PSNR = 10\log_{10} \frac{m \cdot n}{\sum_{i=1}^{m} \sum_{j=1}^{n} [S(i, j) - X(i, j)]^2}. \tag{11.5}$$

The values of PSNR in Fig. 11.2 are obtained according to Eq. (11.5).

To further compare the ABSR and APSR methods under different noise intensities, Fig. 11.3 illustrates the noise reduction performance for several images using both methods. Different variance values of the Gaussian white noise are used to contaminate the images. Once again, the results indicate that the APSR method consistently provides superior visual feedback across all noise levels compared to ABSR.

As shown in Fig. 11.4, we compare the noise reduction performance of the APSR method with four commonly used image denoising methods: the arithmetic mean filter, geometric mean filter, median filter, and Wiener filter. These four conventional methods can be directly implemented using MATLAB functions: the arithmetic mean filter with "imfilter", the geometric mean filter with "exp(imfilter(log(I)))", the median filter with "medfilt2", and the Wiener filter with "wiener2". Compared to these traditional approaches, the APSR method once again demonstrates a clear advantage.

To further analyze the impact of noise and provide both an intuitive and objective comparison, Fig. 11.5 illustrates the dependence of the PSNR value on noise intensity across various image denoising methods. The comparison results are based on six different methods applied to the same noisy image, which is contaminated by Gaussian white noise. To ensure a meaningful evaluation of noise reduction performance, the PSNR value of the original noisy image before processing is also included. According to the curves, the APSR method consistently demonstrates superior noise reduction performance across different noise variance levels.

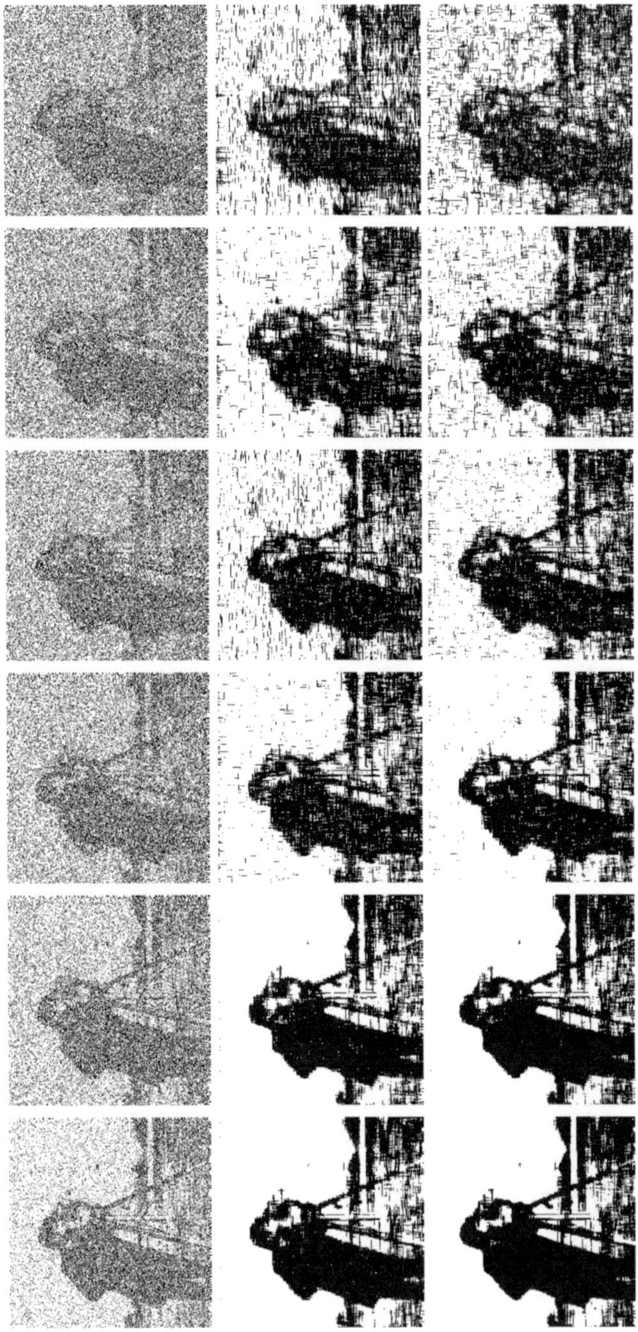

Fig. 11.3 Noise reduction effects by using the ABSR method or the APSR method. The first column on the right shows the noisy images, and the noise parameters (μ, σ) are (0.5, 0.5), (0.5, 1), (0.5, 1.5), (0.5, 2), (0.5, 2.5) and (1, 3) from top to bottom. The first column on the left and the middle column display the denoised images obtained using the APSR and ABSR methods, respectively. From the top row to the bottom row, the PSNR values of the noisy images in the first column on the right are −10.1, −11.5, −13.6, −14.6, −15.2, and −15.6. The PSNR values of the denoised images in the middle column (ABSR method) are 3.9, 1.4, 1.1, −1.6, −4.3, and −4.9, while the PSNR values of the denoised images in the first column on the left (APSR method) are 5.3, 3.3, 2.3, −0.7, −3.2, and −4.3

Fig. 11.4 Noise reduction results obtained using the APSR method and four commonly used denoising methods are presented. From left to right, the first row displays the noisy image with GWN($\mu = 0.5$, $\sigma = 1$), followed by images processed using geometric mean filtering and arithmetic mean filtering, respectively. The second row presents images processed using median filtering, Wiener filtering, and the APSR method, respectively

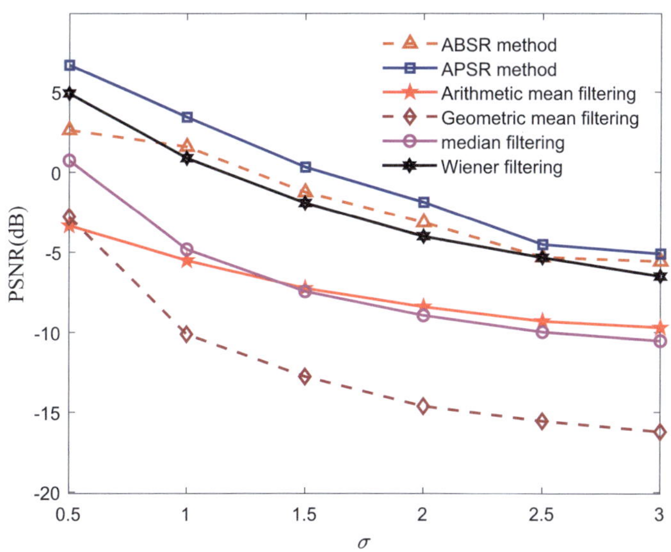

Fig. 11.5 The dependence of PSNR on the noise variance σ under different image denoised methods

To display the noise reduction performance of the APSR method, Fig. 11.6 gives the results obtained when the image is contaminated by different types of noise. Here, we choose the gamma noise, uniform noise, Rayleigh noise, and exponential noise to simulate the noise background of the image. Besides the Gaussian white noise discussed just above, Fig. 11.6 shows the effectiveness of the APSR method in image noise reduction.

In summary, the above examples show the effectiveness of APSR method in image denoising, demonstrating stable performance across both weak and strong noise conditions. Compared to the traditional bistable system, the periodic potential system offers superior processing property. It not only delivers an efficient and high-performance approach for handling heavily noisy images but also offers new insights into image noise dynamics. Additionally, the findings may inspire novel strategies for other image processing applications. Furthermore, for different application scenarios, there are many other types of image noise, such as additive Gaussian noise, Poisson noise, salt and pepper noise, speckle noise, and stripe noise [43–46]. The application of aperiodic stochastic resonance for image denoising under various types of noise remains an important area for further research.

11.2 Image Feature Extraction by Logical Stochastic Resonance

Logical stochastic resonance offers a novel approach to image feature extraction [47, 48]. The LSR theory has been discussed in detail in Chap. 8; here, we focus on its application in image feature extraction. By leveraging noise to enhance signal features, LSR exhibits strong performance in noisy environments and effectively extracts image features even under low SNR conditions [49]. It has shown promise in applications such as underwater target recognition [47] and radar systems [50], improving image resolution, contrast, and detection capabilities while reducing false alarms [51]. However, traditional nonlinear systems are incapable of performing the **XOR** logic operation, which limits the further development and application of logical stochastic resonance.

The LSR phenomenon, driven by random noise, significantly enhances signal detection and processing across diverse fields, including signal processing, image processing, sensor networks, and biomedical applications. However, traditional bistable or multistable potential functions are unable to perform the **XOR** logic operation. This limitation poses a significant obstacle to the practical application of logical stochastic resonance in scenarios that necessitate **XOR** logic functionality.

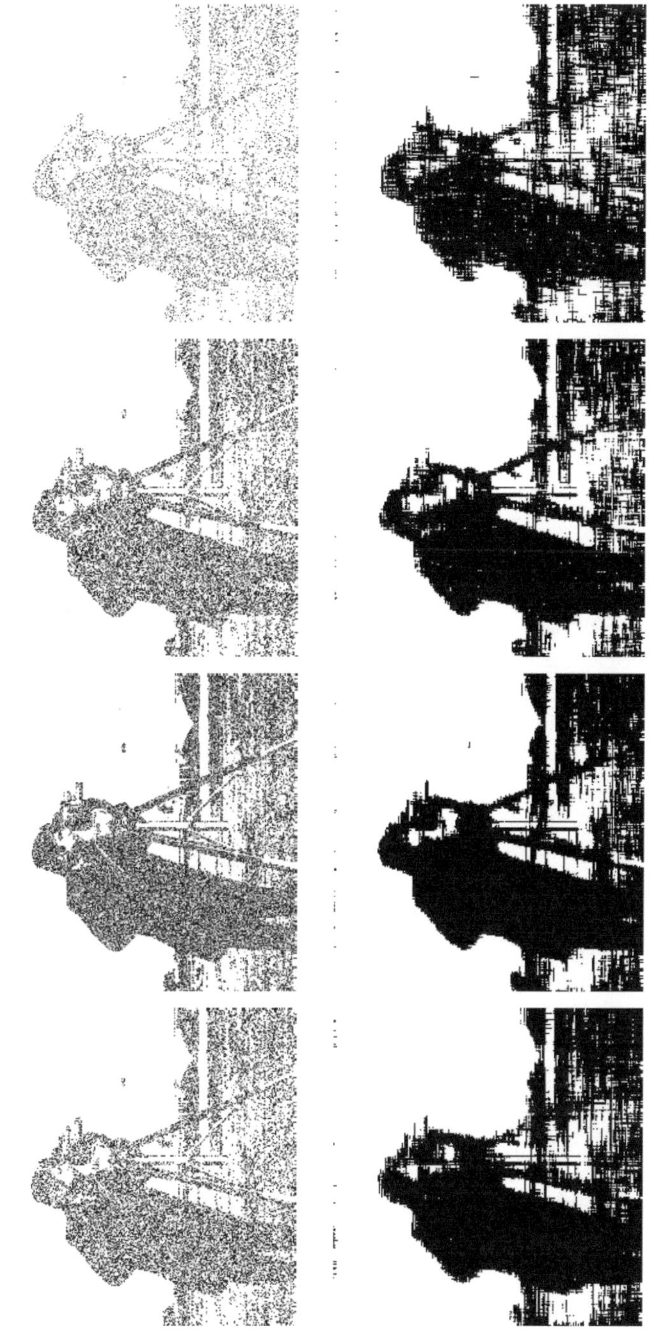

Fig. 11.6 Denoised results of the noisy image contaminated by different types of noise including gamma noise, uniform noise, Rayleigh noise and exponential noise from top to bottom. The right column shows noisy images, while the left column displays the corresponding denoised images by APSR method

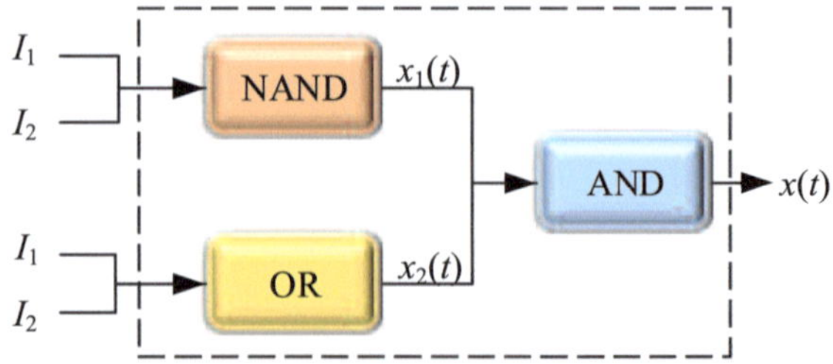

Fig. 11.7 Nonlinear array element producing logical **XOR** stochastic resonance

To achieve stochastic resonance in **XOR** logic operation, nonlinear array elements are constructed using components of a bistable system, i.e.,

$$\begin{cases} \dot{x}_1 = a_1 x_1 - b_1 x_1^3 + r_1 + I_1 + I_2 + \xi(t) \\ \dot{x}_2 = a_2 x_2 - b_2 x_2^3 + r_2 + I_1 + I_2 + \xi(t) \\ \dot{x} = ax - bx^3 + r + x_1 + x_2 + \xi(t) \end{cases} \tag{11.6}$$

In Fig. 11.7, a nonlinear array unit that implements **XOR** logic operations is constructed based on three bistable nonlinear Langevin equations. Perform an **AND** logic operation on the **NAND** logic and **OR** logic of the two inputs to obtain the **XOR** logic of the original input.

The logical truth table corresponding to Fig. 11.7 is presented in Table 1.1. To achieve **XOR** logic, the following steps are carried out:

The **NAND** logic result x_1 and the **OR** logic result x_2 from the two logic inputs I_1 and I_2 are used as new logic inputs. These results are coupled with the nonlinear system to obtain the **AND** logic result x. Therefore, x represents the logical **XOR** of I_1 and I_2.

This procedure overcomes the limitation of bistable and multistable potential functions in realizing **XOR** logic operation, successfully enabling **XOR** operations through the construction of nonlinear array elements. This advancement addresses a broad range of applications. To validate the capability of the nonlinear array unit in generating logical **XOR** operation, the results of each logical modulation are illustrated via the input and output of the nonlinear system. Figure 11.8 displays the intermediate process results for producing logical **XOR** stochastic resonance.

The test image for radiographic detection of printed circuit board welding defects has a resolution of 512×512 and comes from industrial sites. Figure 11.9 illustrates the process of automatically extracting abnormal features from X-ray inspection images using **XOR** logical stochastic resonance with nonlinear array elements.

Initially, $Image_0$ is used as the reference input, which is flattened into a one-dimensional array of 1×512^2. The input I_1 for the nonlinear array element is obtained by applying polarization and variable scaling to $Image_0$. Similarly, input I_2 is processed through the same steps based on nonlinear array elements. Stochastic resonance is achieved by fine-tuning parameters to align the nonlinear system, external inputs, and noise, thereby generating the desired logical output. Finally, the

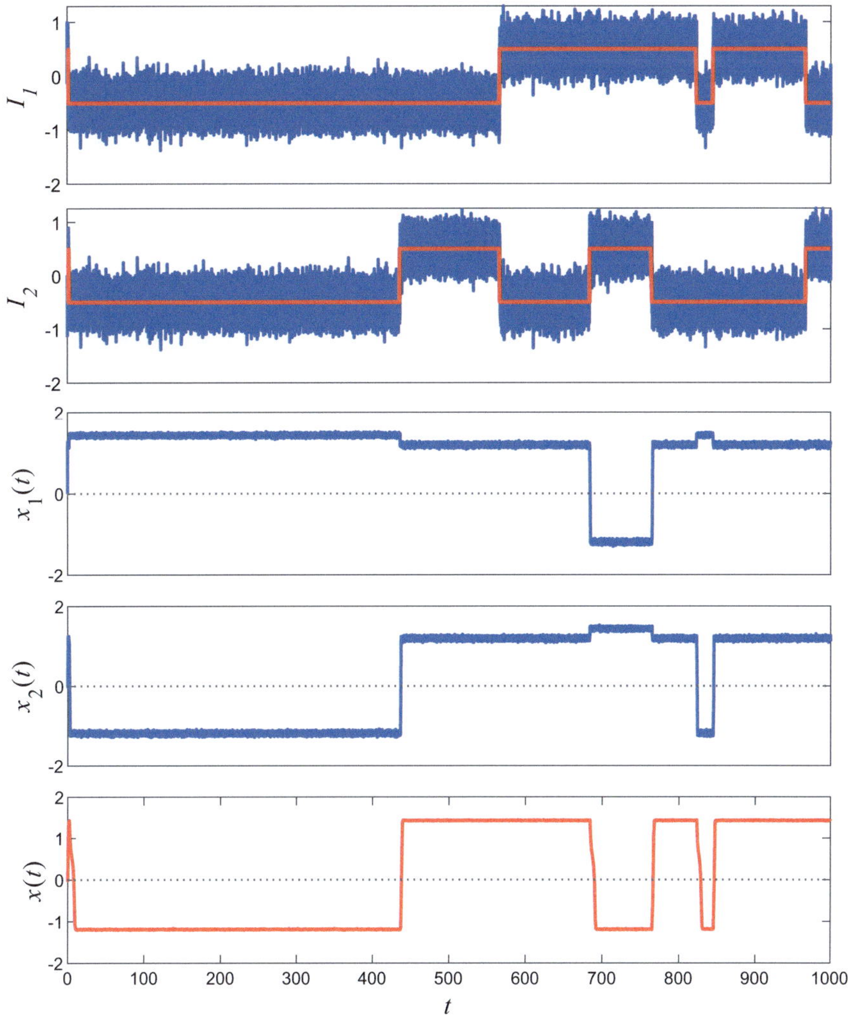

Fig. 11.8 Process of obtaining **XOR** logic using the nonlinear array element. I_1 and I_2 are input signals. $x_1(t)$ is the **NAND** logic operation result. $x_2(t)$ is the **OR** logic operation result, and $x(t)$ is the **XOR** logic operation result obtained by **AND** operations on **OR** and **NAND** logic. The system parameters are $a_1 = 1.1$, $b_1 = 1.1$, $r_1 = 0.4$, $a_2 = 1.1$, $b_2 = 1.1$, $r_2 = -0.4$, $a = 1.1$, $b = 1.1$, $r = 0.4$. I_1 and I_2 are polarized digital signals with random noise intensity $D = 0.01$

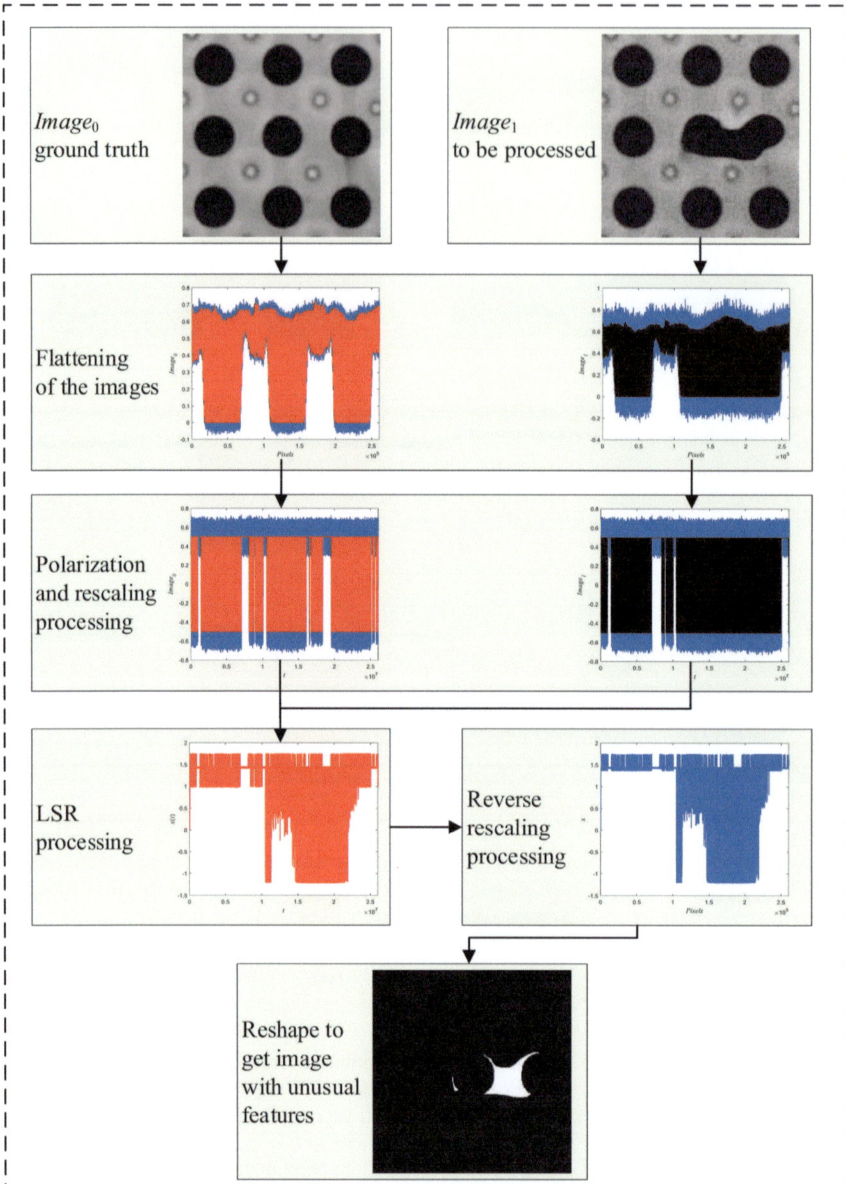

Fig. 11.9 Schematic of the **XOR** logical stochastic resonance method for anomaly extraction in PCB X-ray inspection images using the nonlinear array element

output from the nonlinear array element is transformed back and reshaped to produce a two-dimensional result for abnormal feature extraction.

The printed circuit board frequently house a range of components, from a few to dozens, in various sizes and packages, which can lead to numerous soldering defects. Effective inspection must ensure that no defects are overlooked while maintaining high efficiency, relying significantly on engineering expertise and well-structured procedures. It is essential to analyze the characteristics of different defect types and tailor each inspection step accordingly. Each stage of the printed circuit board soldering quality X-ray inspection process serves distinct objectives and priorities, addressing specific types of defects to achieve comprehensive and accurate evaluations.

Fig. 11.10 illustrates defects such as short circuits, solder slag, oversized solder balls, and missing solder balls, which can lead to product failure. It compares the abnormal feature extraction performance of the **XOR** logical stochastic resonance method with that of the traditional differential morphology method. Traditional methods struggle to clearly delineate abnormal features due to blurred pixel grayscale distributions, resulting in insufficient grayscale differentiation and unclear feature boundaries, especially under noisy conditions. In contrast, the **XOR** logical stochastic resonance method enhances pixel-scale details, effectively reducing noise interference and improving detection accuracy. This method excels in processing image details and extracting features, producing grayscale images with higher contrast and clearer boundaries. Such capabilities are essential for practical applications requiring precise and reliable feature extraction.

To enhance the observation of radiation detection image features, Fig. 11.11 presents a three-dimensional visualization of the two-dimensional image from Fig. 11.10. This three-dimensional subplot reveals the significant effect of noise on the key features, clearly demonstrating the saliency of features and the effectiveness of noise suppression through height variation. In contrast, differential morphology methods show relatively smooth height changes in the three-dimensional images, making it challenging to distinguish between abnormal features and noise. The **XOR** logical stochastic resonance method, however, displays pronounced height variations, effectively highlighting abnormal features and intuitively suppressing noise. This further confirms the superior performance of **XOR** logical stochastic resonance in feature extraction.

The results indicate that **XOR** logical stochastic resonance achieves high detection accuracy for defects like short circuits, welding slag, oversized solder balls, and missing solder balls. Comparative analysis reveals that **XOR** logical stochastic resonance significantly outperforms traditional methods, such as differential morphology. This technique offers a reliable solution for precise feature extraction and automatic batch inspection of printed circuit board welding defects, addressing the increasing demands for quality and efficiency in large-scale electronic device production. In summary, **XOR** logical stochastic resonance introduces a novel approach to image feature extraction in challenging noise environments.

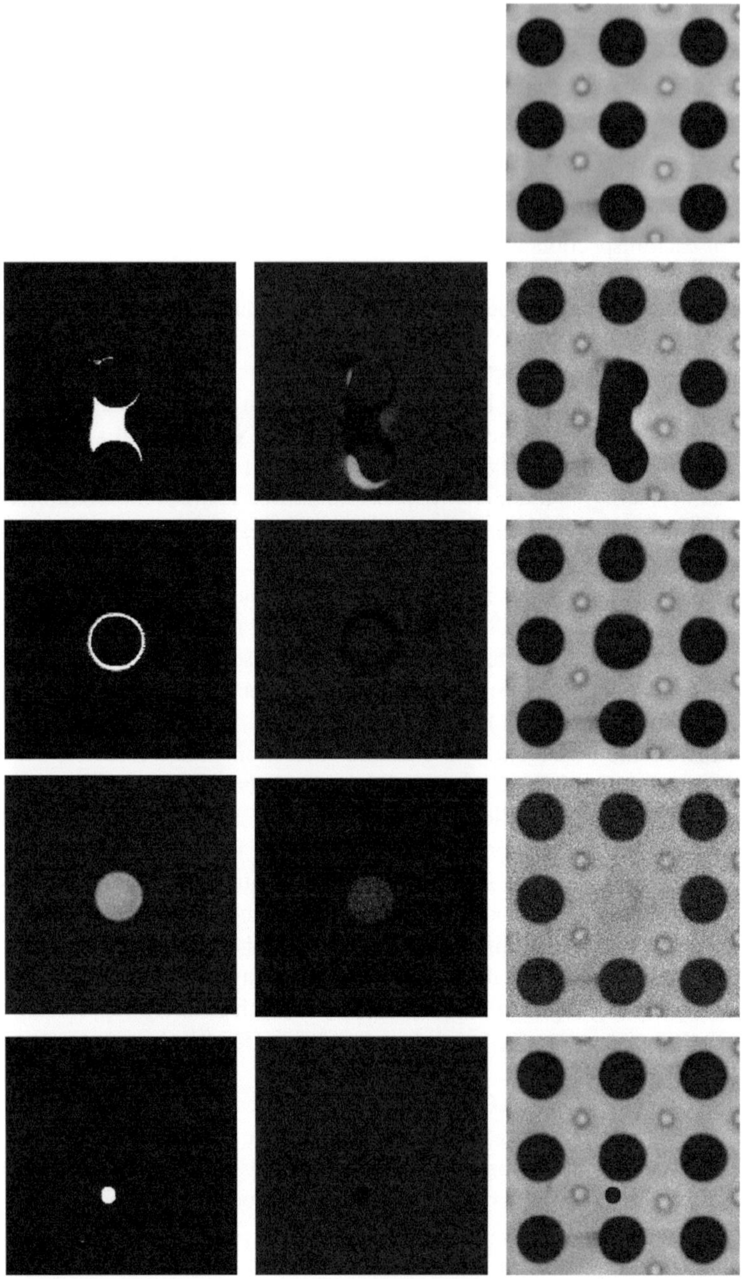

Fig. 11.10 Image feature extraction comparison between traditional differential morphology and **XOR** logical stochastic resonance methods. From top to bottom, the images in the right column are the ground truth $Image_0$, $Image_1$ with solder ball short circuit defects, $Image_2$ with oversized solder ball defects, $Image_3$ with missing solder ball defects, $Image_4$ with solder slag/beads defects. The images in the middle column are detected by traditional differential morphology. The image in the left column are processed by the **XOR**logical stochastic resonance method

Fig. 11.11 Three-dimensional feature extraction results by traditional differential morphology and **XOR** logical stochastic resonance respectively. Images in the right column are the noisy images. Images in the middle column are defected by traditional differential morphology. Images in the left column are the results obtained by the **XOR** logical stochastic resonance method. The defect types are corresponds to that in Fig. 11.10

11.3 Image Contour Perception by Aperiodic Vibrational Resonance

Aperiodic vibrational resonance was first introduced to image processing by Morfu et al. [52, 53]. They also discussed the different effects and interactions of aperiodic vibrational resonance and aperiodic stochastic resonance in image processing. In our former review paper [54], we explicitly stated that the main role of aperiodic vibrational resonance in image processing is for extracting image contour.

We process the image signal by the following procedure. At first, er determine the size of the image (m, n). The noisy image I_N is expressed by the pure image I and the noise N, i.e.,

$$I_{N(i,j)} = I_{(i,j)} + N_{(i,j)}, i = 1, \cdots, m, \ j = 1, \cdots, n, \tag{11.7}$$

where $I_{N(i,j)}, I_{(i,j)}, \sigma_{(i,j)}$ are the grey levels of the noisy image, pure image, and noise of the pixels at coordinates $(i, \ j)$, respectively. The noise $N_{(i,j)} = \sigma\eta_{(i,j)}$, where σ is the noise intensity and $\eta_{(i,j)}$ is the standard Gaussian white noise.

We design the following aperiodic vibrational resonance resonator for the image contour perturbation

$$X_{i,j} = I_{N(i,j)} + B\cos\left(\frac{2\pi ij\Omega \times 0.5(m+n)}{m \times n} + \varphi_{i,j}\right), \quad i = 1 \cdots m, \quad j = 1 \cdots n. \tag{11.8}$$

In Eq. (11.8), $X_{i,j}$ is the output image that processed by the aperiodic vibrational resonance resonator, $B\cos\left(\frac{2\pi ij\Omega \times 0.5(m+n)}{m \times n} + \varphi_{i,j}\right)$ is a cosine function with the initial phase fluctuation. The phase random perturbation $\varphi_{i,j} = \sqrt{2D}\xi(t)$, and $\xi(t)$ is a standard Gaussian white noise. The resonator output $X_{i,j}$ is compared with a threshold value T_h. If $X_{i,j} > T_h$, then $X_{i,j} = 1$, this case indicating the white level. If $X_{i,j} \leq T_h$, then $X_{i,j} = 0$, it representing the black level. Apparently, the resonator output X closely depends on the threshold value T_h. The flow chart of image contour perception by aperiodic vibrational resonance can also be described in Fig. 11.12.

We use the cross-correlation coefficient C_{XI} between the resonator output X and the original pure image to quantify the image contour perception efficiency. We take a weld flaw detection image as an example and discuss the issue in detail in Fig. 11.13. The input weld flaw detection image is shown in Fig. 11.14a. We discuss Fig. 11.13 first. To study the aperiodic vibrational resonance induced by the input image, we first choose different values for the threshold T_h, as shown in Fig. 11.13a. For some cases, such as $T_h = 0.5$, 0.7 and 0.8, the cross-correlation coefficient C_{XI} almost decreases with the signal amplitude B. For other cases, such as $T_h = 0.3$, 1.0, 1.2 and 1.5, the cross-correlation coefficient C_{XI} is a non-monotonic function of the signal amplitude B. There is an appropriate value of B that makes the cross-correlation coefficient C_{XI} achieve the peak. In other words, the aperiodic vibrational resonance is obvious for some cases in Fig. 11.13a. If we fix the amplitude B and make the threshold T_h as a controllable parameter, the change in threshold can also

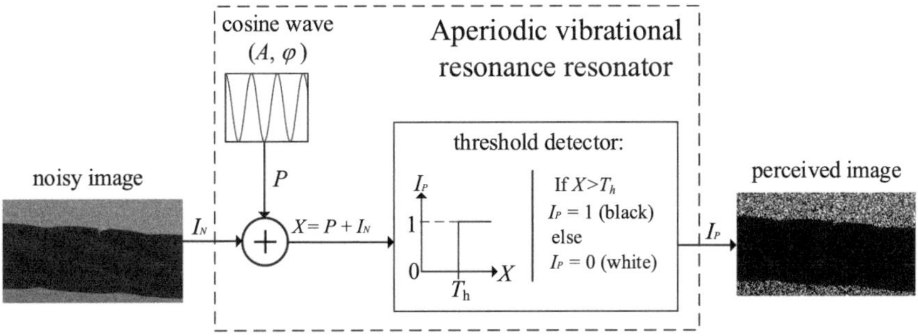

Fig. 11.12 The flow chart of image contour perception by aperiodic vibrational resonance

cause resonance, as presented in Fig. 11.13b. It is particularly evident for the cases of $B = 0$ and $B = 0.3$. There is also an undeniable fact in Fig. 11.13 that when $B = 0$, if an appropriate threshold is chosen, the cross-correlation coefficient C_{XI} may reach its maximum value. This is due to the occurrence of dither phenomenon [55, 56].

The threshold T_h influences the image darker or whiter. To display the image contour perception effect, we provide Fig. 11.14 to visually demonstrate the contour extraction effect of image. The effect of contour extraction varies for different thresholds and B values. On the one hand, it affects the size of the image contour, and on the other hand, it also affects the distribution of black and white points inside the contour. In order to display the image contour clearly, both dithering and aperiodic vibrational resonance have played a positive role.

11.4 Concluding Remarks

In this chapter, we present image processing methods based on aperiodic stochastic resonance, logical stochastic resonance, and aperiodic vibrational resonance. Aperiodic stochastic resonance is primarily used for image denoising, implemented through a periodic potential nonlinear system. When the system output reaches its "'strongest" aperiodic stochastic resonance effect, image noise is maximally suppressed. To further validate this approach, we consider various types of noise in different scenarios. For logical stochastic resonance, specifically **XOR** logical stochastic resonance, we apply this **XOR** logical stochastic resonance can successfully extract much weaker image features. For aperiodic vibrational resonance, we utilize it for image contour perception. By adjusting thresholds and signal amplitudes to induce aperiodic vibrational resonance or dithering, the system output can be optimized to enhance image contour perception at varying levels.

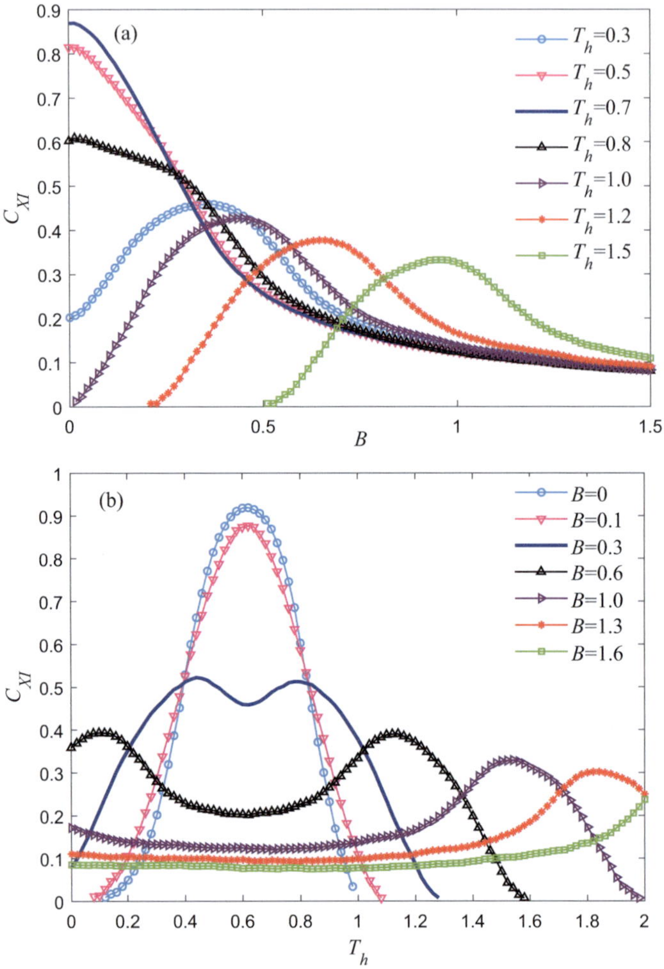

Fig. 11.13 Effects of the signal amplitude B and the threshold value T_h on the cross-correlation coefficient C_{XI} between the output image and the original input image. **a** Dependence of the cross-correlation coefficient C_{XI} on the signal amplitude B under different values of the threshold value T_h. **b** Dependence of the cross-correlation coefficient C_{XI} on the threshold value T_h under different values of the signal amplitude B. The parameters are= 0.1, $D = 10$

Although this chapter explores various applications of aperiodic resonances in image processing, it primarily focuses on monochrome images. Monochrome images can be easily converted into aperiodic binary signals, making them well-suited for the discussed methods. However, extending these approaches to more complex grayscale and color images presents a significant challenge. Developing effective techniques for processing such images based on aperiodic resonances remains a promising area for further research.

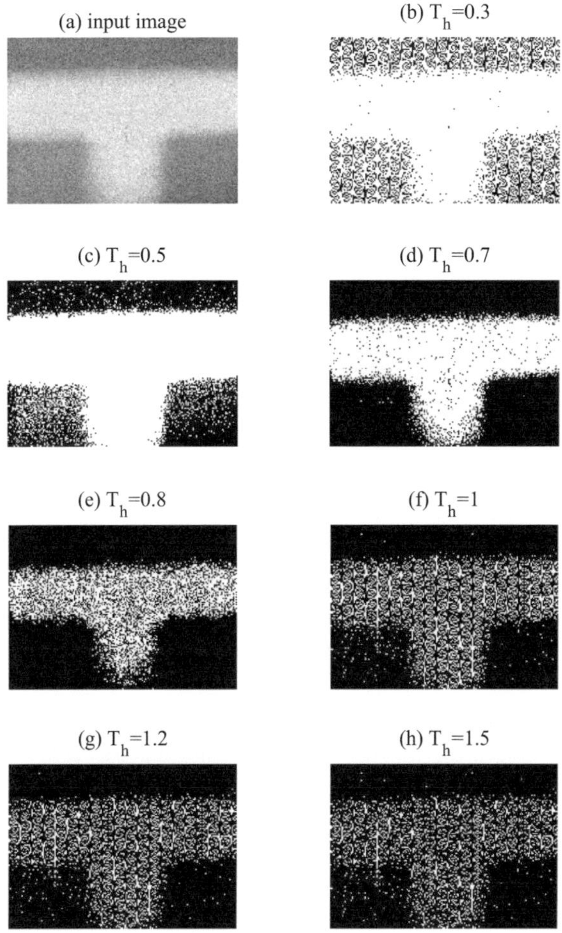

Fig. 11.14 The image contour perception of the noisy image based on the aperiodic vibrational resonance under different threshold values. **a** The input noisy image, **b–h** The contour perception effect of the weld seam image. The parameters are $= 0.1$, $D = 10$, and $B = 0.375$, 0, 0, 0.015, 0.465, 0.66, 0.975 from **b** to **h** in turn

References

1. M. Diwakar, M. Kumar, A review on CT image noise and its denoising, *Biomedical Signal Processing and Control*, 2018, **42**: 73-88.
2. M.A. Habibzadeh, M.R. Ay, A.K. Asl, H. Ghadiri, H. Zaidi, Impact of miscentering on patient dose and image noise in x-ray CT imaging: phantom and clinical studies, *Physica Medica*, 2012, **28**(3): 191-199.
3. B. Li, S. Wang, S. Wang, Defending against noise by characterizing the rate-distortion functions in end-to-end noisy image compression, In *2021 IEEE International Conference on Image Processing (ICIP)* IEEE, 2021, September, pp. 3727-3731.
4. Q. Wu, J. Zhang, W. Ren, W. Zuo, X. Cao, Accurate transmission estimation for removing haze and noise from a single image, *IEEE Transactions on Image Processing*, 2019, **29**: 2583-2597.
5. Y. Himeur, A. Boukabou, Robust image transmission over powerline channel with impulse noise, *Multimedia Tools and Applications*, 2017, **76**: 2813-2835.

6. B. Goyal, A. Dogra, S. Agrawal, B.S. Sohi, A. Sharma, Image denoising review: From classical to state-of-the-art approaches, *Information Fusion*, 2020, **55**: 220-244.
7. Y. Quan, Y. Chen, Y. Shao, H. Teng, Y. Xu, H. Ji, Image denoising using complex-valued deep CNN, *Pattern Recognition*, 2021, **111**: 107639.
8. J. Xiao, R. Zhao, K.M. Lam, Bayesian sparse hierarchical model for image denoising, *Signal Processing: Image Communication*, 2021, **96**: 116299.
9. B. Ghanbari, A. Atangana, A new application of fractional AtanganaCBaleanu derivatives: designing ABC-fractional masks in image processing, *Physica A: Statistical Mechanics and its Applications*, 2020, **542**: 123516.
10. C.S. Asha, M. Singh, S. Suresh, S. Lal, Optimized dynamic stochastic resonance framework for enhancement of structural details of satellite images, *Remote Sensing Applications: Society and Environment*, 2020, **20**: 100415.
11. H. Li, C. Y. Suen, A novel non-local means image denoising method based on grey theory, *Pattern Recognition*, 2016, **49**: 237-248.
12. P. Jain, V. Tyagi, Spatial and frequency domain filters for restoration of noisy images, *IETE Journal of Education*, 2013, **54**(2): 108-116.
13. M.F. Wahab, F. Gritti, T.C. O'Haver, Discrete Fourier transform techniques for noise reduction and digital enhancement of analytical signals, *TrAC Trends in Analytical Chemistry*, 2021, **143**: 116354.
14. A.M. John, K. Khanna, R.R. Prasad, L.G. Pillai, A review on application of fourier transform in image restoration, In *2020 Fourth International Conference on I-SMAC (IoT in Social, Mobile, Analytics and Cloud)(I-SMAC), IEEE* 2020, October, pp. 389-397.
15. H. Joo, J. Park, J. Kim, B.U. Lee, Image noise reduction in Discrete Cosine Transform domain, *IEIE Transactions on Smart Processing and Computing*, 2013, **2**(1): 20-26.
16. C. Wu, X. Ma, W. Wang, Hyperspectral image denoise based on curvelet transform combined with weight coefficient method, *Journal of Intelligent & Fuzzy Systems*, 2019, **37**(4): 4425-4429.
17. A.A. Yahya, J. Tan, B. Su, M. Hu, Y. Wang, K. Liu, A.N. Hadi, BM3D image denoising algorithm based on an adaptive filtering, *Multimedia Tools and Applications*, 2020, **79**: 20391-20427.
18. M. Lebrun, An analysis and implementation of the BM3D image denoising method, *Image Processing On Line*, 2012, **2**: 175-213.
19. M.W. Shi, F. Zhang, S.W. Wang, C.M. Zhang, X.M. Li, Detail preserving image denoising with patch-based structure similarity via sparse representation and SVD, *Computer Vision and Image Understanding*, 2021, **206**: 103173.
20. A. Tavakoli, A. Pourmohammad, 2012, Image denoising based on compressed sensing, *International Journal of Computer Theory and Engineering*, **4**(2): 266-269.
21. K. Zhang, W. Zuo, Y. Chen, D. Meng, L. Zhang, Beyond a gaussian denoiser: Residual learning of deep cnn for image denoising, *IEEE Transactions on Image Processing*, 2017, **26**(7): 3142-3155.
22. C. Tian, Y. Xu, W. Zuo, Image denoising using deep CNN with batch renormalization, *Neural Networks*, 2020, **121**: 461-473.
23. S. Annam, A. Singla, Correlative analysis of denoising methods in spectral images embedded with different noises, 2020 The 6th International Conference on Parallel, Distributed and Grid Computing (PDGC) IEEE, 2020: 318-323.
24. A. Javeed, T. Shah, Attaullah, Lightweight secure image encryption scheme based on chaotic differential equation, *Chinese Journal of Physics*, 2020, **66**: 645-659.
25. Q. Wang, J. Ma, S. Y. Yu, L. Y. Tan, Noise detection and image denoising based on fractional calculus, *Chaos, Solitons & Fractals*, 2020, **131**: 109463.
26. H. Zhang, J. Yu, Y. Ma, Z. Pan, J. Zhao, Image restoration based on stochastic resonance in a parallel array of FitzhughCNagumo Neuron, *Complexity*, 2020, **2020**(1): 8843950.
27. C. Ma, J. Ao, Nonlinear average stochastic resonance for image enhancement, *International Journal of Electrical Engineering & Education*, 2023, **60**(1_suppl): 578-593.

28. Anuranjeeta, S. Sharma, N. Sharma, M. Singh, K. K. Shukla, Enhancement and segmentation of histopathological images of cancer using dynamic stochastic resonance, *International Journal of Medical Engineering and Informatics*, 2020, **12**(2): 180-193.
29. X. Feng, H. Liu, N. Huang, Z. Wang, Y. Zhang, Generation and stability of size-adjustable bulk nanobubbles based on periodic pressure change, *Scientific reports*, 2019, **9**(1): 1118.
30. R. Chouhan, R. K. Jha, P. Biswas, Enhancement of dark and low-contrast images using dynamic stochastic resonance, *IET Image Processing*, 2013, **7**(2): 174-184.
31. M. Shen, J. Yang, W. Jiang, M.A.F. Sanjuán, Y. Zheng, Stochastic resonance in image denoising as an alternative to traditional methods and deep learning, *Nonlinear Dynamics*, 2022, **109**(3): 2163-2183.
32. M. Shen, J. Yang, M.A.F. Sanjuán, Y. Zheng, H. Liu, Adaptive denoising for strong noisy images by using positive effects of noise, *The European Physical Journal Plus*, 2021, **136**:(6): 698.
33. R.K. Jha, P.K. Biswas, B.N. Chatterji, Image denoising using stochastic resonance, in *Proceedings of International Conference on Cognition and Recognition*, 2005, 343-348.
34. R. Chouhan, R.K. Jha, P.K. Biswas, Image denoising using dynamic stochastic resonance in wavelet domain, in *2012 12th International Conference on Intelligent Systems Design and Applications (ISDA) IEEE*, 2012, November, pp. 58-63.
35. S. Jiao, Q. Liu, W. Liu, Image denoising using two-dimensional stochastic resonance in wavelet domain, in *2017 Chinese Automation Congress (CAC) IEEE*, 2017, October, pp. 4489-4494.
36. S. Wang, P. Niu, Y. Li, J. Jia, S. Wang, H. Li, B. Sun, B. Zheng, S. Ximin, Image denoising using adaptive bi-dimensional stochastic resonance system, *Ferroelectrics*, 2023, **609**(1): 148-157.
37. W. Huang, G. Zhang, S. Jiao, J. Wang, Gray image denoising based on array stochastic resonance and improved whale optimization algorithm, *Applied Sciences*, 2022, **12**(23): 12084.
38. V.S. Rallabandi, P.K. Roy, Magnetic resonance image enhancement using stochastic resonance in Fourier domain, *Magnetic Resonance Imaging*, 2010, **28**(9): 1361-1373.
39. D. Dhillon, R. Chouhan, Noise-aided edge-preserving image denoising using non-local means with stochastic resonance, in *2018 IEEE Global Conference on Signal and Information Processing (GlobalSIP)*, 2018, November, pp. 21-25.
40. S. Jiao, J. Shi, Y. Wang, R. Wang, A novel image noise reduction method for composite multistable stochastic resonance systems, *Heliyon*, 2023, **9**(3): e14431.
41. J. Zhang, D. Huang, J. Yang, H. Liu, X. Liu, Realizing the empirical mode decomposition by the adaptive stochastic resonance in a new periodical model and its application in bearing fault diagnosis, *Journal of Mechanical Science and Technology*, 2017, **31**: 4599-4610.
42. K. Liu, Y. Jin, Stochastic resonance in periodic potentials driven by colored noise, *Physica A: Statistical Mechanics and its Applications*, 2013, **392**(21): 5283-5288.
43. C. Boncelet, Image noise models, in *The Essential Guide to Image Processing*, 2009, Academic Press, 2nd edition, pp. 143-167.
44. R. Verma, J. Ali, A comparative study of various types of image noise and efficient noise removal techniques, *International Journal of Advanced Research in Computer Science and Software Engineering*, 2013, **3**(10): 617-622
45. A.K. Boyat, B.K. Joshi, A review paper: noise models in digital image processing, *Signal & Image Processing: An International Journal*, 2015, **6**(2): 63-75.
46. Y. Chang, L. Yan, T. Wu, S. Zhong, Remote sensing image stripe noise removal: From image decomposition perspective, *IEEE Transactions on Geoscience and Remote Sensing*, 2016, **54**(12): 7018-7031.
47. N. Wang, B. Zheng, H. Zheng, B. Yang, When underwater degraded images meet logical stochastic resonance, *Nonlinear Dynamics*, 2018, **94**: 295-305.
48. B. Zheng, N. Wang, H. Zheng, Z. Yu, J. Wang, Object extraction from underwater images through logical stochastic resonance, *Optics Letters*, 2016, **41**(21): 4967-4970.
49. Y. Yao, Cross-correlated sine-Wiener bounded noises-induced logical stochastic resonance, *Pramana*, 2021, **95**(2): 77.
50. Z. Liao, K. Ma, M. S. Sarker, H. Yamahara, M. Seki, H. Tabata, Quadstable logical stochastic resonance-based reconfigurable Boolean operation subjected to heavy noise floor, *Results in Physics*, 2022, **42**: 105968.

51. G. Cheng, W. Liu, R. Gui, Y. Yao, Sine-Wiener bounded noise-induced logical stochastic resonance in a two-well potential system, *Chaos, Solitons & Fractals*, 2020, **131**: 109514.
52. S. Morfu, B.I. Usama, P. Marquié, Perception enhancement of subthreshold noisy image with vibrational resonance, *Electronics Letters*, 2019, **55**(11): 650-652.
53. S. Morfu, B.I. Usama, P. Marquié, On some applications of vibrational resonance on noisy image perception: the role of the perturbation parameters, *Philosophical Transactions of the Royal Society A*, 2021, **379**(2198): 20200240.
54. J. Yang, S. Rajasekar, M.A.F. Sanjuán, Vibrational resonance: A review, *Physics Reports*, 2024, **1067**: 1-62.
55. L. Gammaitoni, Stochastic resonance and the dithering effect in threshold physical systems, *Physical Review E* 52 (1995) 4691-4698.
56. R.A. Wannamaker, S.P. Lipshitz, J. Vanderkooy, Stochastic resonance as dithering, *Physical Review E* 61 (2000) 233-236.

Chapter 12
Bearing Fault Diagnosis by Aperiodic Resonances

In this chapter, we explore the application of aperiodic resonances in bearing fault diagnosis. We begin by introducing the background of bearing fault diagnosis, particularly under time-varying conditions. Next, we describe two experimental platforms used to collect vibration signals from rolling bearings with structural faults. Finally, we focus on feature extraction and enhancement methods for bearing fault diagnosis in time-varying operating conditions. These methods include aperiodic system resonance, aperiodic vibrational resonance, piecewise re-scaled aperiodic stochastic resonance, and real-time re-scaled aperiodic stochastic resonance. Each method, along with its effectiveness in bearing fault diagnosis, has been validated using experimental signals.

12.1 Background of Bearing Fault Diagnosis

Bearings are known as joints of industrial equipment, especially playing an irreplaceable role in rotating machinery. There are many types of bearings, but this chapter only takes the most common rolling bearing as the research background. Due to long-term operation under heavy load, time-varying working condition, and harsh environment, bearing is prone to failure. If the bearing fault diagnosis is not carried out in a timely manner, it will cause equipment accident, and even machine damage and human death. As a result, the bearing fault diagnosis has always been a long-standing and widely studied topic by a various researchers in different disciplines.

There are various methods for bearing fault diagnosis, with mainstream approaches primarily based on deep learning algorithms [1–5] and signal processing techniques [6–12]. Among signal processing methods, aperiodic resonance-based techniques hold a unique position. These include the stochastic resonance method [13–16], vibrational resonance method [17–22], and system resonance method [23, 25], as well as hybrid approaches combining these methods with other techniques [25]. However, the works mentioned in [13–25] primarily focus on periodic characteristic

J. Yang and M. A. F. Sanjuan, *Aperiodic Resonances—Theory and Applications*,
https://doi.org/10.1007/978-981-96-6819-9_12

signals and have only been applied to bearing fault diagnosis under constant operating conditions. In practical scenarios, bearings often operate under time-varying conditions where speed, load, or both fluctuate over time. This results in characteristic signals that present fault features in a non-stationary aperiodic form, rendering the conventional aperiodic resonance methods in [13–25] unsuitable for direct application. There are two approaches to address this issue. The first involves transforming the non-stationary signal under time-varying speed conditions into a stationary signal using computed order tracking [26–29]. This allows for the application of re-scaled stochastic resonance to extract fault features under time-varying conditions. The second approach, which is the focus of this chapter, directly applies aperiodic resonance techniques to extract aperiodic characteristic features [31–33].

12.2 Experimental Platform for Bearing Fault Diagnosis

We have designed two experimental platforms to collect the data and to verify the application of aperiodic resonances in bearing fault diagnosis under time-varying speed conditions. The experimental platform shown in Fig. 12.1 mainly simulates various complex variable speed operating conditions of joint bearing of an industrial robotic arm, while the experimental platform shown in Fig. 12.2 is a conventional bearing-rotor fault simulation experimental platform.

12.2.1 Bearing Fault Diagnosis Platform for the Joint of an Industrial Robotic Arm

A transmission system experimental platform was built for the special motion laws of the joint bearing of an industrial robotic arm, as shown in Fig. 12.1. The experimental platform mainly consists of a signal generator, laptop, 1A206 accelerometer, 5E103 eddy current sensor, 750 watt servo motor, servo driver, RV-20E reducer, magnetic powder brake, NI9234 signal acquisition card, magnetic powder brake controller, and type-6204 faulty bearing. The experimental platform can simulate the motion of any joint of an industrial robotic arm corresponding to the speed characteristics, while considering the torque load, which has certain significance for the fault diagnosis and monitoring of the bearing of an industrial robotic arm joint.

During the experiment, we generated simulated voltage signals using a signal generator and input them into the servo drive to control the forward and reverse rotation of the motor, creating various time-varying speed conditions. The servo drive was set to analog control mode, allowing any simulated voltage within the range of -5V to +5V from the signal generator to be used for control. Positive and negative voltage values corresponded to the forward and reverse rotation of the servo motor, respectively. First, the signal generator produced an analog signal corresponding to

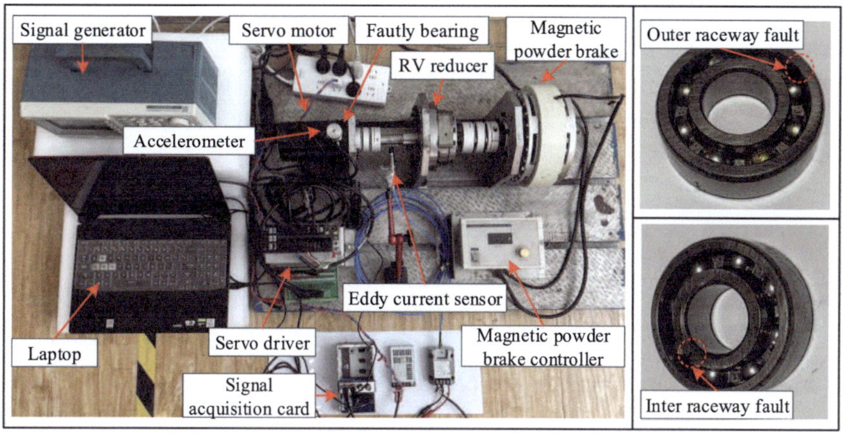

Fig. 12.1 Experimental platform for simulating joint bearing faults of an industrial robotic arm under various time-varying speed conditions

the robot's rotational speed, which was then input into the servo driver to regulate the motor and achieve the desired rotational speed conditions. Next, an accelerometer was used to collect vibration signals during bearing operation. The vibration data were recorded on a computer via a sensor signal acquisition card and an upper computer system, followed by analysis and processing using MATLAB software. The sampling frequency was set to 51,200 Hz. The test bearing used in this experiment was a servo motor end bearing, specifically a 6204 deep groove ball bearing. The structural parameters of this bearing are listed in Table 12.1.

In Table 12.1, D_o, D_i, D_c and D_b are the outer diameter, inner diameter, pitch diameter and rolling element diameter of the bearing, respectively. T is the thickness of the bearing, z is the number of rolling elements, and α is the contact angle.

The theoretical formula for the fault characteristic frequency of a rolling bearing is as follows

$$\begin{cases} f_i = \frac{1}{2} f_r \left(1 + \frac{D_b}{D_c} \cos\alpha\right) z \\ f_o = \frac{1}{2} f_r \left(1 - \frac{D_b}{D_c} \cos\alpha\right) z \\ f_b = \frac{1}{2} \frac{D_c}{D_b} f_r \left(1 - \frac{D_b^2}{D_c^2} \cos^2\alpha\right) \end{cases} \quad (12.1)$$

Here, f_i, f_o and f_b are the theoretical fault characteristic frequency of the inner raceway, the outer raceway and the rolling element, and f_r is the rotational frequency of

Table 12.1 The structural parameters of the 6204-type bearing

Type	D_o	D_i	D_c	D_b	T	z	α
6204	47 mm	20 mm	33.5 mm	8 mm	17 mm	8	0

the bearing. Based on Eq. (12.1), the theoretical fault characteristic frequencies of the 6204-type bearing are $f_i = 4.955 f_r$, $f_o = 3.045 f_r$, and $f_b = 1.974 f_r$. Obviously, when f_r goes over time, the theoretical fault characteristic frequencies also change accordingly. Incidentally, we introduce a new index called as *order*. The order is the ratio of the characteristic frequency to the rotational frequency, i.e.,

$$\begin{cases} O_i = f_i/f_r = \frac{1}{2}\left(1 + \frac{D_b}{D_c}\cos\alpha\right)z \\ O_o = f_o/f_r = \frac{1}{2}\left(1 - \frac{D_b}{D_c}\cos\alpha\right)z \\ O_b = f_b/f_r = \frac{1}{2}\frac{D_c}{D_b}\left(1 - \frac{D_b^2}{D_c^2}\cos^2\alpha\right) \end{cases} \tag{12.2}$$

As a result, the characteristic orders of the inner raceway, outer raceway, and rolling element are 4.955, 3.045, and 1.974, respectively. Notably, even when the rotational speed varies over time, these characteristic orders remain constant. This property makes order analysis a valuable tool for simplifying the processing and extraction of feature frequencies [26–29].

12.2.2 Bearing Fault Diagnosis Platform for a Conventional Bearing-Rotor System

The conventional bearing-rotor fault simulation experimental platform is shown in Fig. 12.2. The experimental platform mainly consists of 5E103 eddy current sensor, 1A206 accelerometer, type-N306E faulty bearing, NI9234 signal acquisition card, 198 BGL series spindle servo motor, laptop, and frequency converter. The frequency converter is used to control the motor speed and transmit force and torque to the

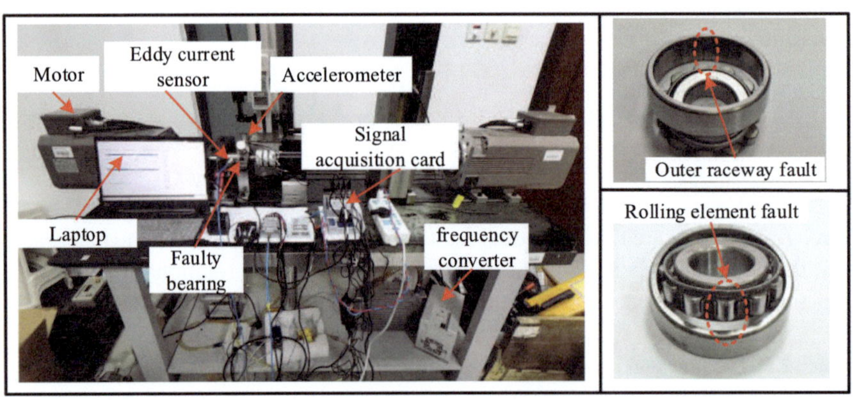

Fig. 12.2 Experimental platform for simulating conventional bearing-rotor faults under various time-varying speed conditions

Table 12.2 The physical parameters of N306E-type bearing

Type	D_o	D_i	D_c	D_b	T	z	α
N306E	72 mm	30 mm	52 mm	10 mm	19 mm	11	0

faulty bearing through the coupling. Analyze the operating status of the bearing by synchronously collecting the vibration signal and speed signal of the shaft. The sampling frequency is set 10240 Hz. The types of bearing faults are scratches on the outer raceway of the bearing (size: 1.5 × 1 mm (width × depth)) and scratches on the rolling elements (size: 1.5 × 1 mm (width × depth)), as shown in Fig. 12.2.

The structural parameters of type-N306E bearing are shown in Table 12.2. According to Eq. (12.1), the theoretical characteristic frequencies of the outer raceway fault, inner raceway fault, and rolling element fault are $4.44 f_r$, $6.56 f_r$, and $5.01 f_r$, respectively.

For experiment platform in Fig. 12.2, the main experimental steps are as follows:

(1) Install the faulty bearing and debug the sensor, signal acquisition card, and laptop, as shown in Fig. 12.2.
(2) Use a frequency converter to control a three-phase motor to produce a speed characteristic of "speed linear rising—steady operating - speed linear decreasing". The eddy current sensor is placed in front of the reference shaft with a fixed frame for collecting phase signal generated by the rotation of the reference shaft. The acceleration sensor is arranged on the bearing seat to collect vibration signals of the bearing.
(3) Collect and save the measured vibration signals and phase signals using a laptop, and analyze the collected signals using softwares MATLAB and LabVIEW.

12.3 Bearing Fault Diagnosis by Aperiodic System Resonance

The aperiodic system resonance method is validated using two sets of experimental signals with complex time-varying speed conditions for enhancing fault characteristic of bearing outer raceway. The two groups of vibration signals are collected from experiment platforms in Figs. 12.1 and 12.2, respectively.

12.3.1 Bearing Fault Diagnosis for the Joint of an Industrial Robotic Arm

The aperiodic system resonance algorithm proposed in Sect. 3.2 is used to analyze the experimental signal from the test platform in Fig. 12.1. The speed satisfies that of a joint of an industrial robotic arm. The maximum speed reaches 1800 revolutions per minute (r/min), and the frequency 30 Hz, corresponding to a fault characteristic frequency 30 Hz $\times$ 3.09 = 92.7 Hz. The measured signal is shown in Fig. 12.3 which shows the collected fault characteristic signal of the joint bearing simultaneously. Within 9 s, the speed condition of the bearing is first in forward rotation with the process "speed linear rising—steady operating—speed linear decreasing", then stop, and goes on in reverse rotation with the same speed varying process, and then moves again and again for the following cycles. Figure 12.3b shows the corresponding time-frequency analysis results. Due to the short acceleration and deceleration time, there may be spectral leakage when using short-time Fourier transform analysis. A weak fault characteristic frequency (93 Hz) can be observed in the steady speed operation stage.

For the convenience of analysis, a segment of the experimental signal (2.8 $\sim$ 4.2 s) is extracted for analysis, and the EMD technique is used to filter out interference signals. The decomposed IMF5 component is chosen as the input characteristic signal for the aperiodic system resonance analysis. Figure 12.4 shows the comparison of the signal performance before and after the aperiodic system resonance processing. Figure 12.4a shows the raw signal with the fault characteristic of the bearing outer raceway, and Fig. 12.4b shows the its enhanced effect by the aperiodic system resonance processing. Compared to the raw signal, the energy value of the STFT spectrum has significantly improved, from 0.025 to 1 in the energy color bar on

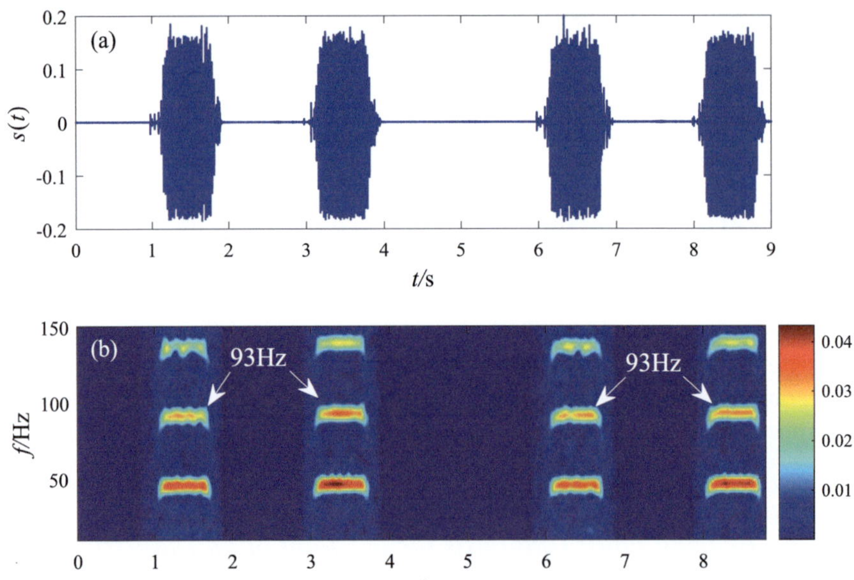

Fig. 12.3 The raw experimental vibration signal of the bearing with fault in the outer raceway, **a** the time domain waveform, **b** the STFT spectrum

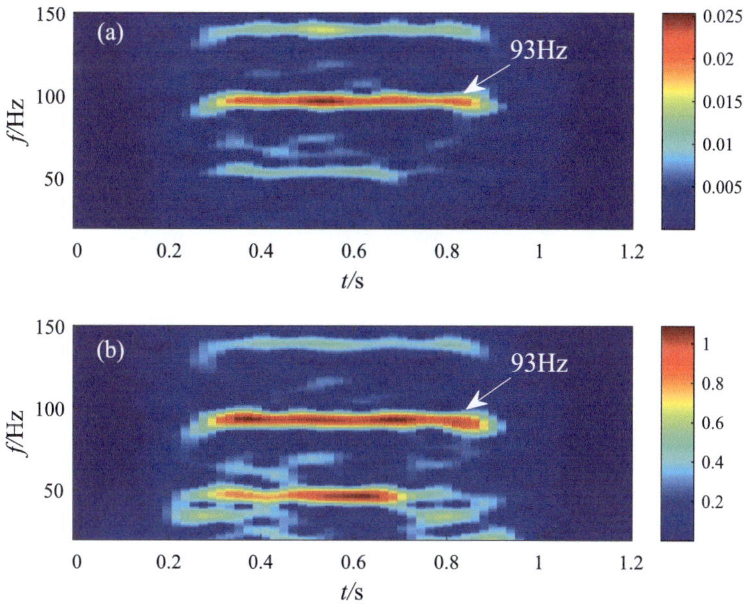

Fig. 12.4 Comparison of the performance before and after the aperiodic system resonance processing, **a** the STFT spectrum of the raw signal collected from the experimental platform in Fig. 12.1, **b** the STFT spectrum of the signal processed by aperiodic system resonance. The parameters are $a_1 = 0.01, b_1 = 0.01, \beta_0 = 500$

the right side of the subplot, indicating the effectiveness of the aperiodic system resonance method in enhancing the weak fault characteristic of a joint bearing.

12.3.2 Bearing Fault Diagnosis for a Conventional Bearing-Rotor System

Collecting the vibration signal of the bearing with fault in outer raceway from the experimental platform in Fig. 12.2, the speed characteristic is still set as the process of "speed linear rising—steady operating—linear decreasing". According to Eq. (12.2) and Table 12.2, the characteristic order of the N306E-type bearing with fault in the outer raceway is 4.44×, corresponding to a fault characteristic frequency $4.44f_r$. The maximum speed of the bearing in the experimental signal is 3000 rpm, and the theoretical fault characteristic frequency is $50 \times 4.44 = 222$ Hz. The time series of the experimental signal is shown in Fig. 12.5, which represents the time-domain waveform of experimental vibration signal corresponding to the bearing with fault in the outer raceway. The amplitude of the signal shows an overall trend of "first small amplitude vibration, then increasing and stabilizing, and finally decreasing vibration", which corresponds to the speed condition of the bearing.

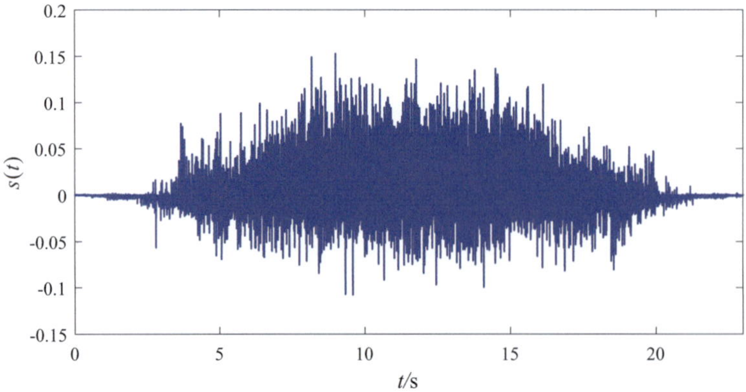

Fig. 12.5 The experimental vibration signal of the bearing with fault in the outer raceway

Fig. 12.6 shows the comparison of the performance before and after the aperiodic system resonance processing. Figure 12.6a presents the characteristic of the raw signal in the STFT spectrum. A weak fault characteristic frequency 224 Hz can be observed (within the allowable error range). Figure 12.6b shows the fault characteristic that obtained by the aperiodic system resonance processing. Compared to the raw signal, the energy value after the aperiodic system resonance processing is improved significantly in the STFT spectrum, increasing from 0.015 to 0.1, indicating the effectiveness of the method in enhancing the weak bearing fault characteristic once again.

12.4 Bearing Fault Diagnosis by Aperiodic Vibrational Resonance

In this subchapter, at first, we still use the data that ever used in Fig. 12.3. Herein, the method and nonlinear model proposed in Sect. 7.2.1 is still used to realize the aperiodic vibrational resonance. A segment of experimental signal ($2.8 \sim 4.2$ s) is extracted for analysis. After the EMD preprocessing, the IMF6 component of the signal is chosen as the input characteristic signal. Figure 12.7 shows the comparison of the effects before and after the aperiodic vibrational resonance processing. Specifically, Fig. 12.7a shows the STFT spectrum of the raw vibration signal of the bearing with fault in outer raceway while Fig. 12.7b shows the STFT spectrum of characteristic signal after it processed by the aperiodic vibrational resonance. Compared to the raw signal, the time-frequency energy value of the processed signal has significantly increased from approximately 0.015 to 1. It indicates the effectiveness of the aperiodic vibrational resonance in enhancing the weak fault characteristic signal of the roll bearing under time-varying speed condition.

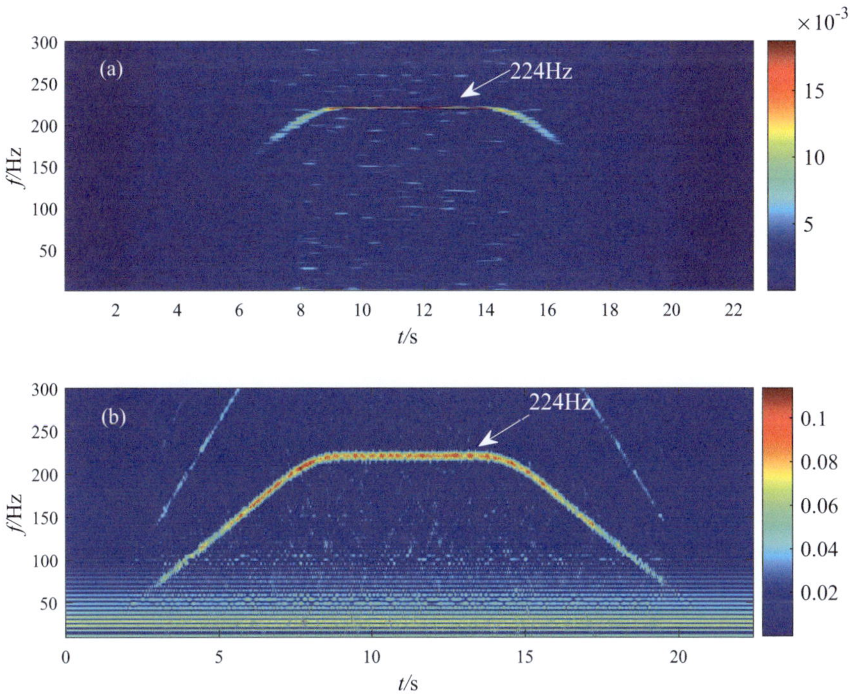

Fig. 12.6 Comparison of the performance before and after the aperiodic system resonance processing, **a** the STFT spectrum of the raw signal collected from the experimental platform in Fig. 12.2, **b** the STFT spectrum of the signal processed by the aperiodic system resonance. The parameters are $a_1 = 0.2$, $b_1 = 0.01$, $\beta_0 = 100$

Now, we process the signal in Fig. 12.5 by the aperiodic vibrational resonance method. Figure 12.8 shows the comparison of the signal performance before and after the aperiodic vibrational resonance. Figure 12.8a shows the STFT spectrum of the raw signal that collected through experiment in the presence of fault in the bearing outer raceway, and Fig. 12.8b shows the STFT spectrum of the output signal that have been processed by the aperiodic vibrational resonance. Compared to the raw signal, the time-frequency energy value of the processed signal has significantly increased from 0.015 to 0.15 in the energy value, as shown in the color bar on the right side of the subplot. The results in this figure indicate the effectiveness of the aperiodic vibrational resonance method in enhancing the weak fault characteristic signal of the bearing under speed time-varying condition once again.

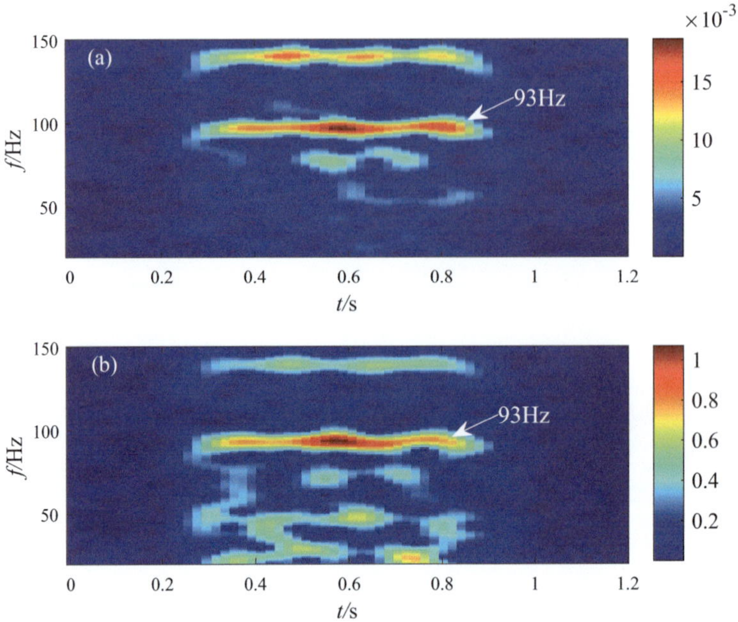

Fig. 12.7 Comparison of performance before and after the aperiodic vibrational resonance processing, **a** the STFT spectrum of the raw signal collected from the experimental platform in Fig. 12.1, **b** the STFT spectrum of the signal processed by the aperiodic vibrational resonance. The parameters are $a_1 = 0.01$, $b_1 = 0.01$, $\beta_0 = 500$, $B = 0.01$

12.5 Bearing Fault Diagnosis by Aperiodic Stochastic Resonance

12.5.1 Piecewise Re-scaled Aperiodic Stochastic Resonance Method

Here, an experiment was still conducted on the bearing fault diagnosis test platform given in Fig. 12.2, and corresponding vibration signals were collected in unit of millivolt (mV). The time-varying speed scheme is as follows: the shaft speed linearly increases from 300 rpm to 1800rpm within 10 s; then, it maintains a constant speed and operation for 16 s; finally, the shaft speed linearly decreases to 300 rpm within 10 s. This process of time-varying speed can simulate the conditions of rotating machinery from start to stop. According to the bearing structural parameters in Table 12.2 and the characteristic frequency formula in Eq. (12.1), the theoretical frequency of the bearing with fault in the outer raceway is 133.27 Hz when the speed is 1800 rpm. Meanwhile, to simulate the noisy background, the intensity of the additional Gaussian white noise is chosen as $D = 300$. The sampling frequency is $f_s = 10240$ Hz, and the number of sampling points is $n = 368640$.

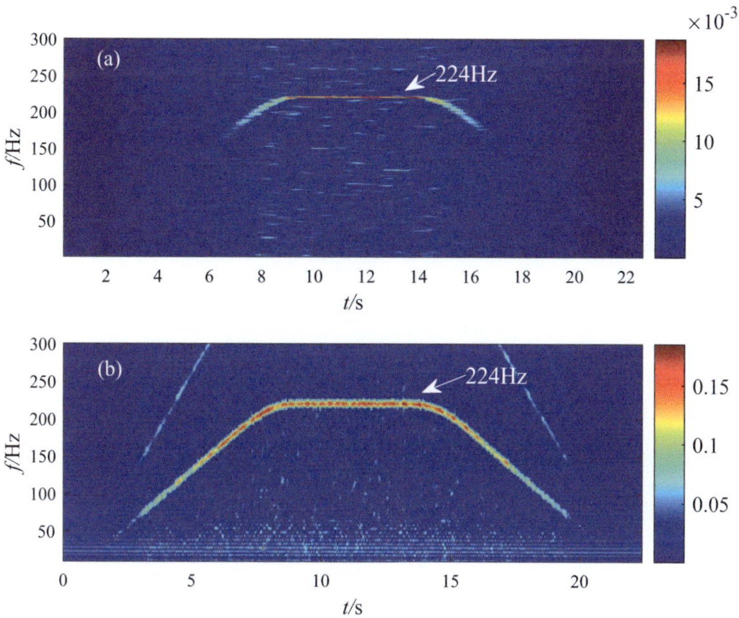

Fig. 12.8 Comparison of performance before and after the aperiodic vibrational resonance, **a** the STFT spectrum of the raw signal collected from the experimental platform in Fig. 12.2, **b** the STFT spectrum of the signal processed by the aperiodic vibrational resonance. The parameters are $a_1 = 0.1$, $b_1 = 0.01$, $\beta_0 = 100$, $B = 0.01$

Under time-varying speed condition, for the bearing with faut in the outer raceway, Fig. 12.9 shows the waveform of the time series and STFT spectrum of the vibration signal. As shown in this figure, the vibration signal contains complex frequency components, and the time-frequency information characterizing the bearing fault is hidden in strong noise, making it difficult to effectively extract the instantaneous frequency curve of the rolling bearing.

To further process the signal by the piecewise re-scaled stochastic resonance method as described in Sect. 5.1. The first step is to deal with the noisy signal by Hilbert demodulation. Then, the preprocessed envelope signal is divided into 18 subsegments. By the quantum particle swarm optimization (QPSO) algorithm, we obtain the optimal system parameters that match each subsegment signal to achieve the expected aperiodic stochastic resonance effect. To better demonstrate this method, herein, we take one of the subsegment envelope signal as an example to illustrate the specific process of extracting and enhancing the weak time-varying characteristic frequency of the damaged bearing from the vibration signal.

As shown in Fig. 12.10, the STFT spectrum of the second subsegment of the vibration signal is displayed. Apparently, the fault feature is completely submerged in the strong background noise. We cannot get any information for the fault. Stated plainly, it is not possible to directly obtain the instantaneous frequency curve of the

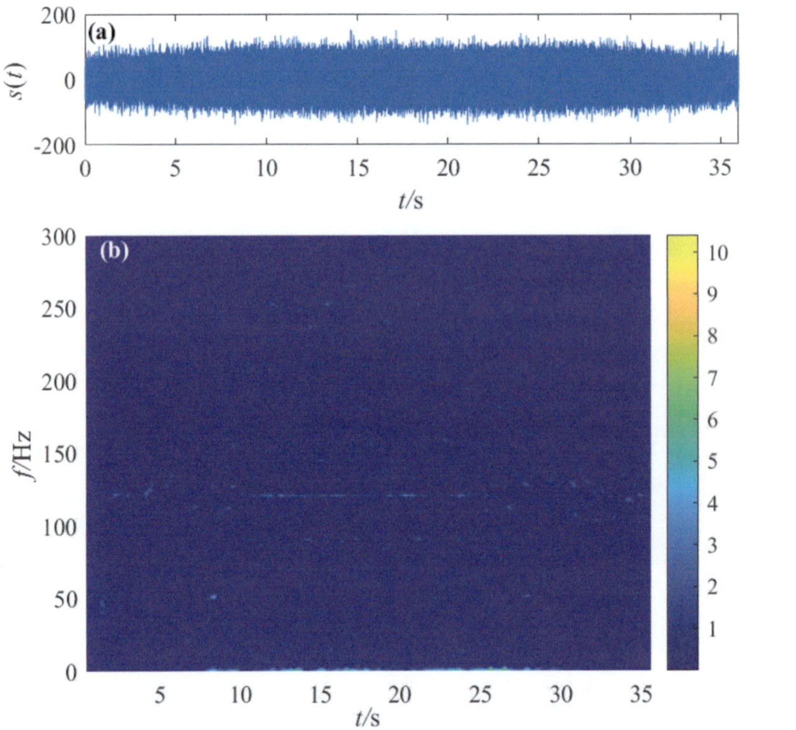

Fig. 12.9 The raw vibration signal of the bearing with fault in the outer raceway under time-varying speed condition, **a** the time domain waveform, **b** the STFT spectrum

second subsegment signal from the figure. To solve the problem, we need to carry out EMD processing to each subsegment signal. This preprocessing can distribute the interference components to different IMF components, playing a preliminary role in eliminating noise. It once again proves the necessity of selecting the optimal IMF component for each subsegment signal.

The second subsegment signal decomposition generated a total of 14 IMF components. In Fig. 12.11, it shows the waveforms of these IMF components in time-domain. It is worth noting that not all IMF components carry the fault characteristic information of the subsegment signal. In fact, some redundant IMF components containing interfering frequency components actually hinders the presentation of the fault characteristic. Therefore, it is necessary to select the optimal IMF component based on the criterion of maximum cross-correlation coefficient. Calculate the cross-correlation coefficient between each IMF component and the original subsegment signal, and select the optimal component which can get the maximum cross-correlation coefficient. The results of the cross-correlation coefficient between the nth IMF and the second subsegment signal are given in Table 12.3. Apparently, IMF8 which make the cross-correlation coefficient achieve the maximum value is chosen as the optimal IMF component replacing the subsegment signal for further processing.

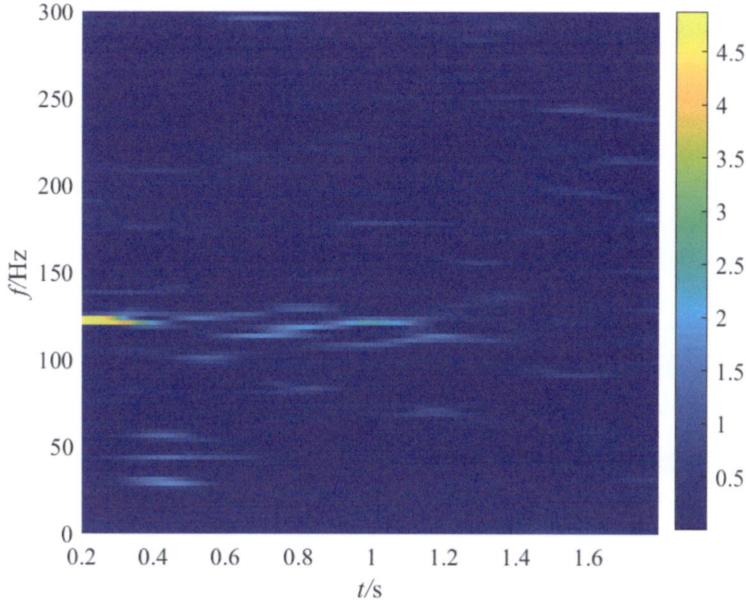

Fig. 12.10 The STFT spectrum for the second subsegment of the bearing vibration signal with fault in outer raceway

After the aperiodic stochastic resonance processing of IMF 8, Fig. 12.12 shows the STFT spectrum of the output. Compared with the result in Fig. 12.10, it depicts the complete instantaneous frequency curve of the second subsegment signal. The result indicates the appearance of the aperiodic stochastic resonance and its application in extracting and enhancing the non-stationary characteristic of the signal with complex frequency components and strong noise.

It is worth noting that the maximum value of the STFT spectrum of Fig. 12.10 is 4.5 dB/Hz, which is reduced to 0.2 dB/Hz as shown in Fig. 12.12 after the aperiodic stochastic resonance processing. However, this does not mean that the fault characteristic of the bearing have been weakened. It is because an increase in the maximum value of the STFT spectrum is not a necessary condition for signal enhancement. The increase in STFT spectrum value may also be caused by noise. The determination of whether the signal has been enhanced can be determined by quantifying the amplification degree of the ratio of the feature signal energy to the surrounding noise energy. The maximum value of the STFT spectrum in Fig. 12.10 corresponds to different noise frequency bands. On the contrary, the energy in Fig. 12.12 is concentrated on the instantaneous frequency curve representing the fault of the subsegment signal. It is obvious that the signal-to-noise ratio in Fig. 12.12 is much higher than that in Fig. 12.10. The result indicates that the proposed method is successful in processing the second subsegment vibration signal of the damaged bearing.

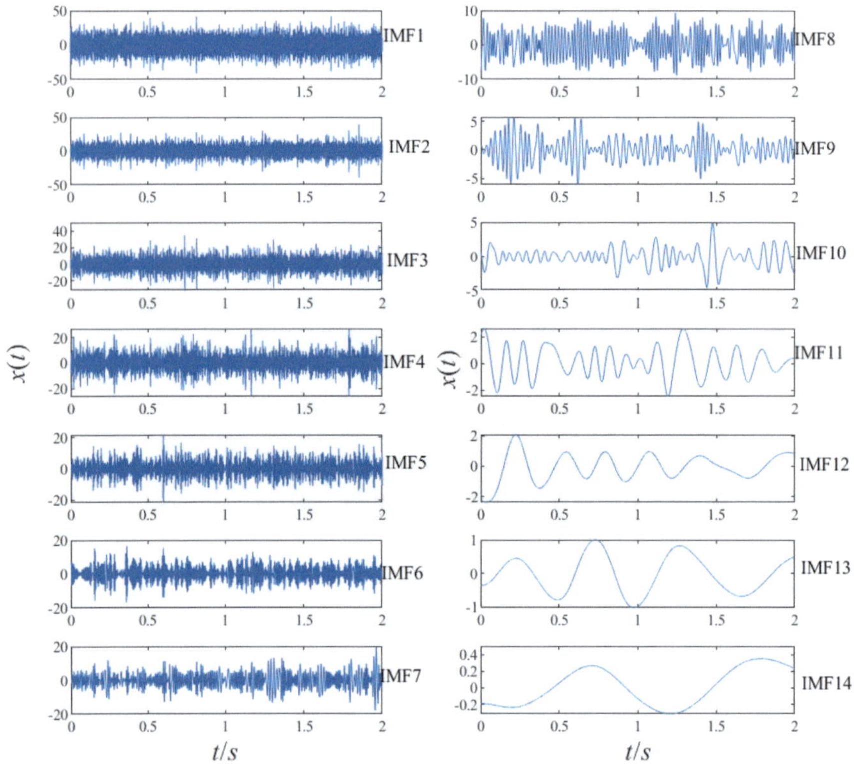

Fig. 12.11 Time domain waveforms of IMFs of the second subsegment vibration signal collected from the experimental platform with bearing fault in the outer raceway

Table 12.3 The cross-correlation coefficient between each IMF and the second subsegment vibration signal collected from the experimental platform with bearing fault in the outer raceway

IMF1	IMF2	IMF3	IMF4	IMF5	IMF6	IMF7	IMF8	IMF9	IMF10	IMF11	IMF12	IMF13	IMF14
0.0911	0.0799	0.0596	0.0844	0.1558	0.2457	0.3345	0.3553	0.0675	0.0028	0.0151	0.0015	0.0009	0.0014

By repeating the above analysis process for all subsegment signals, the expected aperiodic stochastic resonance response for each subsegment signal can be obtained. In order to more intuitively demonstrate the fault characteristic enhancement, the system output of all processed subsegment signals are reconstructed in chronological order. The STFT spectrum of the reconstructed signal is shown in Fig. 12.13. Compared with Figs. 12.9, 12.13 fully displays the nonlinear frequency modulation form characterized by the fault in the bearing outer raceway. The time-frequency curve is clear, and the high-frequency noise is effectively suppressed. The maximum value of the STFT spectrum decreases from 9 in Fig. 12.9 to 0.5 in Fig. 12.13. However, the spectral amplification gain η of the reconstructing characteristic signal is calculated to be 2.8. Once again, it has been proven that it is inappropriate to directly

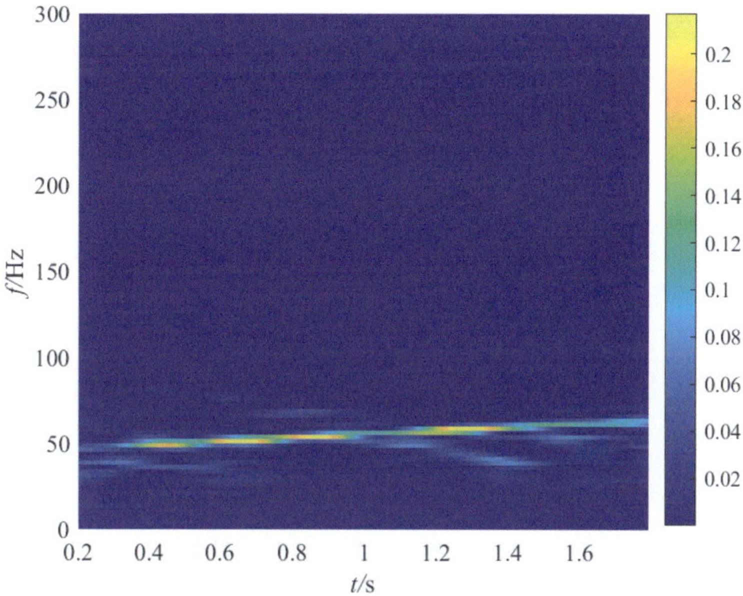

Fig. 12.12 The STFT spectrum of the second subsegment signal processed by the aperiodic stochastic resonance. The parameters are $a_1 = 0.1$, $b_1 = 1.08$ and $\beta = 120$

determine whether a signal is enhanced or attenuated based on the maximum value of the STFT spectrum. In addition, it can be intuitively seen from Fig. 12.13 that the signal-to-noise ratio of the output signal has been improved, which is also a true manifestation of the characteristic signal enhancement.

The results of the processed experimental signals discussed above demonstrate that aperiodic stochastic resonance combined with EMD can effectively enhance the weak fault characteristic of the bearing under the time-varying speed condition. It is an application of the theoretical study of the aperiodic stochastic resonance method in Sect. 5.4.

12.5.2 Real-Time Re-scaled Aperiodic Stochastic Resonance Method

In order to verify the real time re-scaled aperiodic stochastic resonance method that introduced in Sect. 5.1.2 in the application of the bearing fault diagnosis under the time-varying speed condition, we still use the experimental signals collected from the platform shown in Fig. 12.2.

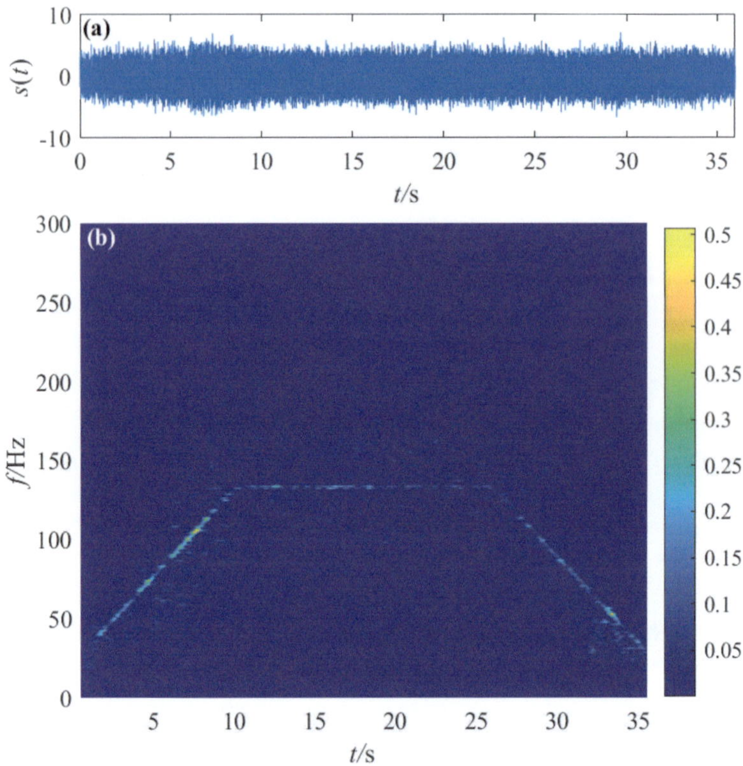

Fig. 12.13 The fault characteristic signal reconstructed from all subsegment signals processed by the piecewise re-scaled aperiodic stochastic resonance method, **a** the time domain waveform, **b** the STFT spectrum

To simulate linear and nonlinear time-varying speed conditions in engineering, the experimental data are collected from two groups. The first group of the data corresponding to the speed linear rising condition, and the second group of the data corresponding to the speed linear rising first and then operating in a steady speed. The instantaneous time-varying speeds of f_{r1} and f_{r2} corresponding to these two groups data are defined in the following equations

$$f_{\mathrm{r1}} = 4.7t + 14.4 \quad t \in [0, 2] \tag{12.3}$$

and

$$f_{r2} = \begin{cases} 5t + 22.5 & t \in [0, 1.5) \\ 30 & t \in [1.5, 2] \end{cases} \tag{12.4}$$

We still use the N306E-type bearing, according to Eq. (12.1) and Table 12.2, corresponding to the two time-varying speed operating conditions in Eq. (12.3) and

Eq. (12.4), the characteristic frequency of the bearing with fault in outer raceway is given respectively in the following two equations, i.e.,

$$f_o = 20.9t + 64 \quad t \in [0, 2] \tag{12.5}$$

and

$$f_o = \begin{cases} 22.2t + 100 & t \in [0, 1.5) \\ 133.3 & t \in [1.5, 2] \end{cases} \tag{12.6}$$

The experimental signal of the rolling bearing collected under linear speed regulation condition is processed by the real-time re-scaled aperiodic stochastic resonance. The raw waveform in time-domain is shown in Fig. 12.14a. There is no obvious fault feature in the time-domain waveform, and the waveform contains a lot of noise interference. Figure 12.14c is the STFT spectrum of the raw signal, with a frequency range of 0–150 HZ, which corresponds to the frequency band of the fault characteristic frequency. However, due to strong noise interference, the time-frequency ridge in the time-frequency diagram is submerged by noise, making it impossible to directly diagnose fault through the STFT spectrum. Figure 12.14b shows the output time series that obtained by the real-time re-scaled aperiodic stochastic resonance. As shown in the subplot, the waveform has obvious regularity and the amplitude is steady at around 0.5. Figure 12.14d shows the STFT spectrum of the output corresponding

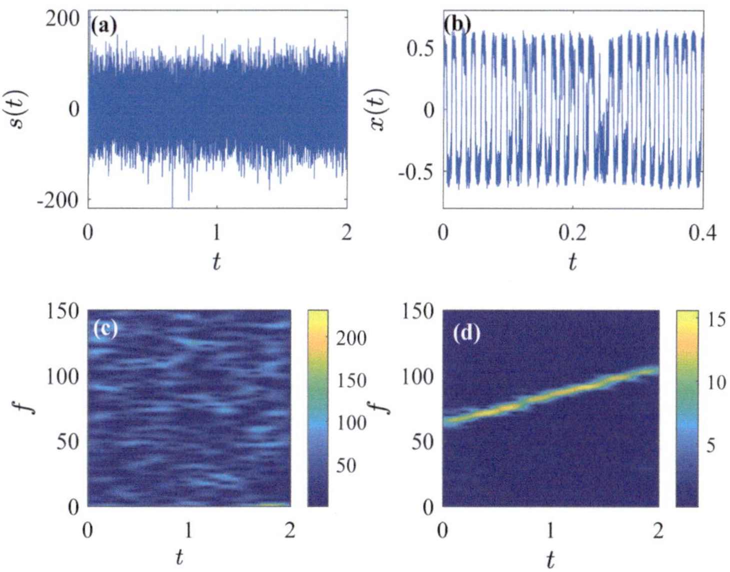

Fig. 12.14 Bearing fault characteristic extraction by the real-time re-scaled aperiodic stochastic resonance method under the linear time-varying speed condition defined in Eq. (12.3), **a** time-domain waveform of the raw vibration signal, **b** time-domain waveform of the processed signal, **c** the STFT spectrum of the raw signal, **d** the STFT spectrum of the processed signal

to Fig. 12.14b, with clear and continuous time-frequency ridge visible in the subplot. The time-frequency characteristic of the ridge line is consistent with theoretical value, indicating the presence of fault in outer raceway of the bearing. The energy of the fault characteristic area in the STFT spectrum is significantly increased, while the noise energy of other frequency area is significantly reduced. The results prove that the real-time re-scaled aperiodic stochastic resonance can extract non-stationary fault feature under strong noise background.

Similarly, the real-time re-scaled aperiodic stochastic resonance is used to process experimental vibration signal of the rolling bearing under nonlinear speed regulation condition, with the fault also in the outer raceway of the bearing. In Fig. 12.15a, it presents the time-domain waveform of the raw signal. There is no obvious fault feature in the waveform, and there is a lot of noise interference existing in the waveform. Figure 12.15c is the STFT spectrum of the raw signal, with a frequency range of 0–150 HZ, which corresponds to the frequency band of the characteristic frequency that characterizes fault in the outer raceway as given in Eq. (12.5). However, due to strong noise interference, the time-frequency ridge in the time-frequency spectrum is submerged by noise, making it impossible to directly diagnose fault through the time-frequency spectrum. Figure 12.15b shows the output waveform of raw signal processed by the real-time re-scaled aperiodic stochastic resonance method, which corresponds to the fault in the outer raceway of the rolling bearing N306E. As shown in the figure, the waveform has strong regularity and the amplitude is stable at around

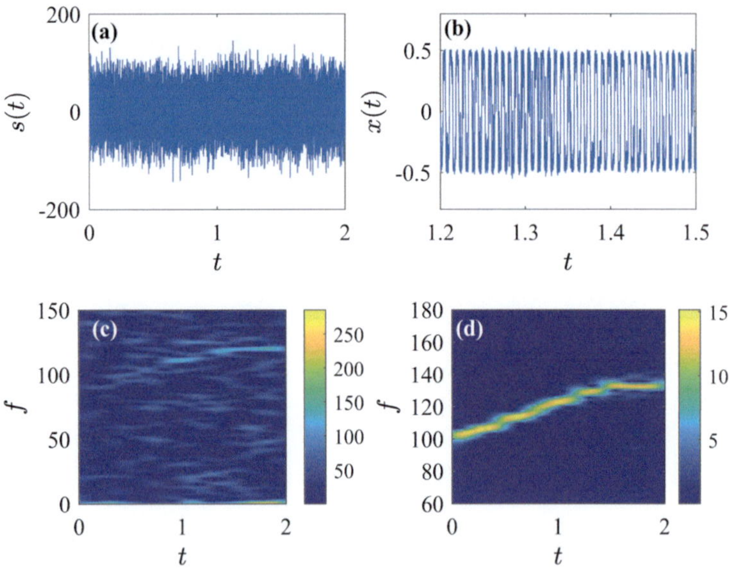

Fig. 12.15 Bearing fault characteristic extraction by the real-time re-scaled aperiodic stochastic resonance method under the nonlinear time-varying speed condition Eq. (12.4), **a** time-domain waveform of the raw vibration signal, **b** time-domain waveform of the processed signal, **c** the STFT spectrum of the raw signal, **d** the STFT spectrum of the processed signal

0.5. Figure 12.15d shows the STFT spectrum of the processed signal, with clear and continuous time-frequency ridge visible in the figure. The time-frequency characteristic of the ridge line is consistent with Eq. (12.6), indicating the presence of fault in the outer raceway of the bearing. The above analysis shows the effectiveness of the real-time re-scaled aperiodic stochastic resonance method in extracting and enhancing the weak fault characteristic under nonlinear speed regulating condition and strong noise background.

12.6 Concluding Remarks

We present the application of aperiodic resonances in bearing fault diagnosis under time-varying speed conditions. The fault characteristic frequency of a bearing typically appears as a non-stationary signal in a frequency-modulated form, making fault diagnosis challenging. Aperiodic resonances offer a novel approach to addressing this issue. To enhance the weak characteristic information of damaged bearings, we introduce the aperiodic system resonance method and the aperiodic vibrational resonance method. Additionally, we utilize aperiodic stochastic resonance combined with the empirical mode decomposition technique to extract unknown characteristic information, enabling bearing fault diagnosis under complex time-varying speed conditions and strong noise backgrounds. The effectiveness of these methods is validated using various sets of experimental data. Although this chapter focuses on diagnosing rolling bearing faults, aperiodic resonance methods can also be applied to other faults occurring under time-varying speed conditions, such as bolt loosening faults, gear faults, and coupling misalignment faults.

References

1. S. Zhang, S. Zhang, B. Wang, T.G. Habetler, Deep learning algorithms for bearing fault diagnostics-A comprehensive review, *IEEE Access*, 2020, **8**: 29857-29881.
2. D.T. Hoang, H.J. Kang, A survey on deep learning based bearing fault diagnosis, *Neurocomputing*, 2019, **335**: 327-335.
3. S. Zhang, S. Zhang, B. Wang, T.G. Habetler, Deep learning algorithms for bearing fault diagnostics-a review, 2019, August,In *2019 IEEE 12th International Symposium on Diagnostics for Electrical Machines, Power Electronics and Drives* (SDEMPED) (pp. 257-263).
4. X. Chen, R. Yang, Y. Xue, M. Huang, R. Ferrero, Z. Wang, Deep transfer learning for bearing fault diagnosis: A systematic review since 2016, *IEEE Transactions on Instrumentation and Measurement*, 2023, **72**: 1-21.
5. R. Liu, B. Yang, E. Zio, X. Chen, Artificial intelligence for fault diagnosis of rotating machinery: A review, *Mechanical Systems and Signal Processing*, 2018, 108: 33-47.
6. A. Rai, S.H. Upadhyay, A review on signal processing techniques utilized in the fault diagnosis of rolling element bearings, *Tribology International*, 2016, **96**: 289-306.
7. Y. Lei, J. Lin, Z. He, M.J. Zuo, A review on empirical mode decomposition in fault diagnosis of rotating machinery, *Mechanical Systems and Signal Processing*, 2013, **35**(1-2): 108-126.

8. Z. Feng, D. Zhang, M.J. Zuo, Adaptive mode decomposition methods and their applications in signal analysis for machinery fault diagnosis: a review with examples, *IEEE Access*, 2017, **5**: 24301-24331.
9. R. Yan, R.X. Gao, X. Chen, Wavelets for fault diagnosis of rotary machines: A review with applications, *Signal Processing*, 2014, **96**: 1-15.
10. Y. Wei, Y. Li, M. Xu, W. Huang, A review of early fault diagnosis approaches and their applications in rotating machinery, *Entropy*, 2019, **21**(4): 409.
11. Z. Feng, Y. Zhou, M.J. Zuo, F. Chu, X. Chen, Atomic decomposition and sparse representation for complex signal analysis in machinery fault diagnosis: A review with examples, *Measurement*, 2017, **103**: 106-132.
12. Z. Shan, Z. Wang, J. Yang, Q. Ma, T. Gong, Novel time-frequency mode decomposition and information fusion for bearing fault diagnosis under varying-speed condition, *IEEE Transactions on Instrumentation and Measurement*, 2023, **72**: 3511210.
13. Z. Qiao, Y. Lei, J. Lin, F. Jia, An adaptive unsaturated bistable stochastic resonance method and its application in mechanical fault diagnosis, *Mechanical Systems and Signal Processing*, 2017, **84**: 731-746.
14. S. Lu, Q. He, J. Wang, A review of stochastic resonance in rotating machine fault detection, *Mechanical Systems and Signal Processing*, 2019, **116**: 230-260.
15. X. Liu, H. Liu, J. Yang, G. Litak, G. Cheng, S. Han, Improving the bearing fault diagnosis efficiency by the adaptive stochastic resonance in a new nonlinear system, *Mechanical Systems and Signal Processing*, 2017, **96**: 58-76.
16. J. Yang, D. Huang, D. Zhou, H. Liu, Optimal IMF selection and unknown fault feature extraction for rolling bearings with different defect modes, *Measurement*, 2020, **157**: 107660.
17. J. Gao, J. Yang, D. Huang, H. Liu, S. Liu, Experimental application of vibrational resonance on bearing fault diagnosis, *Journal of the Brazilian Society of Mechanical Sciences and Engineering*, 2019, **41**: 1-13.
18. L. Xiao, X. Zhang, S. Lu, T. Xia, L. Xi, A novel weak-fault detection technique for rolling element bearing based on vibrational resonance, *Journal of Sound and Vibration*, 2019, **438**: 490-505.
19. J. Li, X. Cheng, S. Zhang, Z. Meng, L. Cao, Fault feature extraction method of rolling bearings based on coupled resonance system with vibrational resonance-assisted enhanced stochastic resonance, *Mechanical Systems and Signal Processing*, 2024, **208**: 111069.
20. J. Li, X. Cheng, J. Peng, Z. Meng, A new adaptive parallel resonance system based on cascaded feedback model of vibrational resonance and stochastic resonance and its application in fault detection of rolling bearings, *Chaos, Solitons & Fractals*, 2022, **164**: 112702.
21. L. Xiao, J. Tang, X. Zhang, T. Xia, Weak fault detection in rotating machineries by using vibrational resonance and coupled varying-stable nonlinear systems, *Journal of Sound and Vibration*, 2020, **478**: 115355.
22. L. Xiao, R. Bajric, J. Zhao, J. Tang, X. Zhang, An adaptive vibrational resonance method based on cascaded varying stable-state nonlinear systems and its application in rotating machine fault detection, *Nonlinear Dynamics*, 2021, **103**: 715-739.
23. S. Zhang, J. Yang, J. Zhang, H. Liu, E. Hu, On bearing fault diagnosis by nonlinear system resonance, *Nonlinear Dynamics*, 2019, **98**: 2035-2052.
24. S. Zhang, Z. Wang, J. Yang, Rolling bearing damage evaluation by the dynamic process from self-induced resonance to system resonance of a Duffing system, *ASCE-ASME Journal of Risk and Uncertainty in Engineering Systems, Part B: Mechanical Engineering*, 2023, **9**(1): 011202.
25. J. Yang, D. Zhou, *Re-scaled Resonance Theory and Applications in Fault Diagnosis*, Beijing: Science Press, 2020, (in chinese).
26. J. Yang, C. Yang, X. Zhuang, H. Liu, Z. Wang, Unknown bearing fault diagnosis under time-varying speed conditions and strong noise background. *Nonlinear Dynamics*, 2022, **107**: 2177–C2193.
27. Z. Wang, J. Yang, Y. Guo, Unknown fault feature extraction of rolling bearings under variable speed conditions based on statistical complexity measures, *Mechanical Systems and Signal Processing*, 2022, **172**: 108964.

28. T. Gong, J. Yang, S. Liu, H. Liu, Non-stationary feature extraction by the stochastic response of coupled oscillators and its application in bearing fault diagnosis under variable speed condition, *Nonlinear Dynamics*, 2022, **108**(4): 3839-3857.
29. Z. Wang, J. Yang, Y. Guo, T. Gong, Z. Shan, Nonstationary feature extraction based on stochastic resonance and its application in rolling bearing fault diagnosis under strong noise background, *Review of Scientific Instruments*, 2023, **94**(1): 015110.
30. T. Gong, Z. Wang, Q. Ma, J. Yang, Motor bearing fault diagnosis in an industrial robot under complex variable speed conditions, *Journal of Computational and Nonlinear Dynamics*, 2024, **19**(2): 021007.
31. Z. Shan, Z. Wang, J. Yang, D. Zhou, H. Liu, Extracting non-stationary signal under strong noise background: Time-varying system analysis, *Journal of Vibration and Control*, 2023, **29**(17-18): 4036-4045.
32. X. Zhuang, C. Yang, J. Yang, C. Wu, Z. Shan, T. Gong, Rolling bearing fault diagnosis by aperiodic stochastic resonance under variable speed conditions, *Fluctuation and Noise Letters*, 2022, **21**(2): 2250015.
33. T. Gong, J. Yang, M.A.F. Sanjuán, H. Liu, Z. Shan, Vibrational resonance by using a real-time scale transformation method, *Physica Scripta*, 2022, **97**(4): 045207.

Chapter 13
Nondestructive Testing of Steel Wire Rope by Aperiodic Stochastic Resonance

This chapter explores the application of aperiodic stochastic resonance in processing magnetic flux leakage (MFL) detection signals for steel wire ropes. The research findings have potential applications in the field of nondestructive testing of steel wire ropes. We begin by introducing the background of nondestructive testing for steel wire ropes. Then, we validate the effectiveness of the aperiodic stochastic resonance method in processing experimental MFL signals. This method not only extracts damage characteristics of the wire rope but also assesses the severity of the damage.

13.1 Background of Nondestructive Testing of Steel Wire Rope

Steel wire rope has light weight, strong load-bearing capacity, and can transmit long-distance loads. It has good flexibility, high tensile and fatigue strength levels, and it is widely used in industries such as port cranes, mine hoisting, cable transportation, elevators, and suspension bridges. Especially in the field of mine hoisting, the steel wire rope is not only affected by the working load during operation, but also by different external environments such as dark and humid conditions, high-altitude falling objects, etc. This often leads to wear, breakage, rust and other damages to the steel wire rope. Moreover, the damage will become increasingly serious over the working time. If the damaged wire rope cannot be replaced in time, it may cause serious accidents [1–3].

The nondestructive testing methods for steel wire rope include electromagnetic testing [4, 5], visual testing [6], acoustic emission testing [7], radiographic testing [8], and eddy current testing [9], information fusion [10, 11], etc. Due to harsh environments, high costs, and low anti-interference capabilities, visual testing method, acoustic emission inspection method, radiographic inspection method, and eddy

J. Yang and M. A. F. Sanjuan, *Aperiodic Resonances—Theory and Applications*,
https://doi.org/10.1007/978-981-96-6819-9_13

current inspection method are still limited in engineering applications and are still studied in the theoretical and laboratory stages. Flaw detection instruments based on the electromagnetic detection method are gradually being applied in the engineering.

The damage characteristics detected using electromagnetic methods primarily manifest as different aperiodic shocks in the signal. These signals are typically weak and often submerged in noise, including strand waves and random vibrations. Therefore, effective denoising preprocessing methods are essential [12]. There are three main sources of vibration noise in steel wire rope damage signals: (1) Instrument-Induced Noise: Portable flaw detection devices inevitably introduce vibration noise due to instrument shaking when operated manually. (2) Structural Vibration Noise: Fixed steel wire rope online monitoring devices detect vibration noise caused by the rope's movement during the operation of large equipment such as mine hoists and elevators. (3) Intrinsic Structural Noise: Steel wire ropes are composed of multiple strands of metal wires twisted into a spiral structure. This unique formation, combined with the presence of dirt, protrusions, and surface irregularities, interacts with the walking wheels of the detection instrument, generating additional vibrations and noise. Beyond these specific wave and vibration noises, background noise from electronic components and environmental factors may also affect the processing of steel wire rope signals.

Due to the complexity of the structure and working conditions of steel wire ropes, various noises are often mixed in when collecting steel wire rope signals, which is a persistent interference to defect signals and affects the identification and judgment of steel wire rope damage signals. Liu et al. proposed an E-index stochastic resonance model and a weak signal detection method [13]. The chaotic characteristics were analyzed through output signals and the performance of the model was simulated and evaluated. The effectiveness of this method was verified by comparing three typical models for detecting defects in steel wire ropes and identifying weak signals through experiments and case studies. Given the superiority of stochastic resonance in processing weak signals with noise, this chapter investigates the identification of steel wire rope damage based on the aperiodic stochastic resonance method.

13.2 Steel Wire Rope Damage Identification by Aperiodic Stochastic Resonance

The signal of a steel wire rope detected by magnetic flux leakage method belongs to typical aperiodic pulse signals similar to that in Eqs. (1.15) and (1.16). We use the following equation to study aperiodic stochastic resonance and to realize the identification of wire rope damage, i.e.,

$$\frac{dx}{dt} = -V'(x) + s(t) + N(t), \tag{13.1}$$

where $V(x) = -ax + \frac{b}{4}x^4$, and $a > 0, b > 0$, which makes the system to be symmetric and monostable. The configuration of the potential function is shown in Fig. A.6 of the Appendix A.2.2. Aperiodic stochastic resonance in Eq. (13.1) has specially advantage in processing the steel wire rope damage signal [14]. In the time domain, the magnetic flux leakage signal of broken wire rope is presented in aperiodic pulse. Due to the harsh working environment of steel wire ropes, the signals collected in practical engineering applications are often submerged in strong background noise. The noise sources mainly originate the strand noise, the vibration noise and the electronic device noise [12, 16–19]. The strand noise produces in the fluctuation of the magnetic flux leakage signal that induced by the uneven and periodical change of the steel wire surface. The vibration noise caused by the shaking of the steel wire rope and the detector. Among these noise sources, the Gaussian white noise dominates the composition. As a result, herein, $N(t)$ is the Gaussian white noise.

We use the signal in Eq. (1.15) as the input characteristic signal, and use the function "awgn" of the MATLAB software to add noise in the signal to simulate the noisy signal of the damaged steel wire rope. Then, we input the noisy signal in turn into Eq. (13.1) to adjust the system parameter a to realize the aperiodic stochastic resonance under different noise intensities. The results are given in Fig. 13.1. We use the local cross-correlation coefficient defined in Eq. (1.18) as the measure to quantify

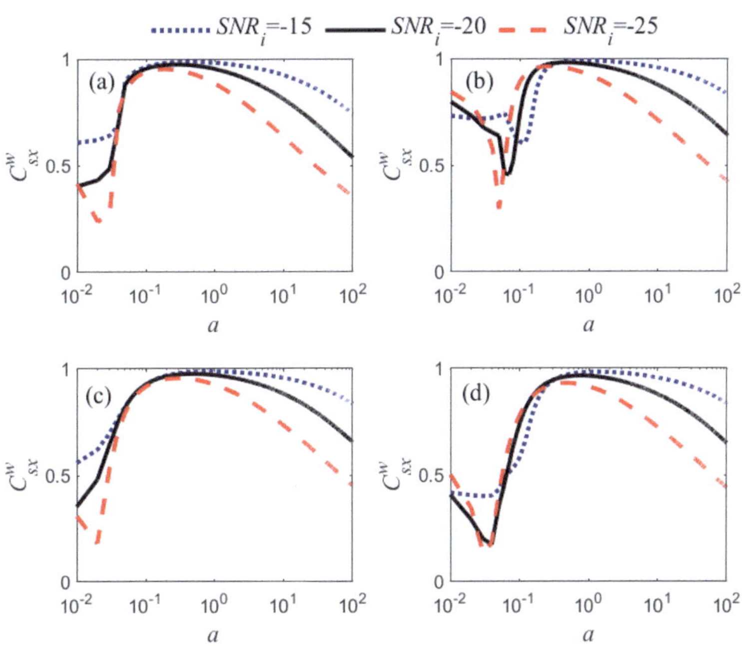

Fig. 13.1 The local cross-correlation coefficient C_{sx}^{w} versus the system parameter a. From **a** to **d**, the curve corresponding to the signal $s_1(t)$, $s_2(t)$, $s_3(t)$, $s_4(t)$, in Eq. (1.15) in turn. The simulation parameter is $b = 0.06$

the feature similarity between the input characteristic signal and the system output. This figure shows the dependence of the local cross-correlation coefficient on the system parameter a for the four types of aperiodic impulse signals under different input signal-to-noise ratios. The curves in Fig. 13.1 indicate the trend of the local cross-correlation coefficient corresponding to a given pulse signal and its output is almost the same change with the system parameter a. Specifically, the local cross-correlation coefficient reaches a maximum value at an appropriate system parameter a. For different types of the pulse signal, the maximal local cross-correlation coefficient may have a slight change. However, it can be determined that the optimal aperiodic stochastic resonance will be realized within approximately similar intervals. Overall, in Fig. 13.1, the curves demonstrate the existing of aperiodic stochastic resonance in the monostable system. It also provides a new approach for handling noisy signal of damaged steel wire rope.

To display the performance of aperiodic stochastic resonance in processing the characteristic signal of the damaged steel wire rope. The simulated waveform in Fig. 13.2a characters the single local flaw (SLF) defect and the waveform in Fig. 13.2b indicates the single loss of metallic cross-sectional area (SLMA), respectively. Selecting one group of experimental signal for steel wire rope with the above typical types of damage to verify the effective of the aperiodic stochastic resonance method in identifying the damage. From Fig. 13.2c, d, it can be seen that the signals corresponding to both the SLF and SLMA cases are completely submerged by the strong noise. In fact, these two subplots are obtained by adding a noise with SNR $= -21$ dB to the subplots in Fig. 13.2a, b respectively. The optimal aperiodic stochastic resonance obtained under the system parameters given in Fig. 13.2e, f. Through the above optimal parameters, the signal with SLF defect or SLMA defect, noise, and asymmetric bias coefficient reach the optimal match, thus obtaining the aperiodic stochastic resonance and achieve the detection of the SLF or SLMA defect of the steel wire rope under strong noise background, as shown in Fig. 13.2e, f.

The experimental signal with multiple local flaw (MLF) defects or multiple loss of metallic cross-sectional area (MLMA) defect is shown in Fig. 13.3a, b, respectively. The signals in Fig. 13.3c, d are submerged by strong noise with SNR $= -21$ dB. In Fig. 13.3e, f, the characteristic signals recovered by the aperiodic stochastic resonance method are presented. As a result, the signal of steel wire rope with multiple damage can also be identified in a strong noise background by the aperiodic stochastic resonance method.

13.3 Steel Wire Rope Damage Degree Judging by Aperiodic Stochastic Resonance

These signals collected from the experimental platform in Fig. 13.4. This test device is also given in [15]. It mainly composed of UGN3503 Hall type steel wire rope flaw detector, steel wire rope sample rope (6×37S-FC), and computer. The discrimination standard for minor damage to steel wire ropes is that the number of broken wires

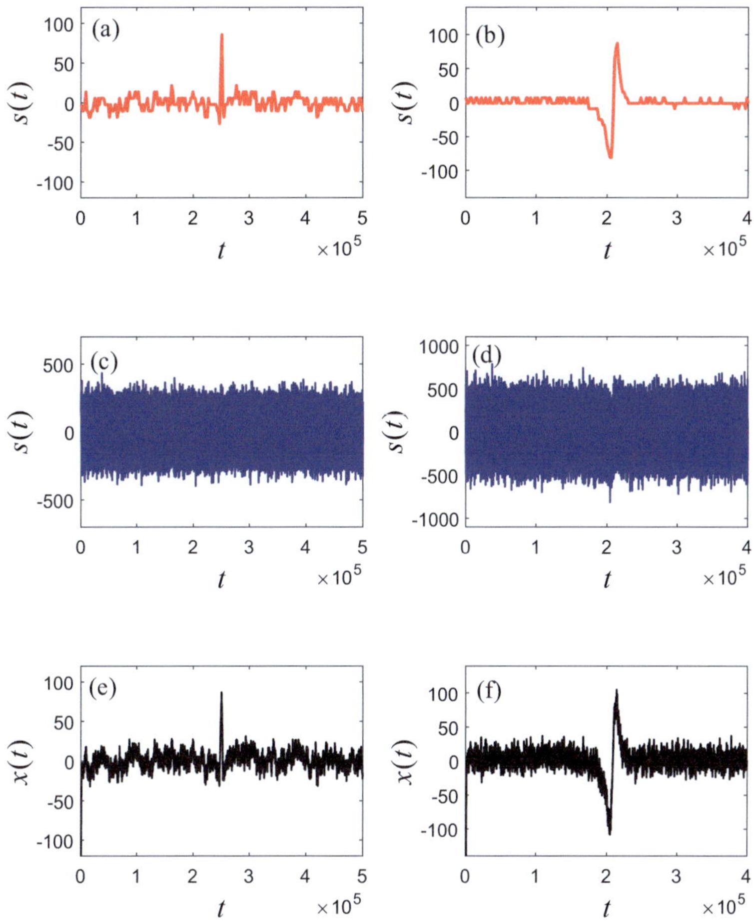

Fig. 13.2 The simulation signal of steel wire rope with SLF or SLMA defect. **a** The pure signal with SLF defect. **b** The pure signal with SLMA defect. **c** The noisy signal with SLF defect. **d** The noisy signal with SLMA defect. **e** The noisy signal in **c** processed by the aperiodic stochastic resonance. **f** The noisy signal in **d** processed by the aperiodic stochastic resonance. The simulation parameters are $b = 0.05$ and $r = 0.8$

within one lay is less than 5% of the total number of wires. The experimental sample rope has a total of 222 wires in one lay, 6 weak damages with 1–6 broken wires were made in different locations of the same steel wire rope.

In Fig. 13.5**a**–**f** the plots show the weak damage signals corresponding to 1–6 broken wires in the steel wire rope. The noise intensity is -40, dB. In this figure, the amplitude of the damage signal gradually increases as the number of broken wires increases. However, due to the strong noise interference, weak damage characteristics are submerged, making it difficult to distinguish the broken degree in the steel wire rope.

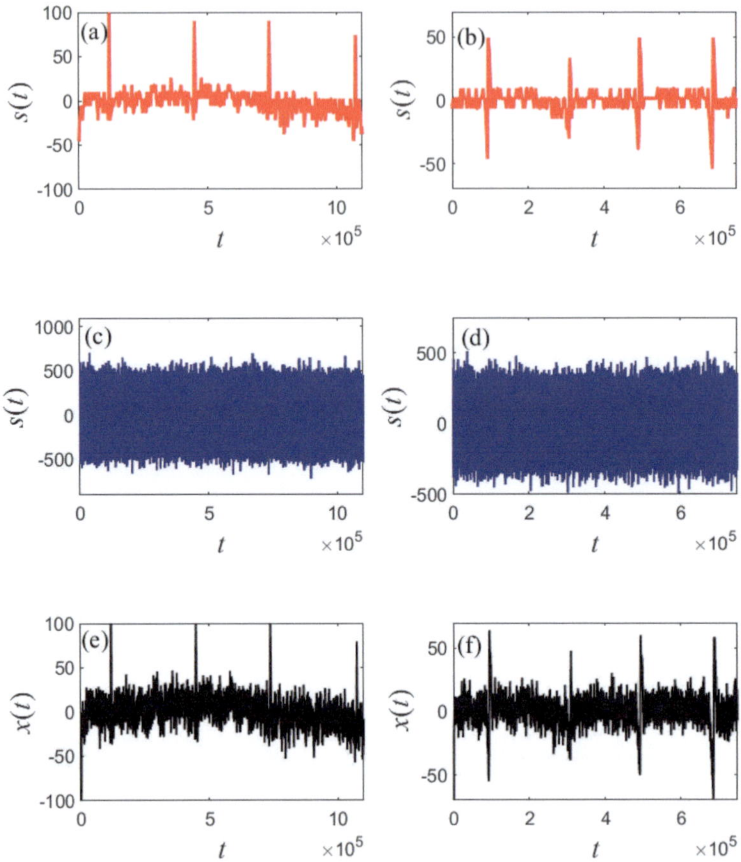

Fig. 13.3 The simulation signal of steel wire rope with MLF or MLMA defect. **a** The pure signal with SLF defect. **b** The pure signal with SLMA defect. **c** The noisy signal with SLF defect. **d** The noisy signal with SLMA defect. The simulation parameters are $b = 0.05$ and $r = 0.8$

The noisy signal, with a noise intensity of -40 dB, with weak damage characteristic is processed by the aperiodic stochastic resonance method, as shown in Fig. 13.6. From the subplots Fig. 13.6a–f, the noise is eliminated when the aperiodic stochastic resonance occurs in the response. In addition, the weak damage characteristic has not only been recovered, but also significantly enhanced. Furthermore, the characteristic amplitude of the weak damage signal increases with the increase of the number of broken wires, and it reflects the trend of damage evolution.

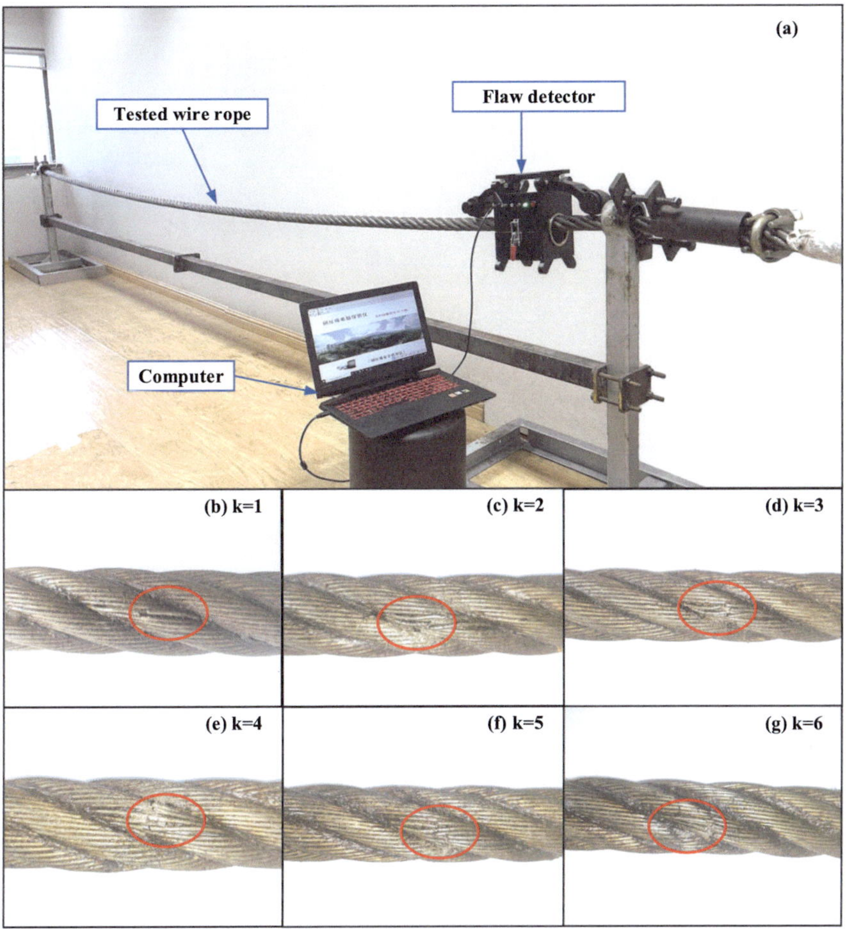

Fig. 13.4 The test rig of nondestructive testing of steel wire rope. **a** The test rig. **b–g** The tested wire rope with k broken wires

13.4 Concluding Remarks

We present an application of aperiodic stochastic resonance for denoising signals in the nondestructive testing of steel wire ropes. Experimental verification demonstrates that aperiodic stochastic resonance processing effectively reduces noise in various damage signals and aids in assessing the severity of steel wire rope damage. However, research on aperiodic stochastic resonance in this field is still in its early stages, and several challenges remain. For instance, due to the considerable length of steel wire ropes, a large volume of data is generated during inspections. Key challenges include developing efficient methods for processing large datasets to enable real-time inspection and analysis, accurately locating damage along the rope,

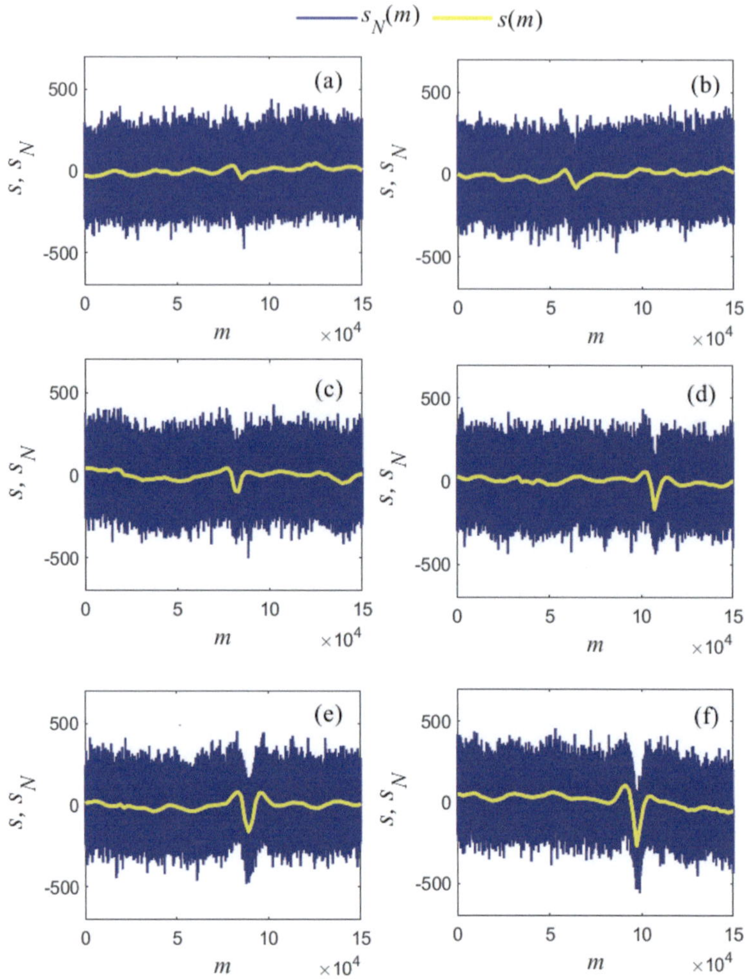

Fig. 13.5 The experimental noisy signal $s_N(m)$ and pure signal $s(m)$ of damaged steel wire rope with different broken wires. **a** 1 broken wire. **b** 2 broken wires. **c** 3 broken wires. **d** 4 broken wires. **e** 5 broken wires. **f** 6 broken wires

and identifying the type and severity of damage in complex noise environments. Addressing these issues is essential for the practical application of this method in real-world engineering scenarios.

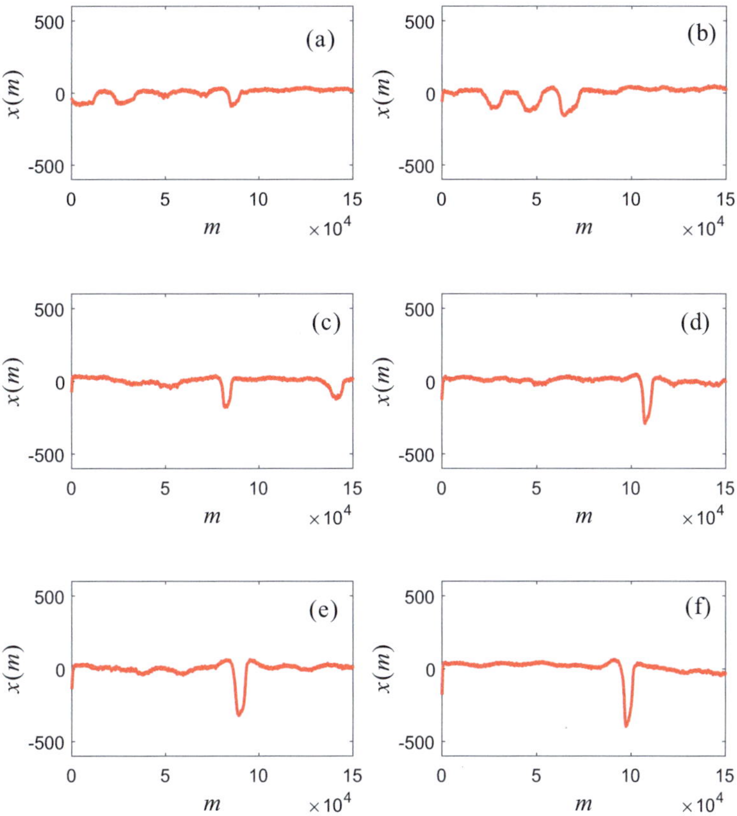

Fig. 13.6 The experimental noisy signal processed by the aperiodic stochastic resonance. **a** 1 broken wire. **b** 2 broken wires. **c** 3 broken wires. **d** 4 broken wires. **e** 5 broken wires. **f** 6 broken wires

References

1. J. Tian, J. Zhou, H. Wang, G. Meng, Literature review of research on the technology of wire rope nondestructive inspection in china and abroad, *MATEC Web of Conferences, EDP Sciences*, 2015, **22**: 03025.
2. K. Kumar, D. Goyal, S.S. Banwait, Effect of key parameters on fretting behaviour of wire rope: A review, *Archives of Computational Methods in Engineering*, 2020, **27**(2): 549-561.
3. P. Zhou, G. Zhou, Z. Zhu, Z. He, X. Ding, C. Tang, A review of non-destructive damage detection methods for steel wire ropes, *Applied Sciences*, 2019, **9**(13): 2771.
4. X.Y. Zhong, X.H. Zhang, Research of non-destructive testing of wire rope using magnetic flux leakage, *Applied Mechanics and Materials*, 2012, **189**: 255-259.
5. J.R. Wait, Review of electromagnetic methods in nondestructive testing of wire ropes, *Proceedings of the IEEE*, 1979, **67**(6): 892-903.
6. L. Shi, J. Tan, S. Xue, J. Deng, Inspection method of rope arrangement in the ultra-deep mine hoist based on optical projection and machine vision, *Sensors*, 2021, **21**(5): 1769.
7. G. Drummond, J.F. Watson, P.P. Acarnley, Acoustic emission from wire ropes during proof load and fatigue testing, *NDT & E International*, 2007, **40**(1): 94-101.

8. P.C. Peng, C.Y. Wang, Use of gamma rays in the inspection of steel wire ropes in suspension bridges, *NDT & E International*, 2015, **75**:80-86.
9. Q. Cao, D. Liu, Y. He, J. Zhou, J. Codrington, Nondestructive and quantitative evaluation of wire rope based on radial basis function neural network using eddy current inspection, *NDT & E International*, 2012, **46**: 7-13.
10. Z. Zhang, J. Zhang, Quantitative nondestructive testing of broken wires for wire rope based on multi-image fusion, *Russian Journal of Nondestructive Testing*, 2022, **58**(7): 643-656.
11. J. Zhang, Z. Zhang, X. Li, B. Liu, Quantitative nondestructive testing for wire rope based on multi-source information fusion, *Journal of Failure Analysis and Prevention*, 2022, **22**(4): 1798-1811.
12. Z. Xu, J. Yang, S. Li, Y. Yang, L. Wang, B. Sun, S. Huang, Research progress in signal processing algorithms for nondestructive testing of steel wire ropes, *Nondestructive Testing*, **45**(11): 72-79. (in Chinese)
13. S. Liu, Y. Sun, Y. Kang, A novel E-exponential stochastic resonance model and weak signal detection method for steel wire rope, *IEEE Transactions on Industrial Electronics*, 2021, **69**(7): 7428-7440.
14. Z. Xu, Z. Wang, J. Yang, M.A.F. Sanjuán, B. Sun, S. Huang, Aperiodic stochastic resonance in a biased monostable system excited by different weak aperiodic pulse signals and strong noise, *The European Physical Journal Plus*, 2023, **138**(5): 1-12.
15. B. Sun, S. Huang, Z. Wang, J. Yang, S. Li, Y. Yang, Stochastic resonance of monostable system induced by strong noise and aperiodic pulse and its application, *Journal of Vibration and Shock*, 2024, **43**(10): 279-284. (in Chinese)
16. F. Pan, Z. Liu, L. Ren, M. Zuo, Adaptive local flaw detection based on magnetic flux leakage images with a noise distortion effect for steel wire ropes, *IEEE Transactions on Industrial Electronics*, 2023, **71**(4): 4120-4129.
17. X. Yan, D. Zhang, S. Pan, E. Zhang, W. Gao, Online nondestructive testing for fine steel wire rope in electromagnetic interference environment, *NDT & E International*, 2017, **92**: 75-81.
18. Y. Yao, G. Li, X. Zhang, X. Teng, M. Huang, Research on wavelet denoising method based on soft threshold in wire rope damage detection, *In 2020 IEEE 8th International Conference on Computer Science and Network Technology (ICCSNT)* 2020, 165-170).
19. S. Huang, Z. Wang, J. Yang, T. Gong, Z. Shan, Y. Yang, Adaptive fast Walsh-Hadamard transform for magnetic flux leakage signal of broken wire damage extraction under noise background, *Nondestructive Testing and Evaluation*, 2025, **40**(2): 564-584

Appendix A
Several Typical Potential Functions

In this appendix, we give several typical potential functions which are used in almost all chapters. In order to facilitate further expansion of the depth of aperiodic resonances, we also mention some other potential functions corresponding to nonlinear models that are not used in this monograph. These models can be further studied by readers to enrich the results of aperiodic resonance researches.

A.1 Bistable Potential Function

A.1.1 Traditional Bistable Potential Function

For the traditional bistable potential function, $V(x) = -\frac{a}{2}x^2 + \frac{b}{4}x^4$, when $a > 0$ and $b > 0$, its configuration is shown in Fig. A.1. It has two stable equilibriums $x_{\pm} = \pm\sqrt{\frac{a}{b}}$ and an unstable equilibrium $x_0 = 0$. This kind of potential function is most considered in this monograph.

A.1.2 Bistable Potential Function with Fractional Exponent Nonlinear Term

For the bistable potential function with fractional exponent nonlinear term, $-\frac{a}{2}x^2 +$ $V(x) = \frac{b}{\alpha+1}|x|^{\alpha+1}$, where $\alpha > 1$ is a positive parameter, the curves of the function under different values of α are shown in Fig. A.2. This kind of bistable potential function is used in Eq. (3.7) of Sect. 3.1.

J. Yang and M. A. F. Sanjuan, *Aperiodic Resonances—Theory and Applications*,
https://doi.org/10.1007/978-981-96-6819-9

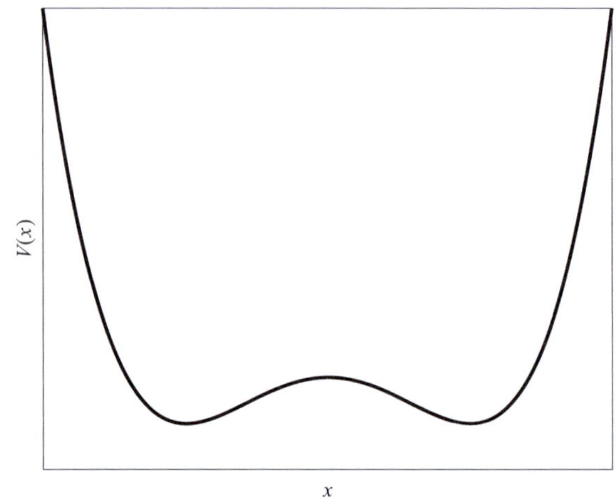

Fig. A.1 The configuration of the traditional bistable potential function $V(x) = -\frac{a}{2}x^2 + \frac{b}{4}x^4$ when $a > 0$ and $b > 0$

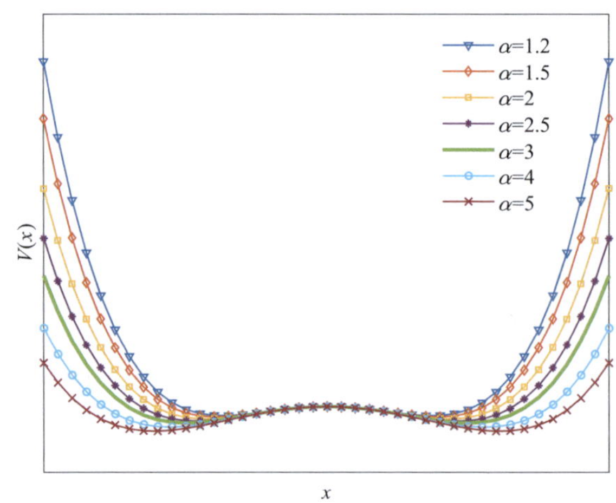

Fig. A.2 The configuration of the bistable potential function $V(x) = -\frac{a}{2}x^2 + \frac{b}{\alpha+1}|x|^{\alpha+1}$

A.1.3 Asymmetric Bistable Potential Function

We give the shape of the asymmetric bistable potential function $V(x) = -rx - \frac{a}{2}x^2 + \frac{b}{4}x^4$ in Fig. A.3 under the precondition of $a > 0$ and $b > 0$. When $r < 0$, the left potential well is high and the right potential well is low. But when $r > 0$, the shape of the potential well exhibits the opposite configuration. Certainly, when $r = 0$, the asymmetric bistable potential function changes to the symmetric case. This kind of potential function is used in Eq. (4.6) of Sect. 4.1.

There is also another form of asymmetric bistable potential function $V(x) = -\frac{a}{2}x^2 + \frac{r}{3}x^3 + \frac{b}{4}x^4$. In Fig. A.4, we give two configurations of the potential function for the case $r < 0$ and $r > 0$ under the precondition $a > 0$ and $b > 0$. Similar to the

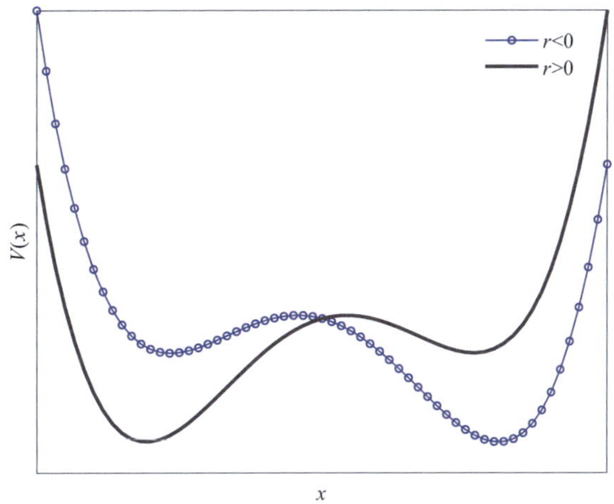

Fig. A.3 The configuration of the asymmetric bistable potential function $V(x) = -rx - \frac{a}{2}x^2 + \frac{b}{4}x^4$ where $a > 0$ and $b > 0$

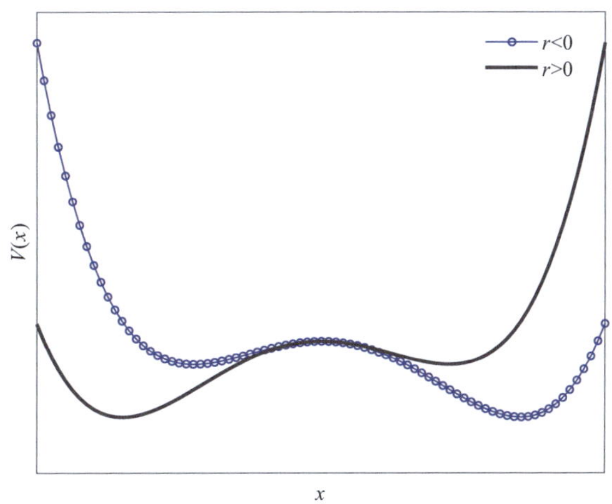

Fig. A.4 The configuration of the asymmetric bistable potential function $V(x) = -\frac{a}{2}x^2 + \frac{r}{3}x^3 + \frac{b}{4}x^4$ where $a > 0$ and $b > 0$

shapes of the potential in Fig. A.3, the positive or negative influences the tilt direction of the potential-well. This kind of potential function is used in Eq. (5.9) of Sect. 5.3.

A.2 Monostable Potential Function

A.2.1 Symmetric Monbostable Potential Function

For the symmetric monostable potential function $V(x) = -\frac{a}{2}x^2 + \frac{b}{4}x^4$, if $a < 0$ and $b > 0$, its configuration is shown in Fig. A.5. It has one stable equilibrium $x_0 = 0$ only.

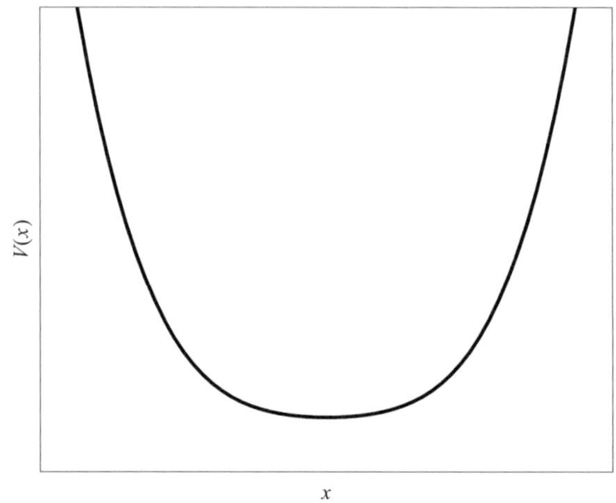

Fig. A.5 The configuration of the symmetric monostable potential function $V(x) = -\frac{a}{2}x^2 + \frac{b}{4}x^4$ when $a < 0$ and $b > 0$

A.2.2 Asymmetric Monbostable Function

In Eqs. (6.10) and (6.11), we investigate the asymmetric monostable potential function $V(x) = \frac{a}{2}x^2 - \frac{r}{3}x^3 + \frac{b}{4}x^4$ and $V(x) = -rx + \frac{a}{2}x^2 + \frac{b}{4}x^4$ respectively. Their configurations are given in Fig. A.6a, b. These two functions are considered in Sect. 6.1.5. In addition, the configuration of the asymmetric monotable potential function $V(x) = -ax + \frac{b}{4}x^4$ with $b > 0$, its configuration is shown in Fig. A.6c. This kind of potential function is used in Eq. (13.1) of Chap. 13 to deal with the aperiodic pulse by the aperiodic stochastic resonance method.

A.3 Tristable Potential Function

Figure A.7. The configuration of the tristable potential function $V(x) = \frac{c}{2}x^2 + \frac{b}{4}x^4 + \frac{a}{6}x^6$ is presented in Fig. A.7, where $a > 0$, $b < 0$ and $c > 0$. This kind of potential function is used in Eq. (4.9) of Sect. 4.1.

A.4 Periodic Potential Function

The configuration of the periodic potential function $V(x) = a\cos(bx)$ is given in Fig. A.8. The periodic potential function is used in Eq. (11.2) of Chap. 11 in the image processing.

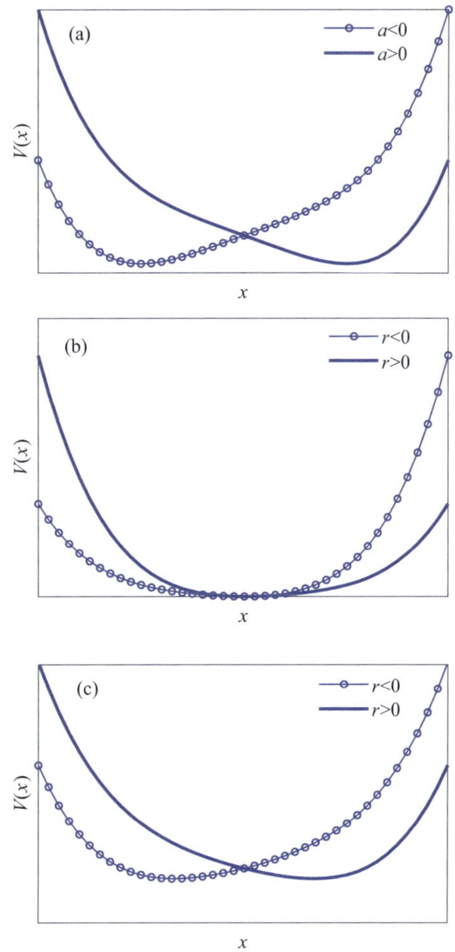

Fig. A.6 The configuration of the asymmetric monostable potential function. **a** $V(x) = \frac{a}{2}x^2 - \frac{r}{3}x^3 + \frac{b}{4}x^4$ with $a > 0$ and $b > 0$, **b** $V(x) = -rx + \frac{a}{2}x^2 + \frac{b}{4}x^4$ with $a > 0$ and $b > 0$, **c** $V(x) = -ax + \frac{b}{4}x^4$ with $b > 0$

A.5 Some Other Typical Potential Functions

The above potential functions only some typical ones used in this monograph. There are still a large number of other nonlinear potential functions worth studying. The studies of aperiodic resonances in these systems are still far from sufficient. We have summarized a series of nonlinear models in our review paper [A1]. In addition, some deformed asymmetric bistable potential [A2], asymmetric tristable potential [A3], we will not elaborate on each one here. It is recommended that readers summarize and conduct research on aperiodic resonances in these systems.

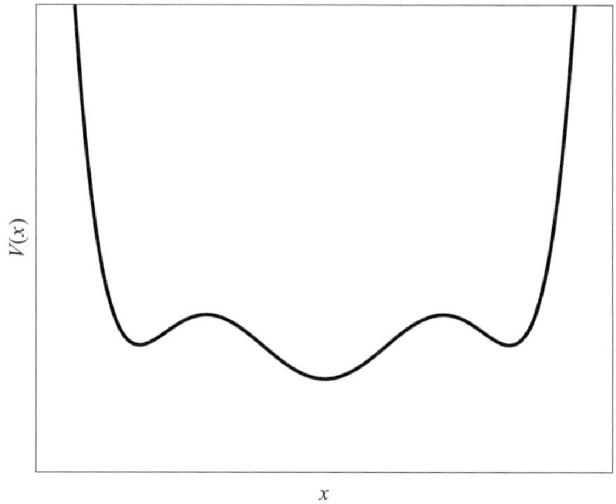

Fig. A.7 The configuration of the tristable potential function $V(x) = \frac{c}{2}x^2 + \frac{b}{4}x^4 + \frac{a}{6}x^6$ with $a > 0$, $b < 0$ and $c > 0$

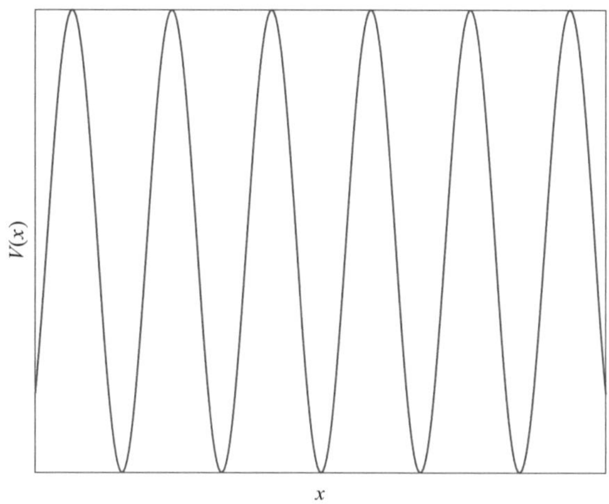

Fig. A.8 The configuration of the tristable potential function periodic potential function $V(x) = a\cos(bx)$

References

A1. J. Yang, S. Rajasekar, M.A.F. Sanjuán, Vibrational resonance: A review, *Physics Reports*, 2024, **1067**: 1-62.

A2. A.M. Tala, Y.W. Ngouongo, G.D. Kenmoé, T.C. Kofané, Ghost stochastic resonance in an asymmetric Duffing oscillator, *Physica A: Statistical Mechanics and its Applications*, 2021, **582**: 126247.

A3. P. Xu, Y. Jin, Stochastic resonance in an asymmetric tristable system driven by correlated noises, *Applied Mathematical Modelling*, 2020, **77**: 408-425.

Appendix B
Several Typical Anharmonic Periodic Signals

In this appendix, we give the expressions and curves of several anharmonic periodic signals. These signals are used in some chapters of the monograph.

B.1 Square Waveform Signal

The expression of a square waveform signal $s_{qua}(t)$ is

$$s_{qua}(t) = \begin{cases} A, & (2n-2)\pi/\omega < t < (2n-1)\pi/\omega, \\ -A, & (2n-1)\pi/\omega < t < 2n\pi/\omega, \end{cases} \quad n = 1,\ 2,\ 3\cdots \tag{B1}$$

Its Fourier series is

$$s_{qua}(t) = \frac{4}{\pi}\sum_{n=1}^{\infty}\frac{1}{2n-1}A\sin[(2n-1)\omega t]. \tag{B2}$$

The time series of the square waveform signal and its Fourier spectrum are given in Fig. B.1.

B.2 Full Sine Waveform Signal

The expression of a full sine waveform $s_{fsi}(t)$ is

$$s_{fsi}(t) = A\left|\sin(\omega t/2)\right|. \tag{B3}$$

J. Yang and M. A. F. Sanjuan, *Aperiodic Resonances—Theory and Applications*,
https://doi.org/10.1007/978-981-96-6819-9

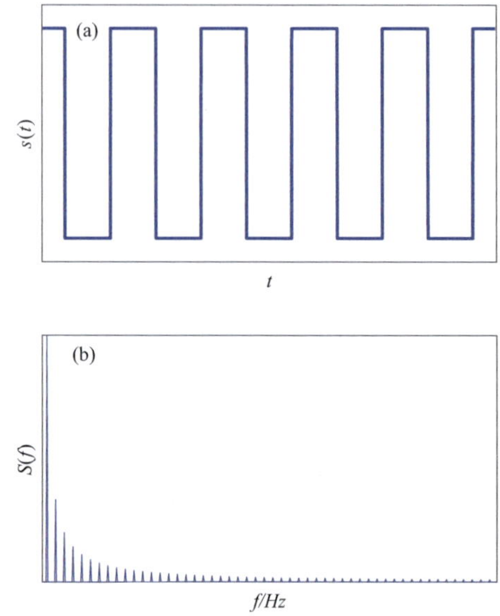

Fig. B.1 The time series and Fourier spectrum of the square waveform signal. **a** Time series, **b** Fourier spectrum

The corresponding Fourier series is

$$s_{fsi}(t) = \frac{2A}{\pi} - \frac{4}{\pi}\sum_{n=1}^{\infty}\frac{n}{4n^2 - 1}F\cos(n\omega t). \tag{B4}$$

The time series of the full sine waveform signal and its Fourier spectrum are given in Fig. B.2.

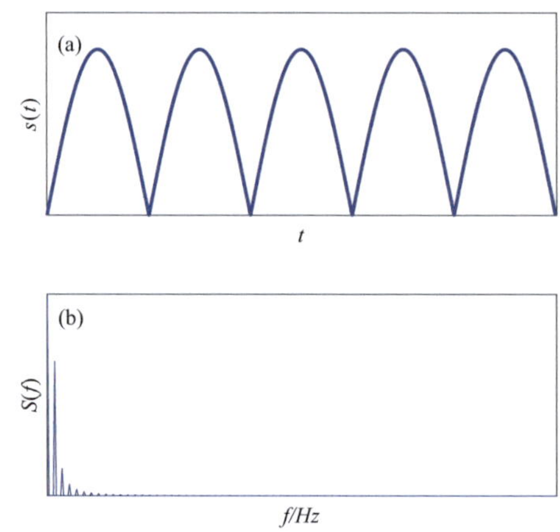

Fig. B.2 The time series and Fourier spectrum of the full sine waveform signal. **a** Time series, **b** Fourier spectrum

B.3 Rectified Sine Waveform Signal

The expression of a rectified sine waveform $s_{rsi}(t)$ is

$$s_{rsi}(t) = \begin{cases} A\sin(\omega t), & (2n-2)\pi/\omega < t < (2n-1)\pi/\omega, \\ 0, & (2n-1)\pi/\omega < t < 2n\pi/\omega, \end{cases} \quad n = 1,\ 2,\ 3\cdots \tag{B5}$$

The corresponding Fourier series is

$$s_{rsi}(t) = \frac{2A}{\pi} - \frac{4A}{\pi}\sum_{n=1}^{\infty}\frac{\cos(2n\omega t)}{(4n^2-1)}. \tag{B6}$$

The time series of the rectified sine waveform signal and its Fourier spectrum are given in Fig. B.3.

B.4 Symmetric Triangular Waveform Signal

A symmetric triangular waveform with period $2\pi/(k\omega)$ is produced by the expression

$$s_{tri}(t) = \begin{cases} -A + \dfrac{2A}{\pi/(k\omega)}[t - 2(n-1)\pi/(k\omega)], \\ 2(n-1)\pi/(k\omega) \le t \le (2n-1)\pi/(k\omega); \\ 3A - \dfrac{2A}{\pi/(k\omega)}[t - 2(n-1)\pi/(k\omega)], \\ (2n-1)\pi/(k\omega) \le t \le 2n\pi/(k\omega), \end{cases} \quad n = 1,\ 2,\ 3\cdots \tag{B7}$$

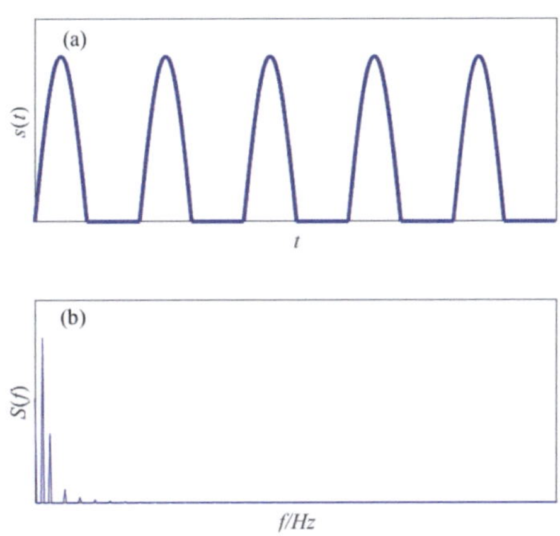

Fig. B.3 The time series and Fourier spectrum of the rectified sine waveform signal. **a** Time series, **b** Fourier spectrum

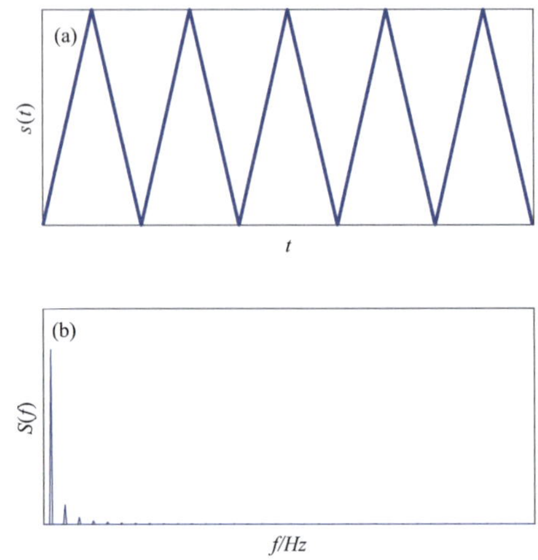

Fig. B.4 The time series and Fourier spectrum of the symmetric triangular waveform signal. **a** Time series, **b** Fourier spectrum

The corresponding Fourier Series is

$$s_{tri}(t) = -\frac{8A}{\pi^2}\sum_{n=1}^{\infty}\frac{1}{(2n-1)^2}\cos(2n-1)k\omega t. \tag{B8}$$

The time series of the symmetric triangular waveform signal and its Fourier spectrum are given in Fig. B.4.

In addition to the four forms of anharmonic periodic signals mentioned above, there are many other forms of this kind of signal that will not be elaborated here.

MIX
Papier aus verantwortungsvollen Quellen
Paper from responsible sources
FSC® C105338

If you have any concerns about our products,
you can contact us on
ProductSafety@springernature.com

In case Publisher is established outside the EU,
the EU authorized representative is:
Springer Nature Customer Service Center GmbH
Europaplatz 3, 69115 Heidelberg, Germany

Printed by Libri Plureos GmbH
in Hamburg, Germany